Hawaii

The Ecotravellers' Wildlife Guide

Other Ecotravellers' Wildlife Guides
Belize and Northern Guatemala *by Les Beletsky*
Costa Rica *by Les Beletsky*
Ecuador and Its Galápagos Islands *by David L. Pearson and Les Beletsky*
Tropical Mexico *by Les Beletsky*

Forthcoming guides
Alaska *by Dennis Paulson and Les Beletsky*
The Caribbean Islands *by H. Raffaele, J. Wiley and L. Beletsky*
East Africa *by C. FitzGibbon, J. Fanshawe and B. Branch*
Florida *by Fiona Sunquist and Les Beletsky*
Perú *by David L. Pearson and Les Beletsky*
Southern Africa *by B. Branch, W. Tarboton and C & T Stuart*

series editor Les Beletsky

On the cover: the Iiwi, one of Hawaii's most striking native birds.

Hawaii
The Ecotravellers' Wildlife Guide

Les Beletsky

Illustrated by:
H. Douglas Pratt (Plates 17–51) and
Colin Newman (Plates 51–83)

Photographs by:
H. Douglas Pratt (Plates 1–16),
William Mull and Pete Oboyski (Plates 84–86) and
Les Beletsky (Habitats)

Contributors:
Paul Banko
Richard Francis
Pete Oboyski
Sarah Reichard
Matti Rossi
Bernd Würsig

ACADEMIC PRESS
SAN DIEGO LONDON BOSTON
NEW YORK SYDNEY TOKYO TORONTO

This book is printed on acid-free paper.

Academic Press
Harcourt Place, 32 Jamestown Road, London NW1 7BY, UK
http://www.academicpress.com

AP Natural World is published by
Academic Press
A Harcourt Science and Technology Company
525 B Street, Suite 1900, San Diego
California 92101–4495, USA
http://www.academicpress.com

ISBN: 0–12–084813–9

Library of Congress Catalog Card Number: 99–68794

A catalogue record for this book is available from the British Library

Typeset by J&L Composition Ltd, Filey, North Yorkshire
Colour Separation by Tenon & Polert Colour Scanning Ltd, Hong Kong
Printed in Hong Kong by Midas Printing Ltd
00 01 02 03 04 05 MD 9 8 7 6 5 4 3 2 1

For Bettie

Contents

Foreword

Throughout the world, wild places and wildlife are dwindling. Their conservation will require ever more intense protection, care, and management. We always value things more when we stand to lose them, and it is perhaps no coincidence that people are increasingly eager to experience unspoiled nature, and to see the great wildlife spectacles. Tourists are increasingly forsaking the package tour and the crowded beach, to wade through jungle streams, to dive on coral reefs, and to track elusive wildlife. But despite its increasing popularity, nature tourism is nothing new, and the attraction to the tourist is self evident – so why should a conservation organization like the Wildlife Conservation Society encourage it?

The answer is that nature tourism, if properly conducted, can contribute directly to the conservation of wild places and wildlife. If it does that, such tourism earns the sobriquet *ecotourism*. A defining quality of ecotourism is that people are actively encouraged to appreciate nature. If people experience wild areas, they can grow to appreciate their beauty, stability and integrity. And only if they do so, will people care about conserving these places. Before you can save nature, people need to know that it exists.

Another characteristic of ecotourism is that people tread lightly on the natural fabric of wild places. By their very definition, these are places with minimal human impact, so people must not destroy or degrade what they come to experience. Tourists need to take only photographs, leave only footprints – and ideally not even that. Wastes and pollution need to be minimized. Potential disturbance to animals and damage to vegetation must always be considered.

The third characteristic, and that which most clearly separates ecotourism from other forms of tourism, is that tourists actively participate in the conservation of the area. That participation can be direct. For instance, people or tour companies might pay fees or make contributions that support local conservation efforts, or tourists might volunteer to work on a project. More likely, the participation is indirect, with the revenues generated by the tourism entering the local economy. In this way, tourism provides an economic incentive to local communities to continue to conserve the area.

Ecotourists thus are likely to be relatively well informed about nature, and able to appreciate the exceptional nature of wild places. They are more likely to travel by canoe than cruise ship. They will be found staying at locally owned lodges rather than huge multi-national hotels. They will tend to travel to national parks and protected areas rather than to resorts. And they are more likely to contribute to conservation than detract from it.

The Wildlife Conservation Society was involved in promoting ecotourism since before the term was generally accepted. In the early 1960s, the Society (then known as the New York Zoological Society) studied how to use tourism to provide revenues for national park protection in Tanzania (then known as Tanganyika).

By the 1970s, the Society was actively involved in using tourism as a conservation strategy, focusing especially on Amboseli National Park in Kenya. The Mountain Gorilla Project in the Virunga mountains of Rwanda, a project started in the late 1970s, still remains one of the classic efforts to promote conservation through tourism. Today the Society continues to encourage tourism as a strategy from the lowland Amazonian forests to the savannas of East Africa.

We at the Wildlife Conservation Society believe that you will find these Ecotravellers' Wildlife Guides to be useful, educational introductions to the wildlife of many of the world's most spectacular ecotourism destinations.

John G. Robinson
Vice President and
Director of International Conservation
Wildlife Conservation Society

• to sustain wildlife • to teach ecology • to inspire care for nature

The mission of the Wildlife Conservation Society, since its founding in 1895 as the New York Zoological Society, has been to save wildlife and inspire people to care about our nature heritage. Today, that mission is achieved through the world's leading international conservation program working in 53 nations to save endangered species and ecosystems, as well as through pioneering environmental education programs that reach more than two million schoolchildren in the New York metropolitan area and are used in 49 states and several nations, and through the nation's largest system of urban zoological facilities including the world famous Bronx Zoo. WCS is working to make future generations inheritors, not just survivors after Bronx Zoo.

With 60 staff scientists and more than 100 research fellows, WCS has the largest professional field staff of any US-based international conservation organization. WCS's field programs benefit from the technical support of specialists based at WCS's Bronx Zoo headquarters in New York. The Field Veterinarian Program sends experts around the globe to assess wildlife health, develop monitoring techniques, and train local veterinarians. WCS's curatorial staff provides expertise in breeding endangered species in captivity. The Science Resource Center helps researchers assess data through computer mapping, statistical treatments, and cutting-edge genetic analysis. The Education Department writes primary and secondary school curricula that address conservation issues and hosts teacher-training workshops around the world.

WCS's strategy is to conduct comprehensive field studies to gather information on wildlife needs, train local conservation professionals to protect and manage wildlife and wild areas for the future, and advise on protected area creation, expansion, and management. Because WCS scientists are familiar with local conditions, they can effectively translate field data into conservation action and policies, and develop locally sustainable solutions to conflicts between humans and wildlife. An acknowledged leader in the field, the Wildlife Conservation Society forges productive relationships with governments, international agencies and local organizations.

To learn more about WCS's regional programs and membership opportunities, please see our pages in the back of this book. And please visit our website at **www.wcs.org**.

Preface

This book and others in the series are aimed at environmentally conscious travellers for whom some of the best parts of any trip are glimpses of animals in natural settings. The purpose of the book is to enhance enjoyment of a trip and enrich wildlife sightings by providing identifying information on several hundred of Hawaii's most frequently encountered land animals, plants, and ocean animals, along with up-to-date information on their ecology, behavior, and conservation. With color illustrations of 18 species of amphibians and reptiles, 21 mammals, and about 135 birds, this book truly includes almost all of the vertebrate land animals that visitors are likely to encounter. In fact, essentially every bird species you might see on the main Hawaiian Islands is illustrated and detailed here. Also included is information on and illustrations of 190 of the most commonly sighted sea creatures seen by divers and snorkelers, plus a chapter on common Hawaiian insects and spiders.

Why write a book for wildlife watchers about a place where most people visit only to lie in the sun, swim in the ocean, and relax? First, there's a growing audience for this kind of book. Visitors to Hawaii increasingly intersperse days spent on the beach with days exploring the islands – taking nature drives, biking back roads, hiking trails, birdwatching. Second, Hawaii's wildlife and plant-life are in many ways unique. Most native Hawaiian species occur nowhere else on the planet and many of them are now quite rare and some are highly endangered – and visitors with an interest in nature want detailed information about these species. And third, public awareness of and education about environmental threats are crucial for conservation; one aim of this educational book, therefore, is to help conservation in a small way.

The idea to write these books grew out of my own travel experiences and frustrations. First and foremost, I found that I could not locate a single book to take along on a trip that would help me identify all the types of animals that really interested me – birds and mammals, amphibians and reptiles, and fish. There are bird field guides, which I've used, but they are often large books, featuring information on every bird species in a given country or region. If I wanted to be able to identify mammals, I needed to carry another book. For "herps" – amphibians and reptiles – another. Thus, the idea: create a single guide book that travellers could carry to help them identify and learn about the different kinds of animals they were most likely to encounter. Also, like most ecotravellers, I am concerned about the threats to many species as their natural habitats are damaged or destroyed by people; when I travelled, I wanted current information on the conservation statuses of animals that I encountered. This book provides the traveller with conservation information on Hawaii in general, and on many of the animal family groups pictured or discussed in the book.

A few administrative notes: The word "Hawaii" refers to both the State of

Hawaii and the largest island in the state; to avoid confusion, in this book I use "Hawaii" for the state and "the Big Island" to refer to the island of Hawaii (a common usage in the islands). Because this book has an international audience, I present measurements in both metric and English system units. By now, you might think, the scientific classification of common animals would be pretty much established and unchanging; but you would be wrong. These days, what with new molecular methods to compare species, classifications of various animal groups that were first worked out during the 1800s and early 1900s are undergoing sometimes radical changes. Many bird groups, for instance, are being rearranged after comparative studies of their DNA. The research is so new that many biologists still argue about the results. I cannot guarantee that all the classifications that I use in the book are absolutely the last word on the subject, or that I have been wholly consistent in my classifications. However, for most users of this book, such minor transgressions are probably irrelevant.

A note on language: The Hawaiian language is written with only 12 letters (a, e, i, o, u, h, k, l, m, n, p, w) but also an apostrophe-like symbol, called the *okina*, or glottal stop – thus the word "Hawai'i" as you often see it rendered. The glottal stop directs you to break the word there and form a new syllable. For instance, as Hawaiians write the word Lanai – Lāna'i – you see more easily that the word should be pronounced not "la-NAI" but "la-NAH-eee." Authors of books about Hawaii face the dilemma of writing with or without the glottal stop (and *macrons*, or emphasis marks, small lines over some vowels, which indicate where to place stresses in words). Because the glottal stop helps break up words with seemingly unending and unpronounceable runs of vowels and more closely reflects the native culture, increasing numbers of authors use it. On the other hand, most visitors to Hawaii don't know what the glottal stop means and its use can be confusing. Except for some local Hawaiian names for animals and plants (in Chapter 3 and in the information opposite the plates), I have opted to write Hawaiian words without the glottal stop and macron. I should note that most Hawaiian words are pronounced pretty much as they look – with the addition of those syllable breaks as indicated by the glottal stops.

I must acknowledge the help of a large number of people in producing this book. First, most of the information here comes from published sources, so I owe the authors of those scientific papers and books a great deal of credit (see References and Additional Reading, p. 203). Many people provided information for or helped in the preparation of this book, including Melinda Renert, Samuel Gon, and Alenka Remec at The Nature Conservancy of Hawaii, Bob Pyle at the Bishop Museum, Leonard Freed of the Unviersity of Hawaii at Manoa, Brenda Becker at National Marine Fisheries Service; and three very helpful biologists at the Pacific Island Ecosystem Research Center located at Hawaii Volcanoes National Park: Paul Banko, Thane Pratt, and Steve Fancy. Special thanks to Richard Francis for writing the chapter on marine life, Sarah Reichard for writing about habitats and vegetation, Pete Oboyski for the chapter on insects, Matti Rossi for the essay on lava flows, Paul Banko for the essay on Hawaii forest birds, Bernd Würsig for the essay on dolphins, and David Pearson for letting me use some of his writing on birds and mammals. Also thanks to Douglas Pratt for reading and commenting on most of the book's chapters, which greatly improved the book, to Kerri Mikkelsen for reading several chapters, and to the artists who produced the wonderful illustrations, Douglas Pratt (amphibians, reptiles, birds, and mammals) and Colin Newman (marine life), and to photographers Douglas Pratt (plants), Bill Mull (insects), Pete Oboyski (insects), and Jack Jeffrey (cover photo). Also thanks to the Burke Museum at the University of Washington for kindly allowing me access to its facilities, and to my editor at Academic Press, Andrew Richford and his assistant, Samantha Richardson.

Please let me know of any errors you find in this book. I am also interested in hearing your opinions on the book, suggestions for future editions, and of your experiences with wildlife during your travels. Write care of the publisher or e-mail: ECOTRAVEL8@aol.com

Chapter 1

Ecotourism and Hawaii

Why Visit Hawaii?

Why should you, an ecotraveller, visit Hawaii? You plucked this wildlife guidebook from a bookstore's shelf, so chances are that when travelling you are interested in seeing natural areas and wildlife, and in pursuing nature activities – hiking, birding, and the like. But isn't Hawaii a beach resort destination, the type of place where 99.87% (author's estimate) of the 6 million annual visitors never venture far from beachfront hotels and time-share condominia, and the average 5-day package tourist spends 4.3 of those days lying motionless on hot sand not more than 10 m (33 ft) from the Pacific Ocean? And even should Hawaii have a few nice natural sights to behold (and I assure you, it does), and some wildlife you cannot see elsewhere, why choose it as a travel destination over other places, other USA states, other countries? And finally, why take leave of a great big solid continent – Europe, say, or the secure-feeling mainland mass of the USA – or even one of the large major islands of Japan, and travel a great distance in too-narrow airline seats to a pack of tiny specks of land set essentially in the very middle of the world's largest body of water – volcanic specks at that, specks that, even as you read this page, are the site of innumerable tons of super-hot molten rock volcanically spewing onto the land's surface and into the sea? Why indeed? Here's a partial list of reasons:

- For hikers and birders: *Kauai's Kokee State Park* and its *Alakai Swamp Trail*. This is simply one of the more beautiful spots on Earth. As I research these travel books, I cover a lot of ground, sometimes at a rapid pace, often visiting in a day what a typical traveller might take in over two or even three days. But every once in a while I'm stopped in my tracks; I start hiking somewhere and realize that the scenery, the habitats, or the wildlife is so stunning that I don't want to leave. Such a place is Kokee State Park. You stare at the sweeping views, the matchless high-elevation habitats, you walk the Pihea Trail, the Alakai Swamp Trail (p. 36), almost in awe; and you don't want to leave, because you know that you might not be back for a long time and it's just so incredibly beautiful. To top it off: this is one of the best places to see some of Hawaii's native forest songbirds – small brightly colored birds, many of which are highly endangered, that live nowhere else in the world. Similarly, Maui's Haleakala National Park and the Nature Conservancy's adjacent Waikamoi Preserve present, for the biologically inclined, gorgeous high-elevation tropical forest habitat and forest birds found nowhere else.

- For adventurers: *Kauai's Na Pali Coast Trail* (p. 38) provides magnificent coastal hiking along lushly vegetated green cliffs that in places plunge vertically 700 m (2300 ft) or more into the sea. The trail is globally recognized for its marvelous views and stunning natural scenery, and for the opportunity it affords experienced hikers to walk along narrow cliff-top paths, sheer drop-offs to the crashing surf far below on one side, a huge wilderness area of steep slopes, deep gulches, and impenetrable vegetation on the other.
- For those geologically inclined: The amazing sights of the Big Island's *Hawaii Volcanoes National Park* would by themselves make a trip to the Hawaiian Islands worthwhile. The place is quite simply unlike anything else you have ever seen. Volcanic landscapes of terrible destruction, haunting views across huge, flat, pitch-black recent lava flows, hikes across still-steaming but temporarily-restful volcanic craters, an active volcanic vent chugging out lava, and all-but-deserted trails through high-elevation native forests make this site one of the world's premier natural attractions.

Add to this partial list of wonderful natural areas the following considerations, and you will quickly see why the Hawaiian Islands are a prime ecotravel destination: (1) many of the best ecological sites are easily accessible; (2) reaching Hawaii, for many people, particularly from the mainland USA, is relatively inexpensive, and accommodations in Hawaii span the range of prices; (3) Hawaii is part of a fairly stable, democratic country, with relatively little social unrest, and where English is commonly understood. Owing to these reasons, it is worth your while to travel to Hawaii specifically to hike, birdwatch, and otherwise enjoy its natural attractions. But even if your main objective is the beach, you can still spend a portion of your stay hiking and visiting parks and nature preserves (in fact, many package trips include a rental car – so you must leave the beach at some point and go somewhere else!).

Hawaiian Tourism and Ecotourism

Hawaii is, of course, one of the capitals of mass tourism. People come mainly for the sun, for beach and water activities, perhaps for the scenic coastal vistas, and, largely, to "escape" or "get away" from the stresses and routine of their everyday lives. Most visitors come from the western part of the mainland USA, but large numbers come from the rest of the USA, Canada, Japan, and other Pacific Rim countries. For residents of the USA mainland, an obvious attraction is to be able to sample a bit of the tropics while remaining within the psychologically secure confines of the USA.

Tourism began in Hawaii probably about 1867, with the first regularly scheduled steamship service from the North American mainland. Interestingly, early visitors came not for the sun and beach, but to view the volcanic activity of

Kilauea on the Big Island. By the 1930s, Hawaii was widely promoted in the USA as a "Paradise of the Pacific," where the climate was June-like in December, a place of rainbows and fiery-orange sunsets amid tropical splendor. About 30,000 people per year were visiting Hawaii when, in 1941, the entry of the USA into World War II largely halted tourism for four years. But during the war hundreds of thousands of USA military personnel passed through or were stationed in Hawaii, giving the islands great exposure; in later years, many returned for vacations. Hawaii, a USA possession since 1898, became the 50th state in 1959. The number of tourist arrivals grew at an average rate of 15% or more annually during a 40–year period beginning at the end of World War II, with about 50,000 arrivals in 1951, more than 300,000 in 1961, almost 2 million in 1971, and 4 million in 1981. Today, with roughly 6.5 million visitors per year, tourism is by far the state's largest industry, providing about a third of income (the military and agriculture are the state's second- and third-largest income producers).

Hawaiian ecotourism is still relatively undeveloped, but interest in it is growing. Changing attitudes of both travellers and state authorities are sparking the upsurge in interest. Travellers increasingly want a variety of activities available to them during their holidays – days on the beach interspersed with hiking, birding, visiting parks, etc. Hawaii state planners for decades have been trying to direct and redistribute tourism. When tourism on Oahu during the 1960s led to explosive growth and development on that island, policy was to try to channel new tourism development to the other main islands (the "neighbor" islands). Now, with many of the main islands already dense with coastal beach resorts, and much of the islands' natural coastal habitats altered by these resorts and other types of development, many believe that it would be beneficial if some of the future growth of tourism were directed toward non-coastal areas; among other directions, toward ecotourism – having tourists utilize inland sites such as national parks, nature preserves, wildlife refuges. The islands already have a good number of excellent ecotravel sites that are easily visited (see Chapter 4). Some tour agencies are starting to specialize in bringing tourists to natural areas for cycling, birding, or hiking. And the media, doing their jobs, have begun to take notice of the subtle change in emphasis. For instance, *New York Times* travel section articles during the past few years have detailed such ecotravel activities as snorkeling and diving off Maui and Kauai, walking the lava trails of the Big Island's Hawaii Volcanoes National Park, and hiking Kauai. Not to worry, though, ecotravel is still lightly represented in Hawaii, and people visiting specifically for nature activities are few and far between. At most parks, if you are willing to move a few paces from your vehicle, you can get far away from the crowd. (Contact Hawaii Visitors Bureau for information on parks, tours, eco-tour operators: 808–923–1811, 800–353–5846; www.visit.hawaii.org)

Ecotourism and Its Importance

People have always travelled. Historical reasons for travelling are many and varied: to find food, to avoid seasonally harsh conditions, to emigrate to new regions in search of more or better farming or hunting lands, to explore, and even, with the advent of leisure time, just for the heck of it (travel for leisure's sake is the definition of tourism). For many people, travelling fulfills some deep need. There's something irreplaceably satisfying about journeying to a new place: the sense of being in completely novel situations and surroundings, seeing things never before encountered, engaging in new and different activities.

During the 1970s and 1980s, however, there arose a new reason to travel, perhaps the first wholly new reason in hundreds of years: with a certain urgency, to see natural habitats and their harbored wildlife before they forever vanish from the surface of the earth. *Ecotourism*, or *ecotravel*, is travel to destinations specifically to admire and enjoy wildlife and undeveloped, relatively undisturbed natural areas, as well as indigenous cultures. The development and increasing popularity of ecotourism is a clear outgrowth of escalating concern for conservation of the world's natural resources and biodiversity (the different types of animals, plants, and other life forms found within a region). Owing mainly to the actions of people, animal species, plant species, and wild habitats are disappearing or deteriorating at an alarming rate. Because of the increasing emphasis on the importance of the natural environment by schools at all levels and the media's continuing exposure of environmental issues, people have an enhanced appreciation of the natural world and increased awareness of global environmental problems. They also have the very human desire to want to see undisturbed habitats and wild animals before they are gone, and those with the time and resources increasingly are doing so.

But that's not the entire story. The purpose of ecotravel is actually twofold. Yes, people want to undertake exciting, challenging, educational trips to exotic locales – wet tropical forests, wind-blown deserts, high mountain passes, mid-ocean coral reefs – to enjoy the scenery, the animals, the nearby local cultures. But the second major goal of ecotourism is often as important: the travellers want to help conserve the very places – habitats and wildlife – that they visit. That is, through a portion of their tour cost and spending into the local economy of destination countries – paying for park admissions, engaging local guides, staying at local hotels, eating at local restaurants, using local transportation services, etc. – ecotourists help to preserve natural areas. Ecotourism helps because local people benefit economically as much or more by preserving habitats and wildlife for continuing use by ecotravellers than they could by "harvesting" the habitats for short-term gain. Put another way, local people can sustain themselves better economically by participating in ecotourism than by, for instance, cutting down rainforests for lumber or hunting animals for meat or the illicit exotic pet trade.

Preservation of some of the world's remaining wild areas is important for a host of reasons. Aside from moral arguments – the acknowledgment that we share the Earth with millions of other species and have some obligation not to be the continuing agent of their decline and extinction – increasingly we understand that conservation is in our own best interests. The example most often cited is that botanists and pharmaceutical researchers each year discover another wonder drug or two whose base chemicals come from plants that live, for instance, only in tropical rainforest. Fully one-fourth of all drugs sold in the USA come from natural sources – plants and animals. About 50 important drugs now manufactured come from flowering plants found in rainforests, and, based on the number of plants that have yet to be cataloged and screened for their drug potential, researchers estimate that at least 300 more major drugs remain to be discovered. The implication is that if the globe's rainforests are soon destroyed, we shall never discover these future wonder drugs, and so will never enjoy their benefits. Also, the developing concept of *biophilia*, if true, dictates that, for our own mental health, we need to preserve some of the wildness that remains in the world. Biophilia, the word recently coined by Harvard biologist E. O. Wilson, suggests that because people evolved amid rich and constant interactions with other species and in natural habitats, we have deeply ingrained, innate tendencies to affiliate with other species and actual physical need to experience, at some level, natural habitats. This instinctive, emotional attachment to wildness means that if we eliminate species and habitats, we shall harm ourselves because we shall lose things essential to our mental well-being.

How Ecotourism Helps

To the traveller, the benefits of ecotourism are substantial (exciting, adventurous trips to stunning wild areas; viewing never-before-seen wildlife); the disadvantages are minor (sometimes, less-than-deluxe transportation and accommodations that, to many ecotravellers, are an essential part of the experience). But what are the real benefits of ecotourism to local economies and to helping preserve habitats and wildlife? Because Hawaii is part of the USA, as an ecotravel destination it is quite different – economically – from many other major ecotourism sites, which are often located in poorer and relatively undeveloped regions of the world. In theory, the pluses of ecotourism are considerable, particularly to sites in developing countries:

(1) Ecotourism benefits visited sites in a number of ways. Most importantly from the visitor's point of view, through park admission fees, guide fees, etc., ecotourism generates money locally that can be used directly to manage and protect wild areas. Ecotourism allows local people to earn their livings from areas near their homes that have been set aside for ecological protection. Providing jobs and encouraging local participation is essential because people will not want to protect the sites, and may even be hostile toward them, if they formerly used them (for farming or hunting, for instance) to support themselves but are no longer allowed such use. Finally, most ecotour destinations are in rural areas, regions that ordinarily would not warrant much attention, much less development money, from central governments for services such as road building and maintenance. But all governments realize that a popular tourist site is a valuable commodity, one that it is smart to cater to and protect.

(2) Ecotourism benefits education and research. As people, both local and foreign, visit wild areas, they learn more about the sites – from books, from guides, from exhibits, and from their own observations. They come away with an enhanced appreciation of nature and ecology, an increased understanding of the need for preservation, and perhaps a greater inclination to support conservation measures. Also, a percentage of ecotourist dollars are usually funnelled into research in ecology and conservation, work that will in the future lead to more and better conservation solutions.

(3) Ecotourism can also be an attractive development option for developing countries. Investment costs to develop small, relatively rustic ecotourist facilities are minor compared with the costs involved in trying to develop traditional tourist facilities, such as modern beach resorts. Also, it has been

estimated that, at least in some regions, ecotourists spend more per person in the destination countries than any other kind of tourists.

Ecotravel Ethics

A conscientious ecotraveller can take several steps to maximize his or her positive impact on visited areas. First and foremost, if travelling with a tour group, is to select an ecologically committed tour company. Basic guidelines for ecotourism have been established by various international conservation organizations. These are a set of ethics that tour operators should follow if they are truly concerned with conservation. Travellers wishing to adhere to ecotour ethics, before committing to a tour, should ascertain whether tour operators conform to the guidelines (or at least to some of them), and choose a company accordingly. Some tour operators in their brochures and sales pitches conspicuously trumpet their ecotour credentials and commitments. A large, glossy brochure that fails to mention how a company fulfills some of the ecotour ethics may indicate an operator that is not especially environmentally concerned. Resorts, lodges, and travel agencies that specialize in ecotourism likewise can be evaluated for their dedication to eco-ethics. Some travel guide books that list tour companies provide such ratings. The Ecotourism Society, an organization of ecotourism professionals, may also provide helpful information (USA tel: 802–447–2121; e-mail: ecomail@ecotourism.org; www.ecotourism.org).

Basic ecotour guidelines, as put forth by the United Nations Environmental Programme (UNEP), the International Union for Conservation of Nature (IUCN), and the World Resources Institute (WRI), are that tours and tour operators should:

(1) Provide significant benefits for local residents; involve local communities in tour planning and implementation.
(2) Contribute to the sustainable management of natural resources.
(3) Incorporate environmental education for tourists and residents.
(4) Manage tours to minimize negative impacts on the environment and local culture.

For example, tour companies could:

(1) Make contributions to the parks or areas visited; support or sponsor small, local environmental projects.
(2) Provide employment to local residents as tour assistants, local guides, or local naturalists.
(3) Whenever possible, use local products, transportation, food, and locally owned lodging and other services.
(4) Keep tour groups small to minimize negative impacts on visited sites; educate ecotourists about local cultures as well as habitats and wildlife.
(5) When possible, cooperate with researchers; for instance, Costa Rican researchers are now making good use of the elevated forest canopy walkways in tropical forests that several ecotourism facility operators have erected on their properties for the enjoyment and education of their guests.

Committed ecotravellers can also adhere to the ecotourism ethic by patronizing lodges and tours operated by local people, by disturbing habitats and wildlife as little as possible (including fish and other coral reef wildlife, not to mention the coral reef itself!), by staying on trails, by being informed about the historical and present conservation concerns of destination countries, by respecting local cultures and rules, by declining to buy souvenirs made from threatened plants or animals, and even by actions as simple as picking up litter on trails.

Chapter 2

Hawaii: Natural History, Geography, and Climate

Essentials of Hawaiian Natural History (or, What Every Ecotraveller Should Know About the Place)

If you were interested in animal and plant life but had never visited small oceanic islands, you would quickly notice, upon your arrival in a place such as Hawaii, that something was wildly amiss. First of all, the flora and fauna would be a lot less diverse than what you were accustomed to back home. There would be many fewer species, with fewer groups represented; for instance, there may be no frogs or snakes, or even no amphibians or reptiles. Second, a significant proportion of species on the islands would be unique to the place; they would occur on that island or group of islands and nowhere else on the planet. Third, if the island is populated, or has a history of human habitation, many of the native plants and animals would probably be rare and even threatened with extinction (and, almost certainly, some would have recently become extinct). Hawaii, owing to its location and history, exemplifies these traits.

For an ecotraveller, understanding some of the basics of island biology will strongly enhance any Hawaiian trip. Four main features of island biology to keep in mind are listed here and then explained in more detail. I discuss mainly animals, but the same forces and processes affect plants.

(1) Because of the mid-ocean locations of islands such as Hawaii, only a few land creatures successfully colonize them.
(2) Owing to the presence on freshly formed islands of habitats empty of or only lightly filled with competitors, the colonizing animals change, or evolve, rapidly into new types, so that many of the species seen today on islands occur nowhere else.
(3) These new island species are in a sense "protected" by their continued geographic isolation, and the lack of contact with mainland species that such isolation brings.
(4) Exposure of island plants and animals to species accidentally or purposefully brought to the islands by people often leads to rapid declines in the populations of the native species, causing many to go extinct. Much of the conservation effort in Hawaii today is concerned with reducing the exposure of native organisms to the human-brought, or non-native, species, and trying to reverse some of the population declines that the exposures created.

Geographic isolation and the colonization of islands. The main feature of the Hawaiian Islands determining the species that naturally occur there is their geographic isolation. Few terrestrial sites on Earth are more isolated. The Hawaiian Island chain arose relatively recently (starting perhaps 6 million years ago) in the middle of the ocean owing to underwater volcanic activity and is now more than 3000 km (1850 miles) from the nearest island neighbors (the Marquesas), more than 4000 km (2500 miles) from North America, and more than 6000 km (3700 miles) from Japan; from a conventional mainland perspective, the islands truly sit in the middle of nowhere. Geographic position largely determines which species occur naturally on the islands because only those animals and plants that could cross thousands of kilometers of open ocean by themselves made it to the islands and had a chance to become established (see Close-up, p. 100). Some birds? Yes, a few species made the crossing and became established. Terrestrial mammals? No – that's a long swim. The flying mammals? Yes, apparently one bat species landed on the islands, survived, and prospered. Snakes? No. Frogs? No. Plants with tiny wind-borne seeds or with seeds that birds transported in their digestive tracks or in the mud on their feet? Yes, many. Freshwater fish? No way. You get the idea – it's a very special, very select group of organisms that you find on isolated islands such as Hawaii. There are fewer types of animals and plants naturally represented on islands for the simple reason that many types, because of their biology (their size, their method of reproduction, the way they locomote, etc.), or owing to chance, never made it to the islands' shores, or, if they did, never became successful enough to become firmly, permanently established.

The high degree of endemism among island species. The selective process of island colonization explains the relative lack of *biodiversity* (the different types of animals, plants, and other life forms found within a region) on isolated islands such as the Hawaii group. But why should these islands have large numbers of *endemic* species – those that occur nowhere else (see Close-up, p. 67)? If these islands are colonized primarily by species from far-away mainland sites, shouldn't they share all or most of their species with the mainland areas? The answer is that after a small subset of mainland species finally reach small, isolated islands, the physical setting and the fact that the islands started as lifeless mid-ocean volcanic rock, exert powerful fast-acting forces on the organisms, leading to rapid changes that create new species. In effect, evolution is speeded up, a single successful colonizing species sometimes rapidly diverging into several new species as, with little or no competition, it fills previously empty habitats and ways of life, or "ecological niches." The result (known in evolutionary biology circles as an *adaptive radiation*) is a set of new species, all of which differ slightly from the initial colonizer, and which are unique to their new island homes. (See p. 54 for a bit more detail.) On the Hawaiian Islands, which have a great variety of environments, or habitat types, into which new species can spread, the consequences of these processes are profound: 98% of the naturally occurring bird species, 95% to 99% of the insects, and up to 90% of the flowering plants, for example, are endemics.

Island species are protected by their isolation, but highly vulnerable when placed in contact with non-native species. Unfortunately, the small-island isolation that permits rapid evolution of many unique species also makes them highly vulnerable to environmental threats and extinction. Simply put, these island species are protected for many generations in a world of low competition for resources (because of the naturally low number of species), and in a world lacking many of the dangers that other, similar species on the mainland face routinely. But if exposed to stiff competition for resources, many native island species would rapidly lose out, populations would crash, the species would become rare, then threatened, then extinct. If exposed to previously unknown dangers – predation dangers, diseases – the island species, lacking natural defenses, would quickly succumb. So, island species of animals (and plants) are fairly safe, as long as new species that are competitors for food or space, or are predators, or bring diseases, do not colonize the islands.

Great portions of native Hawaiian wildlife have recently died out as a result of exposure to non-native species and the threat is ongoing; people are largely to blame. Native Hawaiian species have been disappearing at an alarming rate ever since people reached the islands, bringing new species of plants and animals with them. Species transported by people long distances to sites they would probably never reach on their own are said to be *introduced* to the new area. People may, for various reasons, intentionally introduce organisms. But many others, such as rats and mice, ants and cockroaches, are accidentally introduced when, for example, they are transported in food-storage areas of ships. As a result of such introductions, and the spread of the introduced, or *alien*, species, it is estimated that about half of Hawaii's native bird species (at least 40 species) became extinct between the time Polynesian people reached the islands perhaps 1600 years ago and the time Europeans arrived, 200 years ago; that half the remaining bird species have died out since the Europeans arrived; and that half the still-living bird species are currently threatened with imminent extinction. The same could be said of native land snails and insects. At least 10% of native Hawaiian plants have become extinct during the past 1600 years, about 275 species in the last 200 years, and many are now endangered.

How do alien species negatively affect native species, causing their populations to crash and eventually causing their extinctions? Aliens can harm natives directly or indirectly (see Chapter 5 also). Wild domestic cats, introduced to the islands as pets, and mongooses (p. 172), introduced to control rodents in agricultural fields, directly harm native bird species via predation. They kill many ground-nesting birds, including Nene (Hawaiian Goose), Hawaiian Duck, Hawaiian Coot, Hawaiian Stilt, and shearwaters, both adults and young (mongooses also eat eggs). There are many types of indirect harmful influences, such as:

(1) Aliens compete with natives for the same food. For example, the now-ever-present tiny alien songbird, the Japanese White-eye (p. 145), competes for fruit, nectar, and insects with such Hawaiian native birds as Elepaio and Amakihi.
(2) Introduced pigs and goats eat or otherwise damage or destroy plants that some native forest birds depend on. Several recent extinctions of Hawaii's *honeycreeper* species, and significant reductions in others, are thought to have been contributed to by the destruction by pigs and goats of nectar-producing plants.
(3) Alien species often bring with them diseases for which native species have no defenses or immunity. Along with people came mosquitos, which brought diseases such as *avian malaria*. Several native bird species probably succumbed to this disease and even today, many species that in the past occurred at a range of elevations, now are restricted to areas above 1000 to 1200 m (3300 to 4000 ft), where the relative coolness eliminates the mosquito threat.

The reason that island species are so sensitive to alien introductions, and can be so quickly harmed and endangered, is that having lived in isolation for thousands of years, in environments free from many competitors, free from many kinds of predators, and free from many diseases, they never evolved appropriate defenses or behaviors to deal with dangerous situations, or, if their colonizing ancestors had such defenses, they were lost through generations of non-use.

Knowing how alien species harm natives permits conservation biologists to begin trying to reverse some of the aliens' harmful effects, and to try to prevent future introductions; indeed, Hawaiian conservation biology concentrates, in one sense or another, on the relationships between native and alien species, and on increasing the former and decreasing the latter (see Chapter 5). In fact, a result of all the research and media emphasis in the islands on the harm done to native species by aliens is that, increasingly, in a morality play of nature, native species have come to be regarded as "good" plants and animals, with aliens playing the part of evil-doers. Visitors to the islands are exposed to Hawaii's ecological vulnerabilities even before landing there. A state Department of Agriculture form that tells of the dangers of introducing more non-native species into Hawaii must be filled out while en route, declaring the kinds of agricultural items, plants, or animals you are carrying.

A chief concern is snakes. The Hawaiian Islands are essentially snake-less (p. 100) and will try to remain that way. Snakes would be dangerously efficient predators on bird eggs and young, and on adult birds sitting on their nests. Snakes becoming firmly established on the islands, biologists generally agree, would cause ecological Armageddon for many of the few surviving native birds. The most immedate fear is that the Brown Tree Snake (*Boiga irregularis*), a large (to 2.5 m, 8 ft, long), night-active, mildly venomous native of other parts of the Pacific, will reach Hawaii and cause the same kind of devastation to ecological communities that it caused recently on Guam, where, as one conservation biologist puts it, the snake "stands accused of exterminating all the forest birds" (Meffe and Carroll 1997). The snake had been introduced to Guam in the 1940s. During the past 20 years, several of the snakes have been found in Hawaii alive or dead on airport runways, probably stowaways on airplanes, most likely from Guam. The state Department of Agriculture has a special phone number for people to report possible sightings (808–586–PEST), and local people, through school education programs, know that, at least in the case of Hawaii, the only good snake is a dead snake or a far-away snake. Snake-sniffing dogs routinely patrol commercial and military airports in Hawaii, seeking slithering serpents. Interestingly, state brochures on the dangers of the snake reaching Hawaiian shores emphasize that, in addition to the harm the snake would cause to native wildlife, a major problem would be expensive electricity outages, as the snakes climb utility poles and short-circuit electrical equipment (apparently a common occurrence on Guam).

In summary, and something to keep in mind as you roam Hawaii's natural areas, every terrestrial animal (and plant) species you see on your trip is in some respects ecologically special: it's a native, engaging in ongoing competition with aliens or otherwise fighting off their threats; or it's an *invader*, an introduced species, relatively recently arrived, and struggling to compete, adapt and expand into new areas.

The Main Hawaiian Islands: Geography and Features

The Hawaiian Islands (Map 1, p. 13) constitute an *archipelago*, a large group of islands. The archipelago runs from northwest to southeast in a chain approximately 2500 km (1550 miles) long, straddling the Tropic of Cancer, the northernmost limit of the tropics (p. 73). Most of the islands in the northwestern portion of the chain (called, logically enough, the Northwestern Hawaiian Islands, or the Leeward Islands), including such out-of-the-way spots as Laysan Island, French Frigate Shoals, Midway, and Kure Atoll, are very small places, barely rising above the waves. They are known mainly to birdwatchers interested in seeing the seabirds, such as albatrosses and shearwaters, that nest there, and to former military personnel, who manned bases there during and after World War II. Heavy reference books cite 132 islands in the Hawaiian chain, but most of these are small reefs, *atolls* (coral reef islands that enclose shallow lagoons), and *seamounts* (underwater mountain peaks) that barely break the surface of the ocean; only about 30 can be considered real islands. The State of Hawaii officially includes all of the Hawaiian Island chain, but the Northwestern islands are actually administered by the USA federal government, in the guise of the US Fish & Wildlife Service, as a wildlife refuge; they provide breeding habitat for sea turtles, seabirds, and Monk Seals, among other creatures. There are no full-time human residents on these small islands except the occasional scientist or government employee. The Hawaiian Islands in total encompass an area of 16,640 sq km (6425 sq miles) of usually dry land. The Northwestern chain, however, occupies less than 1% of this area, while more than 99% of it is contained within what are generally referred to as the eight

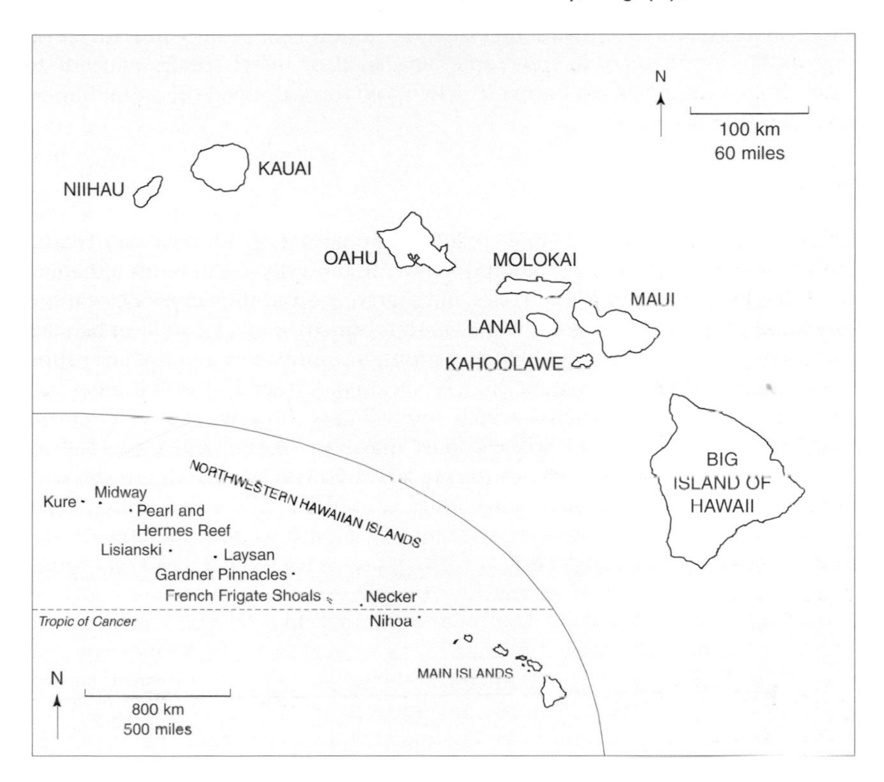

Map 1 The main Hawaiian Islands and (inset) the positions of the main islands and the Northwestern Hawaiian Islands with respect to the Tropic of Cancer.

main islands, which lie at the southeastern end of the chain, south of the Tropic of Cancer.

The eight main islands include two very small ones that few non-residents visit. Niihau, west of Kauai, with an area of about 180 sq km (70 sq miles), has a few hundred full-time residents and is essentially a preserve dedicated to the preservation of native Hawaiian culture. Kahoolawe, just south of Maui, about 116 sq km (45 sq miles) in area, is uninhabited, as befits its recent history as a military bombing practice area. Of the six islands that tourists regularly reach (The Big Island of Hawaii, Kauai, Lanai, Maui, Molokai, and Oahu), Lanai is the smallest (at 360 sq km, 140 sq miles), the least-visited, and the least treated in this book. The reason I give Lanai short shrift is that there's little of interest there from an ecotravel perspective; very little is left in the way of native Hawaiian habitats. Lanai, known as the Pineapple Isle, was essentially a Dole pineapple plantation from the 1920s through the 1980s, but today no pineapples are grown there commercially, and the island is promoted as an upscale beach resort. There are a few small sections of undisturbed native forest on the island (for instance, The Nature Conservancy's Kanepuu Preserve in the western part of the island, which protects about 240 hectares, 590 acres, of dry forest), but much of the land now consists of old, overgrown pineapple fields. There are no birds there that you cannot see on some of the other main islands (in fact, only two natives occur on Lanai, the

Short-eared (Hawaiian) Owl and Apapane). So, unless you, as an ecotraveller seeking wildlife, want to try to spot some land snail or insect species endemic to Lanai, there's not much reason to go. Skip it and instead spend the extra time on Kauai and the Big Island.

Big Island

Geography. The Big Island (Map 2, p. 32) is the largest of the Hawaiian Islands (10,450 sq km, 4034 sq miles), and probably the most diverse in terms of habitat types, landscapes, and wildlife. The island's large size and the range of elevations provided by its five visible volcanoes (a sixth is buried), converge to form habitats that include seacoast, low and high elevation rainforests, low and high elevation dry forests, subalpine regions, and high elevation desert scrub. The volcanoes that constitute the Big Island, each of which you will pass and gawk at as you tour the island, are: Mauna Loa (the world's most massive volcano, rising to 4169 m, 13,677 ft, above sea level, and constituting about 50% of the island; it is still considered active, having erupted many times since 1832, and a massive eruption could threaten Hilo with lava flows), Mauna Kea (with its peak the highest spot in the Hawaiian Islands, at 4205 m, 13,796 ft, above sea level; it's also the world's highest island peak and, if measuring from the sea floor, the highest spot on Earth, and it makes up about 23% of the island); Hualalai (2521 m, 8271 ft), Kohala (1670 m, 5480 ft), and Kilauea (1248 m, 4093 ft). The Kilauea volcano, which actually formed on the southeast slope of Mauna Loa, is currently considered the world's most active volcano, in almost perpetual eruption since 1983.

History, Economy, Eco-attractions. The Big Island was probably the first of the chain settled by Polynesians, more than 1000 years ago. Today, about 140,000 people reside on the island. Main industries are tourism, cattle ranching, and agricultural products that include macadamia nuts ("the world's richest tasting, most expensive nut"), fruit, Kona coffee, and flowers – especially orchids. Main tourist attractions are the Kona Coast (which is the dry and sunny *leeward*, or *kona*, side of the island; its western coast), especially between the Kailua-Kona area and Kawaihae (Map 2, p. 32), where self-contained upscale coastal resorts offer sun, beach, golf and water sports (including snorkeling and diving); the rustic, cattle-ranch-containing "Old Hawaii" of the Waimea area and Kohala Peninsula in the island's northern reaches; and Hawaii Volcanoes National Park in the southern portion of the island (which includes sections of Mauna Loa and Kilauea volcanoes). For the wildlife enthusiast, the Big Island's big attractions are the sites that provide access to higher-elevation forests, such as Hawaii Volcanoes National Park, the Saddle Road/Mauna Kea Forest Reserve area, and Hakalau Forest National Wildlife Refuge. It is at these protected sites, along with a few others, that several native bird species, highly endangered and endemic here, are making last stands – Palila and Hawaiian Hawk, for instance. The Big Island is also one of the few places in the state to see Nene (Hawaiian Goose), at the national park and higher spots on the Mauna Kea and Hualalai volcanoes. If pressed to rate the islands as ecotravel destinations and for their scenic beauty, I would say the Big Island is tied for first place with Kauai. (Big Island Visitors Bureau, Hilo, 808–961–5797; Kona, 808–329–7787; State Parks District Office, 808–974–6200)

Kauai

Geography. Kauai, essentially circular in shape, has an area of about 1445 sq km (558 sq miles), making it the fourth largest of the main islands (Map 3, p. 36). Kauai is formed of a single volcano, Waialeale, with top peaks called Waialeale (1569 m, 5148 ft) and nearby Kawaikini (1598 m, 5243 ft). It is the oldest of the main islands geologically, and the northernmost of the group. Owing to its northerly location and so greater exposure to moisture-bearing trade winds, Kauai has more rainfall, on average, than the other main islands and, indeed, some areas, such as the Alakai Swamp, are reputed to be some of the wettest places on the planet (occasionally with more than 1520 cm, 600 in, of rain in a year!). The Alakai Swamp, 25 sq km (10 sq miles) of mud, bogs, low-stature high-elevation rainforest, stunning landscapes, and famous for the native birds that call it home, sits on a high plateau that stretches out beneath the Waialeale summit, at elevations of 1000 to 1500 m (3300 to 5000 ft). Waimea Canyon, about 23 km (14.5 miles) long, 760 m (2500 ft) deep, and resembling a not-too-miniature Grand Canyon, lies west of the Swamp. The gorge was carved out over the past few million years by the 32 km, 20 mile, Waimea River. Much of the inland portion of the island is rather inaccessible, a wilderness forest reserve. But beautiful beaches are easily reached in several parts of the island, the breathtaking Na Pali Coast on the island's northwest side is reachable by hiking trail or helicopter, and some of the best sights, such as Waimea Canyon and Alakai Swamp, are a short car-ride and hike away.

History, Economy, Eco-attractions. Kauai, known touristically as the Garden Isle, is usually considered the most visually appealing of the main islands, with its green coastal cliffs that in places plunge more than 760 m (2500 ft) to the sea, lush, forested mountains and valleys, waterfalls, and long, wide white-sand beaches. About 60,000 people live along the coast, distributed in several small towns, mostly in the eastern and southern sections. Major sectors of the economy are tourism, retailing, and farming, including sugar, guava, papaya, coffee, and macadamia nuts. Kauai is probably the least developed of the main islands and this, together with its rich scenic beauty, make it the island of choice for outdoors-active visitors to Hawaii. The major eco-attractions are the Na Pali Coast (or Kalalau) Trail, along the steep cliffs above the island's northwestern coast; Waimea Canyon State Park, in western Kauai, with its trails and overlooks; and Kokee State Park, north of Waimea Canyon, with its splendid scenery, trails, and access to the Alakai Swamp. Kauai is also a main Hawaiian destination for birders because the island now supports the greatest number of surviving native bird species and, with its Kokee State Park trail system, permits good access to areas where some of these birds can be seen. Kauai appears to be mongoose-free, the only main island so blessed, and to the degree that mongooses have been responsible for declines and eliminations of some native birds on the other islands (p. 173), Kauai may increasingly become the island-of-last-resort for several bird species. On the Ecotravellers' Wildlife Guide quality-of-enjoyment scale, Kauai is tied for first place with the Big Island. (Hawaii Visitors Bureau, Kauai, 808–245–3971; State Parks Division Office, 808–274–3444)

Maui

Geography. Maui, the second largest of the main islands (1888 sq km, 729 sq miles), takes the form of a smaller circular land mass, called West Maui, and a much larger circular area, perhaps three times larger, East Maui, separated by a flat, narrow isthmus (the Isthmus of Maui, Map 4, p. 41). Two volcanoes form the island. The West Maui Volcano, with its peak at Puu Kukui, 1764 m (5788 ft) above sea level, constitutes West Maui, and Haleakala, with a summit at 3055 m (10,025 ft), constitutes East Maui. The isthmus, a low-lying valley-like region between the high, rugged West Maui Mountains and the dominating height and massive bulk of Haleakala, is responsible for Maui's nickname, the Valley Isle. The windward, eastern slopes of the volcanoes, fissured with deep valleys, support lush green rainforests and form a rugged, rocky coastline. The sunny western coast, along the leeward slopes of the volcanoes, provides most of the broad white-sand beaches adored by sun-worshiping tourists. Much of inland West Maui is occupied by the West Maui Mountains Forest Reserve, and is mostly inaccessible. East Maui, however, sports Haleakala National Park and some adjacent reserves, which provide good access to some stunning inland habitats, including high-elevation wet and dry forests, and to bleak volcanic landscapes.

History, Economy, Eco-attractions. Maui today has the reputation as Hawaii's second-most-developed and second-most-touristy island, after Oahu. Tourism grew rapidly on Oahu during the 1950s and early 1960s, so much so that, with Waikiki over-crowded, the state decided to shift some of the tourists, some of the development, some of the wealth, to other parts of Hawaii. Maui was selected for new beach resorts – mostly in the Lahaina/Kaanapali area of West Maui and the Kihei area of East Maui (Map 4, p. 41). Between 1970 and 1980, tourism exploded on Maui, the island becoming the second-most popular destination in the state, the resident population doubling. The population is now at about 110,000, half the people living in the side-by-side towns of Wailuku and Kahului. The island's economy depends on tourism and agriculture, including sugar, pineapple, and cattle. After heyday years of being considered "the real Hawaii" by many promoters and tourists, meaning that it provided a finer, slower Pacific paradise experience than Oahu's too-commercial Waikiki, Maui has now, in some respects, joined Oahu; many now consider it over-developed, plagued with heavy car traffic and too many humans. But it is still a beautiful place, with some wonderful eco-attractions. Foremost are Haleakala National Park, a cool, high-elevation site with extensive trails across the volcanic landscapes of the huge Haleakala Crater as well as through high-elevation forested areas; and the adjacent Nature Conservancy Waikamoi Preserve, with trails through several habitats, including highland wet forest, cloud forest, and highland dry forest. Both the park and the preserve protect some of Hawaii's least disturbed natural areas, and are excellent places to catch glimpses of some of Hawaii's native forest songbirds and others, such as Nene (Hawaiian Goose). Some of the most scenic drives anywhere can be taken along the coastal road to Hana, on East Maui, and around the rough northeast coast of West Maui. The island also represents the best place on dry Hawaiian land to watch offshore Humpback Whales; the mammalian leviathans winter in the area and often give birth and nurse their young off Maui's western coast. Consider Maui as priority number three in the list of must-visit islands for the ecotraveller, after Kauai and the Big Island. (Maui

Visitors Bureau, 808–244–3530; 800–525–6284; State Parks District Office, 808–984–8109)

Molokai

Geography. Except for tiny Lanai, Molokai is the smallest of the six main islands, roughly rectangular (Map 5, p. 45), with an area of about 674 sq km (260 sq miles). Three volcanoes form the island: East Molokai constitutes the eastern portion of the rectangle, with a highest point at Kamakou Peak (1515 m, 4970 ft, above sea level); West Molokai, the western third of the rectangle, with its high point at Puu Nana (421 m, 1381 ft); and the small volcano forming the low Kalaupapa Peninsula, which juts northwards from the island's north coast (maximum elevation 123 m, 405 ft). Much of the southern shore of Molokai is low and flat, essentially a coastal plain, whereas much of the northern coast consists of steep cliffs, some more than 1000 m (3300 ft) high, plunging to the sea. Main beach sites are in the northwestern and western portions of the island. Much of the difficult-to-access eastern portion of the island is mountainous and constitutes the Molokai Forest Reserve.

History, Economy, Eco-attractions. Molokai is a small, lightly-touristed, rural island, with a single aggregation of humans and buildings sufficiently dense that it could really be called a town (Kaunakakai), a couple of out-of-the-way tourist developments, and a few nice ecotravel destinations. The 7000 or so Molokaiians make their living mostly from agriculture and tourism. The long, broad, mostly deserted white-sand beaches on the island are its primary tourist attraction. The Nature Conservancy's Kamakou Preserve, in the island's eastern highlands, has amazing mountain vistas that appear intermittently through the ever-present mist, and hiking trails through beautiful habitats such as high elevation wet forests, bogs, and shrublands. Unfortunately, owing probably to the island's small size and destruction of much of its native habitats for cattle ranching and crop agriculture from the mid-1800s to mid-1900s, the few endemic songbirds that would have inhabited Kamakou's forests are mostly gone now, the last individuals of at least two species, the Molokai Creeper and Molokai Thrush, dying out just recently (the creeper was last seen in 1963, the thrush in 1991). The preserve, however, still contains many threatened or endangered native plants, as does the Nature Conservancy's Moomomi Preserve, a coastal beach dune reserve on the island's northwest shore. A seasoned ecotraveller might rate Molokai fourth on the priority list of islands to visit. (Molokai Visitors Association, 800–800–6367; within Hawaii, 808–553–3876 or 800–553–0404)

Oahu

Geography. Oahu, the political and economic center of Hawaii, is an irregularly shaped island of approximately 1550 sq km (600 sq miles), with the state's capital and largest city, Honolulu, lying along its southeastern coast. Oahu is the third largest island in the chain (Map 6, p. 48). It is formed of two volcanoes that now make up two sets of mountains: the Waianae Range, running northwest to southeast in the island's western sector, with the island's highest point at Kaala Peak (1226 m, 4025 ft); and the Koolau Range, in the eastern sector, running roughly parallel to the Waianae Range, with its highest peak at Konahuanui (960 m, 3150 ft). Both ranges contain rugged areas of largely inaccessible mountainous terrain with high ridges above narrow, steeply formed verdant valleys. The low-lying region between mountain ranges (the Schofield Plateau and adjacent coastal plain), including Honolulu, is the most developed area, and where most island residents live. The windward (eastern) side of the island is the wetter side, the leeward (western) side is the sunnier, warmer side.

History, Economy, Eco-attractions. Oahu is mostly known for its beach resorts at Waikiki, for being the main Hawaiian island in terms of people, politics, and commerce, and for being the most developed of the islands, but it also has a number of easily reached, excellent eco-attractions, some of them right in the Honolulu area. Historically, the low-lying Pearl Harbor area on the island's south shore was a place for residents of the islands to congregate for ceremonies and festivals; consequently, the island's nickname is "The Gathering Place." Most believe that too many have now gathered here. The island's population is nearing a million people, with almost half of them concentrated in the Honolulu area. Residents work in the tourism industry, for the federal and state governments (especially the USA Department of Defense), in business wholesaling and retailing, and in agriculture – pineapples and coffee are main crops. The island, and particularly the Honolulu/Waikiki region, has all the problems associated with cramming huge numbers of people into a small area – dense housing, high-rise buildings, heavy vehicular traffic, multi-lane freeways, ugly urban landscapes. Still, Oahu is where millions of tourists arrive each year in the state and where many of those people either spend their entire visits (mostly package tours to Waikiki) or at least a portion of their trips. Despite the crowds, a good time can be had on the island, which sports some great beaches, some of the world's most challenging surfing and wind-surfing shorelines, diving and snorkeling spots (with Hanauma Bay, near Waikiki, a popular tourist snorkeling site, having what must be the world's most over-watched fish) and cultural sites such as the heavily visited war memorials at Pearl Harbor Naval Base. There are ways to access the wild, higher-elevation parts of Oahu's mountains (p. 48) and try to see some of Hawaii's endemic forest songbirds. And there are some nice forest trails (p. 49) near Honolulu that allow visitors to hike to overlooks that provide wonderful views of some of the island's magnificent mountain scenery. Still, by reason of its heavy development and large population, I would say Oahu ranks last among the five larger islands as an ecotravel destination. (Hawaiian Visitors Bureau, 808–924–0266; State Parks Division Office, 808–587–0300.)

Climate and Best Time to Visit

The Hawaiian climate is tropical, the days usually warm and pleasant year-round. Average daily temperatures differ in winter and summer by only a few degrees. Breezy winds are common all year, often providing a degree of relief from a sometimes overly hot sun. Remember that, in contrast to temperate regions, where season largely determines temperature, in tropical Hawaii, elevation has the most important effect – the higher you are, the cooler you will be. For instance, on the Big Island in August, the average daily temperature in coastal Hilo, at sea level, is

26° C (76° F); at Kilauea in Hawaii Volcanoes National Park, at 1210 m (3970 ft), it is 18° C (64° F); and at the Mauna Loa Observatory, at 3400 m (11,150 ft), it is 8° C (47° F).

Rain is frequent, and annual rainfall totals distressingly high in some areas of great interest to ecotravellers (at Kauai's Kokee State Park, for instance). But keep in mind that, even during the rainier part of the year, winter, seldom does it rain all day: a lot of the rain falls as brief showers. On trails in higher-elevation areas (that is, almost anywhere away from the beach), passing showers can soak you at almost any time of year, any time of day. Generally, the *windward* coasts of the islands, the sides where the moisture-bearing winds first contact land (generally the east coasts), are wetter and rainier; and the *leeward* coasts, the opposite sides of the islands (the west coasts), are warmer, drier and sunnier.

There really is no "best" time to visit Hawaii – it's great all year. It's a little hotter there in summer, a little wetter in winter. Find a period when low airfares are offered, and go. If you want to try to avoid the biggest crowds, don't visit from November 15 through January 15, the period encompassing the big USA winter holidays.

Midway Atoll

A fairly new ecotravel destination in the region is Midway Atoll National Wildlife Refuge. (An *atoll* is a coral reef island that encloses a shallow lagoon.) Midway Atoll (Map 1, p. 13), in the far reaches of the Northwestern Hawaiian Islands, 2200 km (1300 miles) from the main Hawaiian Islands, consists of three small, sandy islands, only one of which (Sand Island) is large enough and rises far enough above the waves to support a permanent human colony. Until recently, it was a military base; the US Navy moved out of Midway in 1996 (after an extensive environmental clean up) and the US Fish and Wildlife Service (FWS) moved in. Now officially a wildlife refuge, Midway provides protected feeding, breeding, and roosting habitat for numerous types of animals, including seabirds, sea turtles, and monk seals. FWS contracts with a private company to operate tourist facilities, with profits helping to defray the costs of running and maintaining the refuge. Visitors, either on tours or independently, fly from Oahu and land on the 2.5-km-long (1.5-miles) island for week-long or half-week stays in a hotel that still resembles the military barracks it once was. The tourist population is limited to 100, but usually it is much smaller. Available activities are snorkeling and diving the reef, watching feeding Green Sea Turtles in the lagoons, keeping a look-out for monk seals (a small population resides there) and spinner dolphins, sport-fishing, touring in rented motorized golf-carts and, of course, the primary reason most people go, seabird-watching. About 15 species occur at Midway, including albatrosses, tropicbirds, noddies, shearwaters, petrels, and boobies. The world's largest Laysan Albatross breeding colony (about 400,000 pairs) occurs there. Aside from seabirds, however, the only other avian life is a few species of shorebirds, the Common Myna (which may or may not have found the island on its own), and Island Canaries (which, the story goes, were first brought to the island by navy wives, to keep them company when their husbands were at sea; but, because the small yellow birds have been on Midway since 1911, the story cannot be true). Best weather occurs from May to October (cooler and windier in other months);

Laysan and Black-footed Albatrosses breed from November through June; albatrosses are largely absent from July through October. (Midway Atoll, Hawaii office: 808–245–4718; US Fish and Wildlife Service: 808–541–1201; 808–541–2749.)

Chapter 3

Habitats and Common Vegetation

by Sarah Reichard
Ecosystem Science Division
University of Washington, Seattle

- Remoteness
- Geology
- Climatic Extremes
- Topography
- Vegetation Types
- Environmental Close-up 1. Lava Flows: When Volcanoes Alter the Real Estate Market (by Matti Rossi)

Hawaii's native vegetation, simply put, is like no other. The islands' remoteness, geology, climatic extremes, and varied topography have worked together to produce a grouping of native plants, or *flora*, not found anywhere else in the world. Such species are said to be "endemic" (see Close-up, p. 67). With 89% of its native species being endemic, the rate of plant endemism is higher in Hawaii than anywhere else in the world. Unfortunately, these species are often naturally rare, and rampant residential and commercial development and the introduction of aggressive non-native species have increased their rarity. Along with having high rates of endemism, Hawaii also has the most species of threatened plants in the USA. Seeing these rare plants, or even the more common of the approximately 1000 native species, often requires effort on the part of the visitor.

Most of the plants that the average visitor will see are not native, especially in lowland wet forests. Early Polynesian settlers introduced a few non-native species, and Europeans have introduced thousands of plant species in the 200 years or so since their first arrival in the islands. Many introduced plants are not a problem, but many others are extremely aggressive and significantly alter the vegetation where they invade. For instance, a vine known as banana poka (*Passiflora mollisima*; Plate 10) covers trees on several islands, competing with the trees beneath the vines for light and probably increasing storm damage to the trees because of the vines' weight. Strawberry guava (*Psidium cattleianum*; Plate 10) actually poisons other plants because of chemicals found in its decomposing leaves. Other introduced species may actually change ecosystems. Firetree (*Myrica faya*; Plate 9) has nodules on its roots containing bacteria that change nitrogen in the air to a form that the plant can use. Nitrogen is also leached into the surrounding soil. A recent study found that recent lava flows that are invaded by firetree contain several times the amount of nitrogen than flows without firetree. The "extra" nitrogen causes early successional native trees to grow and die faster, leading to changes in the rate of succession into more mature forest. Biologists now recognize that aggressive non-native species, such as the ones mentioned here, are one of the most serious threats to native plants.

Remoteness

The Hawaiian Islands are the most remote island chain in the world. As volcanic islands, they rose from the sea as barren rocks. All species had to arrive there by water, wind, or wing (see Close-up, p. 100). Because of Hawaii's remoteness, relatively few species probably came by wind and, if ocean currents in the past were as they are today, probably few species came by sea. In fact, the greatest number of plant seeds may have arrived with passing migratory birds, either within bird digestive tracts, or stuck to wings or feet. Ray Fosberg, a noted expert on the Hawaiian plants, hypothesized that the islands' native flora has derived over millions of years from the successful establishment of only 272 plant species. Most of Hawaii's plants have affinities to other South Pacific species, so that region was probably the major source of the seeds that colonized Hawaii. A somewhat ironic twist is that some of these plants, which apparently have super powers of dispersal (successfully colonizing Hawaii and other remote island groups), often lose their ability for long-distance dispersal after they arrive. Through evolution, they may lose special mechanisms that aid in dispersal. For instance, fruits in the genus *Bidens* (Aster family) have barbs that attach to fur, wings, and clothing when found anywhere else in the world, but in Hawaii they lack barbs.

Geology

The Hawaiian Islands are actually the tops of large volcanoes. The basaltic rock that forms the islands is gradually broken down by weathering to produce soil; in high rainfall areas the weathering occurs more rapidly. These soils are conspicuously red (sand and some chemicals leach out and what remains is iron and aluminum oxides) and are generally low in nutrients, although plants do grow in them. In areas with high rain and poor drainage, bogs may form. Dead organic matter accumulates in these bogs and forms mats on which vegetation often grows. In drier areas, the breakdown of the lava rock occurs more slowly, and large, well drained cinder fields are formed. The various soil types support different kinds of plants.

Climatic Extremes

Visitors love Hawaii for its mild tropical climate but, in reality, the climate is quite varied. Because the mountains can be very high, temperatures at the top may be cold, the summits even receiving snow. Rainfall also varies extensively, even over very short distances. The reason for this mainly relates to topography: when moisture-heavy trade winds hit the mountains they either go around them or rise over them. As the air rises it cools, losing its moisture. On the islands' *windward* sides (where winds first contact land) rainfall can be quite high, especially in the mountains. The *leeward* sides (the opposite sides) get much less rain because the air that passes over the mountains has already lost its moisture. These leeward *rain shadow* areas can be quite extensive – such as over much of the western portion of the Big Island.

Topography

Hawaii's volcanoes, slowly weathering, are not smooth, classic-looking volcanic cones. Rather, their varied topography includes steep slopes, sharp ridges, and deep gulches, or valleys. These "rough" topographical features can act as barriers to plant breeding. Hence, it is very common to find plants of the same species that live fairly close to each other geographically but cannot exchange genes because pollinators won't move between them (across the ridges or gulches) and their seeds cannot disperse that far. In some groups of Hawaiian plants this *reproductive isolation* has led to related but different species evolving in each separate valley. The topography also contributes to the variety of weather, as described above.

Vegetation Types

Lowland Wet Forests

Lowland wet forests are found on all the main islands and can be considered to extend from about 100 to 900 m elevation (300 to 3000 ft; and even a bit higher in some cases. Note: Using this book's definitions of the altitudes at which various animal species occur, given on p. 77, these "lowland" wet forest types can occur at low and middle elevations.). Rainfall in these areas can be as high as 500 cm (200 in) per year. Most of these wet forests have been highly altered by human activity and most of the plants and animals found there now are actually non-native invasive species. Feral pigs (p. 174) are one of the most destructive elements (pp. 57, 175). They dig up plants such as tree ferns, killing the plants but also disturbing the soil, which makes it easier for non-native plants to become established.

Lowland wet forests are characterized by dense growth and rotting logs. The canopy can be quite high, up to 40 m (130 ft). Usually there is a well developed understory. Trees in these forests generally support many *epiphytes* (plants, such as ferns, mosses, and orchids, that grow on other plants) and many also support vines. Among the common species found in these forests are the versatile 'ōhi'a lehua (*Metrosideros polymorpha*; Plate 9), tree ferns (*Cibotium* spp.; Plate 15), koa (*Acacia koa*; Plate 6), *Psychotria* spp. (Plate 12), and kawa'u (*Ilex anomala*; Plate 1). Common non-native species include African tulip tree (*Spathodea campanulata*; Plate 3), golden pothos (*Epipremnum pinnatum*; Plate 13), shoebutton ardisia (*Ardisia elliptica*), rose apple (*Syzygium jambos*; Plate 16), and mountain apple (*S. malaccense*; Plate 10).

Lowland Dry Forests

It is hard to say what Hawaiian lowland dry forests were like before humans arrived on the scene. They have been heavily disturbed for more than a thousand years; Polynesians burned them, and now they are often the sites for resorts and golf courses. They occur from about 300 to 900 m (1000 to 3000 ft) elevation on the leeward sides of the larger islands, in rainforests that have well drained, rocky soils, and over much of some smaller islands such as Lanai and Niihau. They are generally very open forests and the canopy is low. They have a less layered structure than lowland wet forests. Tree species are generally not as tall as in lowland wet forests, there is a strong shrub component, and there are few herbaceous species, ferns, or mosses. The most accessible and least disturbed places to see these forests are on some of the southern slopes of East Maui (Auwahi).

Hawaii's dry forests tend not to get much attention from the average visitor, but plant species here may be more interesting than in any other habitat type in the islands. About 22% of the native species occur in this zone, including the ever-present 'ōhi'a lehua (*Metrosideros polymorpha*; Plate 9), the native persimmon, or lama (*Diospyros sandwicensis*; Plate 5), sandlewood species (*Santalum* spp.; Plate 12), the beautiful wiliwili (*Erythrina sandwicensis*; Plate 6), and the native cotton, or ma'o (*Gossypium tomentosum*; Plate 16). Unfortunately, non-natives are also common, such as koa haole (*Leucaena leucocephala*; Plate 7), Brazilian pepper (*Schinus terebinthifolius*; Plate 1), and fountain grass (*Pennisetum setaceum*; Plate 14). The last species jeopardizes dry forests because of its tendency to burn easily; native Hawaiian plants are not adapted to survive these kinds of fires.

Middle and Higher Elevation Forests

Middle elevation forests are generally wet, but the rainfall is not quite high enough to produce a true "wet forest." The canopy is closed and is usually not higher than about 25 m (80 ft). There is a well developed shrub layer. There are also middle elevation forests on the dry sides of the islands, but they have been almost all converted to pasture for grazing.

Middle elevation forests are among the most diverse and rich habitats in the islands, and many of Hawaii's very rare native species are found in this zone. The older islands of Kauai and Oahu have the best examples of these forests. Interesting native species found in these forests include pukiawe (*Styphelia tameiameiae*; Plate 5), 'ōhelo (*Vaccinium reticulatum*; Plate 5), maile (*Alyxia oliviformis*; Plate 1), 'ie'ie (*Freycinetia arborea*; Plate 13), māmane (*Sophora chrysophylla*; Plate 14), and, of course, 'ōhi'a lehua (*Metrosideros polymorpha*; Plate 9), as well as other *Metrosideros* species. Open areas may be filled with 'uluhe (*Dicranopteris linearis*; Plate 15), a sprawling, thorny fern. Non-natives are also present, especially strawberry guava (*Psidium cattleianum*; Plate 10), banana poka (*Passiflora mollisima*;

Plate 10), the weed *Buddleia asiatica*, thimbleberry (*Rubus rosifolius*; Plate 11), and kahili ginger (*Hedychium gardnerianum*; Plate 14).

A particular kind of native forest, largely composed of only two tree species, is known as māmane-naio forest. These forests are characteristic mostly of some higher-elevation sites on Maui and the Big Island (the other islands not rising high enough to support the habitat type), and the two brushy trees – māmane (*Sophora chrysophylla*; Plate 14) and naio (*Myoporum sandwicense*; Plate 9) greatly predominate here.

High Elevations Above Treeline

Above treeline in Hawaii, mostly on Maui's Haleakala and the Big Island's Mauna Kea, Mauna Loa, and Kilauea, are striking subalpine scrub habitats dominated by grasses and low-stature shrubs. Among the common shrubs are pukiawe (*Styphelia tameiameiae*; Plate 5), ´ōhelo (*Vaccinium reticulatum*; Plate 5), and Maui na'ena'e (*Dubautia menziesii*; Plate 2). True alpine zones exist around the top sections of Hawaii's highest peaks – Haleakala and the Big Island's Mauna Kea, Mauna Loa, and Hualalai – and it is largely in these cold, rocky, windy areas with little soil that Hawaii's famed and endangered ball-of-spikes silversword (*Argyroxiphium sandwicense*; Plate 2), a member of the sunflower family (Asteraceae), occurs.

Forest Edge and Streamside

Forest edges and habitats along rivers and streams are generally disturbed and usually contain many non-native plants. Some species seen in these habitats are shoebutton ardisia (*Ardisia elliptica*), kahili ginger (*Hedychium gardnerianum*; Plate 14), strawberry guava (*Psidium cattleianum*; Plate 10), and 'uluhe fern (*Dicranopteris linearis*; Plate 15). *Eucalyptus* trees (Plate 9) are also found in these areas.

Open Habitats

Open grass areas and shrublands are found on the leeward sides of all the main islands and also in the Big Island's southern reaches. Many of the plants in these areas, subject to frequent burning, have now been replaced by non-natives. Some non-native species that invade these areas, such as fountain grass (*Pennisetum setaceum*; Plate 14), create a large flammable biomass and are actually adapted to burn. Early Hawaiians burned these areas to induce the growth of native pili grass (*Heteropogon contortus*; Plate 14), which they used for thatch. However, fountain grass now dominates these grasslands to a great degree. Some native shrubs are still found in the rocky soils of these habitats, including lama (*Diospyros sandwicensis*; Plate 5), naio (*Myoporum sandwicense*; Plate 9), māmane (*Sophora chrysophylla*; Plate 14), and 'ulei (*Osteomeles anthyllidifolia*; Plate 16); non-native species such as koa haole *Leucaena leucocephala* (Plate 7) are also common. In moister areas, 'uluhe fern (*Dicranopteris linearis*; Plate 15) may be dense.

Freshwater

There is relatively little freshwater habitat in Hawaii. There are some small spring-fed pools and streams and there are drainage areas that may have high water after rainfalls. Many streamside sites have been converted to taro patches (*Colocasia esculenta*; Plate 13). In wet gulches there is generally an understory of ferns and mosses, as well as species of *Clermontia* (Plate 3) and *Cyanea* (Plate 3). There are also wet grasslands and sedge communities that tend to grade into wet shrubland. In high elevation areas where rainfall exceeds drainage, bogs may form. There are several bogs on Maui, but Hawaii's largest is probably the Alakai Swamp on Kauai, where there are extensive hummocks of shrubs, sedges, and grasses, as well as ʻōhiʻa lehua trees (*Metrosideros polymorpha*; Plate 9). High elevation bogs are also places to find greenswords and some silverswords (*Argyroxiphium* spp.; Plate 2).

Saltwater/Marine

Habitats from sea level to 300 m (1000 ft) around the islands are generally known as the *coastal strand*. Plants that live here must be capable of tolerating salt in both the ground water and the air. These species are mostly incapable of terrestrial dispersal and so their seeds are generally dispersed by currents and waves. Many of these areas have dunes, created through shoreline processes. Because of very frequent trade winds, most of the plants here tend to be windswept and low. The dominant dune shrub is naupaka (*Scaevola sericea*; Plate 8). There may be non-native trees such as ironwood (*Casuarina equisetifolia*; Plate 4), tree heliotrope (*Tournfortia argentea*; Plate 3) and Indian almond (*Terminalia catappa*; Plate 4). There may also be herbs such as pohinahina (*Vitex rotundifolia*) and nanea (*Vigna marina*). On drier sides of the islands there may be small trees such as kiawe (*Prosopis pallida*; Plate 7) and koa haole (*Leucaena leucocephala*; Plate 7), and pili grass (*Heteropogon contortus*; Plate 14). In moister sites there may be hala (*Pandanus tectorius*; Plate 14) and coconut (*Cocos nucifera*; Plate 13).

Environmental Close-up 1 Lava Flows: When Volcanoes Alter the Real Estate Market

by Matti Rossi
Department of Geography
University of Turku, Finland

Lava flows are the most important geological elements in Hawaii – all of the Hawaiian islands were "built" by lava flows. When layers of lava flows accumulate during thousands of years of eruptive activity, shallow-sloped volcanic edifices, or *shield volcanoes*, form. The Big Island of Hawaii consists of five shield volcanoes, two of which, Mauna Loa and Kilauea, are at present very active. Hualalai shield volcano (on the island's leeward, or kona, side) is not active but has erupted during historical times and may thus erupt again in the near future.

Lava flows form during eruptions in which molten rock pours out of a crater or fissure. Lava-forming eruptions are relatively mild in explosivity compared with the powerful eruptions that take place, for instance, in the Cascades region in the northwestern USA (think Mt. St. Helens in, oh, about 1980). The mild character of the Hawaiian eruptions follows from the great fluidity of Hawaiian lavas; in other words, the lava is not very thick, or viscous. The viscosity of lava is primarily governed by two factors: chemical composition and temperature.

Silica is a chemical substance that usually makes lava flows sticky. Hawaiian lava flows, however, are relatively low in silica. The eruption temperature of Hawaiian lava is high, typically 1100 to 1200 °C (2000 to 2200 °F). Lava flows at these temperatures are very fluid compared with those lavas that are initially a few hundred degrees cooler. The gas content of Hawaiian lava flows is low and because of the great fluidity of the lava, gas is released easily and so violent explosions are not common. However, if molten lava meets a water source such as ground water or sea water, vaporization of the water may lead to hazardous explosions. There are several examples of these so-called *hydromagmatic explosions* in Hawaii, such as those sometimes taking place in the summit area of Kilauea and at the ocean entries of the lava flows.

There are primarily two types of lava flows in Hawaii, *pahoehoe* and *aa*, which differ in their appearance and mechanism of flow. They are easily distinguished by their surface roughness: pahoehoe is smooth, ropy and gently hummocky, whereas aa is rough, covered with surface rubble and blocks of various sizes.

Pahoehoe lava flows advance more slowly than aa lava flows. Pahoehoe forms in eruptions where the rate of release of lava from the crater is low. Pahoehoe lava flows consist of small advancing *lava lobes* on which a crust forms very rapidly. Lava continues to flow within this insulating crust and remains hot and fluid. Gradually a network of lava streams forms under the crust and the subsurface pathways may widen into *lava tubes*. As the eruption proceeds, lava tubes grow longer and wider. In long eruptions lasting several weeks to several months, very large lava tubes may form, carrying lava to the advancing lava flow fronts. Because of the insulating crust and a well developed network of lava tubes, pahoehoe lava generally has potential to move over great distances, up to several tens of kilometers, or miles. Sometimes the roofs of the lava tubes collapse and hot lava is exposed, allowing volcanologists to examine the rate of lava flow and the temperature of the lava in the tubes. Often, all the hot lava drains out of a lava tube at the end of an eruption, leaving an elongated cave such as the famed, oft-visited, Thurston Lava Tube in Hawaii Volcanoes National Park.

Aa lava forms in eruptions where the rate of lava release at the vent is higher than that for pahoehoe lava formation. The margins of the lava flow are thinner than the central parts and therefore they cool and stagnate more rapidly. Soon a central lava channel forms within the flow. The amount and speed of lava in the central channel is high and the flow front moves relatively fast. Therefore, the lava crust breaks continually, exposing the hot interiors of lava to the atmosphere and rapid cooling takes place. The flow becomes viscous and sticky and generally does not have the ability to flow as far as pahoehoe flows. After the eruption, a typical aa lava flow has a surface of broken lava blocks and a central lava channel.

Many eruptions in Hawaii, such as the latest Pu‘u O‘o – Kupaianaha eruption, are long and complex and may last years or even decades. During the eruptions, many craters may be active and the rate of lava production may vary greatly. As a result, complex lava flow fields form that consist of many overlapping lava flows, with both pahoehoe and aa lava flow features.

After an eruption, a cooled lava flow is subject to atmospheric processes such as wind, rain and temperature changes, which cause weathering. The surface of the lava flow changes chemically and is fragmented physically. Rock breakage and the resulting gradually increasing release of chemical nutrients allow plants to occupy the originally hostile environment on the surface of cooled lava flows.

The rate of plant community development on cooled lava varies according to temperature, amount of rainfall, and roughness of the lava flow. In Hawaii, on east-facing hot and moist slopes, the reoccupation of plant life is more rapid than on west-facing drier slopes; similarly, on higher slopes, where temperatures are low, plant communities develop more slowly than on lower slopes. Therefore, in different locations, lava flows of the same age may look rather different in terms of degradation and plant life. Rough aa lava flows often catch wind-blown sand and organic waste, such as leaf litter, between blocks of lava. Similarly, the small cracks on the surface of pahoehoe flows may catch enough sediment to allow plants to lay roots on the tiny pockets of loose soil. Thus, wind-blown debris may play a significant role in the early succession of plant communities on lava flows. The first plants to occupy barren Hawaiian lava fields are brackens (ferns) and Ohia trees. Under hot and wet conditions, a Hawaiian lava flow only a few hundred years old may already support a dense tropical forest.

Chapter 4

Parks, Preserves and Getting Around

Getting Around in Hawaii

Driving Here and There. As you might predict for a heavily touristed part of the USA, main roads on the islands are generally kept in good shape. Except on busy, densely populated Oahu, which has some multi-lane freeways, most main roads are one lane in each direction. In certain areas at certain times – particularly near heavily visited tourist sites – this leads to very slow or stopped traffic. Oahu roads can have huge traffic jams, and it's better not to be driving anywhere near Honolulu during morning or afternoon rush hours (evening hours in and around Waikiki can also be unpleasant). Also, even when Oahu's main highways are lightly trafficked, there can be long lines of cars on the narrow two-lane coastal roads. This is especially true on weekend afternoons, when large fractions of the local population congregate at popular beach spots for family gatherings, parties, etc. Signs along main roads and in cities and towns are good and, in general, it's easy to get around. The James Bier maps, one for each of the main islands, sold at many stores, are excellent navigation aids. Here are just a few things to keep in mind:

(1) Some of the driving to more popular sites is along narrow roads that twist and turn, and that climb or descend steeply – the roads up to Maui's Haleakala National Park, and then up to Haleakala's summit, being good examples. A consequence is that sometimes on a map it looks as if the driving time between two points is 30 minutes, but the actual trip duration, given your slow speed as you drive through hairpin turns and switchbacks, is double or even triple that. Also, on some of these steep narrow roads, there are no guard-rails, so you are on your own. Some roads start out normally but then become narrow,

twisty and extremely dangerous – so narrow, that it's all but impossible to turn around. For instance, the road around West Maui is driven by many tourists, and it's a worthwhile drive. But the road narrows in places so that it's a single-lane thoroughfare – this with traffic in two directions going around blind curve after blind curve. I drove it during a rainstorm, came around a curve, only to be confronted with a 2-m-high (6-ft) boulder, apparently loosened from above by the rain, sitting smack in the middle of the road. The best advice: drive slowly and *take it easy*. Remember that many of the drivers on the road with you are tourists out for joy-rides and have no idea where they're going or what they're doing – watch them like a hawk and be careful.

(2) Gasoline prices, in comparison with those in the rest of the USA, are quite high – something about having to ship the stuff in from afar. On the other hand, the islands are small, and you will need to try fairly hard on a brief trip to any one island to go through a single tank of gas.

(3) If you want or need a 4-wheel-drive vehicle (for instance, you will need one on Molokai to visit the Kamakou Preserve), make rental reservations for them far in advance (particularly for weekend/holiday travel); they are limited in number on the smaller islands.

(4) Language. I was always dimly aware that I had trouble differentiating many Hawaiian place names; so many look alike. The cause finally became apparent to me during my research for this book: the Hawaiian language alphabet has only 12 letters, mostly, it usually seems, vowels (the lack of consonants suggests they may want to borrow some from one of those eastern European languages that seem to be chronically over-supplied). Generally, however, unless you're interested in languages and go out of your way to learn some Hawaiian, you won't need to know many Hawaiian words. However, a few are very important. Directions in Hawaii are often given using the words *mauka*, meaning inland, toward the mountain, away from the ocean, and *makai*, meaning toward the ocean. For instance, directions might take the form of "proceed 3 miles to where Route 81 ends, then turn mauka." *Kapu* is a Hawaiian word meaning prohibited, taboo, or sacred, and it is sometimes used on signs to mean No Trespassing (a surfeit of large, vocal, anti-tourist dogs will also keep you out of many inhabited areas).

(5) Eco-hint #742: GO EARLY! To enhance your eco-pleasure, I cannot stress this enough. Regular tourists rise at 08:00 or 09:00, have a leisurely breakfast, then head out for a day's activities. Ecotourists (this means you!), rise at 06:00 (yes, that's six in the morning) and get a 2- or 3-hour headstart on everyone else. The earlier morning hours are better times to see wildlife, especially birds, it's of course cooler, and you avoid most other people. In the early morning, you will enjoy a few hours of blessed isolation, even at popular sites. It's something of a perverse pleasure to be returning to your car after a relatively cool 3-hour morning hike only to meet groups of tourists just leaving the parking lot to start hiking beneath the blazing sun.

Step-by-step directions to many of the sites mentioned here can be found in some general travel guide books to Hawaii, and sometimes in more specialized travel books such as *Enjoying Birds In Hawaii*, by H. D. Pratt, and *Adventuring in Hawaii*, by R. McMahon.

Hiking advice. When hiking in higher-elevation areas, in the mountains (that is, on almost any Hawaiian hike not along a beach), it's wise to bring along very light raingear or a tiny umbrella; rainshowers often come and go all day. Also prepare for strong, bright sun (sunglasses, sunscreen, hat, etc.), and pack along lots of water. Although many people wear their beach-resort footwear on hikes, a pair of lightweight hiking boots is smarter. Trails are often rocky, wet and slippery, muddy, or all three, and the added traction and ankle-support provided by hiking boots can be important for safety. You will also find a stout walking stick helpful; find or make one per island and keep it stored in your rental car trunk.

Descriptions of Parks, Preserves, and Other Eco-sites

The parks and preserves described below were selected because they are the ones most often visited by ecotravellers or because they have a lot to offer. The animals profiled in the color plates are keyed to parks and preserves in the following way: the profiles list the islands (BIG = Big Island, KAU = Kauai, LAN = Lanai, MAUI = Maui, MOLO = Molokai, OAHU = Oahu, MID = Midway) on which each species is likely to be found, and the parks listed below are arranged by island. Park locations are shown in Maps 2 through 6. If no special information on reaching parks and on wildlife viewing is provided for a particular park, you may assume the site is fairly easy to reach and that wildlife can be seen simply by walking along trails, roads, beaches, etc. Tips on increasing the likelihood of seeing mammals, birds, reptiles, or amphibians are given in the introductions to each of those chapters.

Big Island (Map 2)

Saddle Road and Mauna Kea State Park. Saddle Road crosses old (and new!) lava flows as it bisects the island, running from Hilo to the Waimea region, with massive Mauna Kea looming to its north and Mauna Loa to the south. Even just driving the road, as it cuts past hard black lava fields below the towering volcanoes, is worthwhile. Rental car companies forbid taking their precious vehicles here, but the road is paved and adequately maintained and the rule is widely flouted; mine was certainly not the only rental car on the road. There are trails over lava flows to explore (Habitat Photo 2), a high-elevation astronomical observatory and visitor center to see, and a state park in which to rest and picnic (and where else can you pass a military base where signs along the main road say "Caution: Overhead Artillery.") Coming from Hilo, the trails begin on the south side of Saddle Road; the Powerline Road/Trail (a jeep track) starts at about mile-marker 21.5, and the Puu Oo trail (foot trail) at mile 22.5. You can park on the side of the road and hike over the rocky lava surface (this stuff tears up sneakers in a hurry – use hiking boots) to *kipukas*, forested islands on older lava flows that stick out of the broad newer lava flows; they were spared by chance when the molten lava swept around them. Some small forest birds, such as Apapane, Common Amakihi, Omao, and Hawaii Elepaio are common in the kipukas, and Hawaiian Hawks frequent the area. It's an interesting region, but it's often like

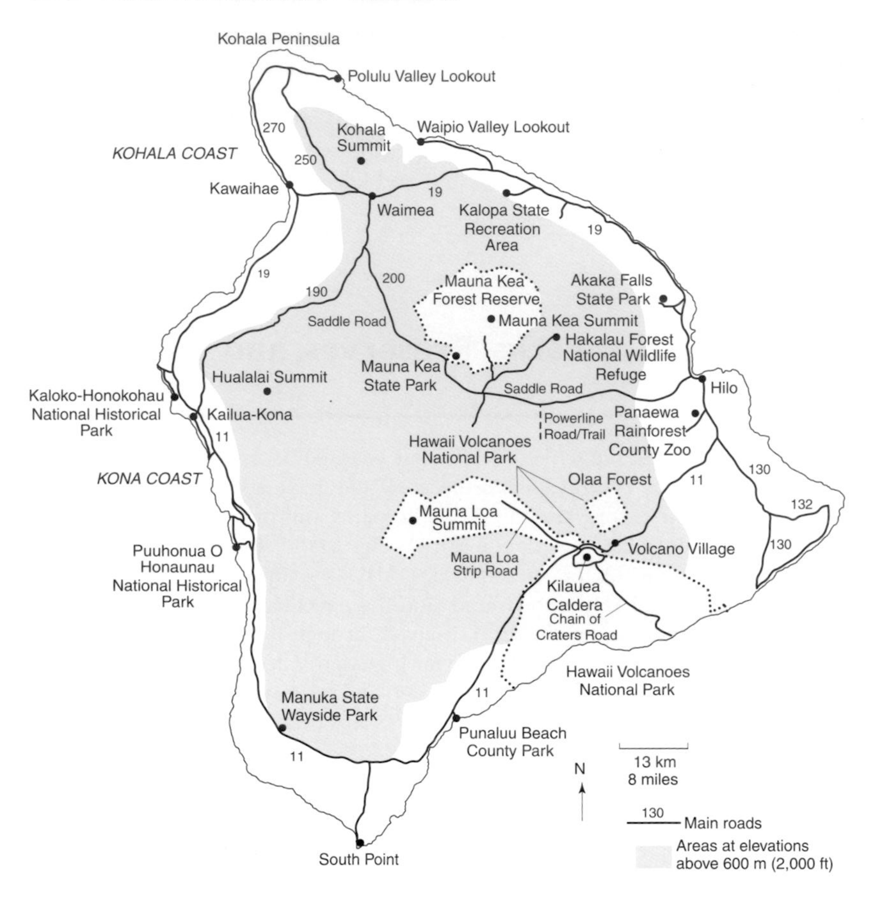

Map 2 The Big Island of Hawaii.

hiking in a hot gravel-pit; a world of gray and black rock. The Puu Oo Trail crosses into the Kipuka Ainahou Nene Sanctuary, and eventually merges with the Powerline Trail.

At the 28-mile marker on Saddle Road, a marked or unmarked good paved road leads northwards to the Mauna Kea Summit area, visitor center, and observatories. About 6.5 km, 4 miles, up the road you cross a cattle guard and pass into the Mauna Kea Forest Reserve (Habitat Photo 1). There are places soon after to pull off the road. Among the tree groves and scrub vegetation here, some native forest birds, such as Common Amakihi and Iiwi, can usually be found, as well as ground birds such as Chukar, pheasants, and francolins; Hawaiian Hawks cruise the area. You can continue up the road to a visitor center; a strenuous trail to the top of Mauna Kea (4205 m, 13,796 ft) starts near the visitor center. Returning to Saddle Road and heading west, the road crosses smooth black lava flows and at about the 35–mile marker you arrive at Mauna Kea State Park, a good place to rest, picnic under shade trees, and birdwatch (Common Amakihi, white-eyes, doves, finches, among others).

Hakalau Forest National Wildlife Refuge. This huge, largely wilderness refuge, on Mauna Kea's eastern (windward) slopes is mostly closed to the public. One section, the Maulua Tract, is open for birding, hiking, and pig-hunting on weekends, by special permission (contact the refuge office in Hilo at 808–933–6915). About 45 bird species occur in the refuge, including 13 native (7 of them endangered, including Akiapolaau, Akepa, Hawaii Creeper, Hawaiian Hawk, and the occasional Nene), 13 migratory, and 20 introduced. At least 12 species of endangered plants occur here. The refuge is about 16 km (10 miles) north of Saddle Road, along a 4-wheel-drive gravel road. (Area: 13,240 hectares, 32,700 acres. Elevation: about 730 to 1950 m (2400 to 6400 ft). Habitats: High-elevation rainforest, scrub forest, bogs, pastureland.)

Hawaii Volcanoes National Park. Simply one of the most visually arresting places you will visit in your lifetime, this park is a must-see. In addition to esthetics, this is a wonderful place for volcanological and biological reasons. Aside from the volcanoes and sprawling lifeless lava flows, the eerie dead scenery, etc., the park is an absolute ecotravel treasure. There are about 240 km, 150 miles, of well maintained trails in the park, including ones that descend into and cross huge, still-steaming volcanic craters (for instance, the popular Kilauea Iki Trail). But much of the trail system networks through native Hawaiian vegetation, including tropical rainforest (Habitat Photo 4). And owing to the fact that most of the park's visitors are car-tourists who refuse to venture more than a kilometer (half-mile) from their cars, forest trails here are often deserted. Also, much of the park is at sufficiently high elevation to take the edge off the hot Hawaiian sun; it's even cool in places. Some native forest birds are easy to see here, even without venturing far from your car. Apapanes, for instance, are all over the place (they are like zebras in Africa: great to see for the first time but, owing to their great numbers, you tire of them in about a half-hour).

Crater Rim Drive, onto which you pass as you enter the park, circles in 18 km, 11 miles, the huge Kilauea Caldera and has various points of interest along its length: viewpoints, volcanic formations, some trail access, visitor center, museum. (Nenes are sometimes sighted in this area. Last time I was there an entire parking area was closed because, according to a large orange sign, the site was now "Nene Territory," the geese having decided to nest there.) Turning from Crater Rim Drive, Chain of Craters Road descends for 39 km, 24 miles, losing 1130 m, 3700 ft, in elevation, as it runs to the island's coast. Aside from the scenery, many people make the drive to get as close as they can to the point where Kilauea's molten lava falls into the sea. The road ends where recent lava flows engulfed it. I hate to break the news, but although most people enter the park believing they will surely get near actively flowing lava (as seen on TV!), regular visitors are often kept many miles from hot lava. In fact, in 1999, when this book was written, Kilauea's lava was oozing out of a vent (called Puu Oo) on the side of the volcano and essentially flowing underground, through lava tubes, to emerge by the sea. All that can be seen from the ocean-front viewing area at the terminus of Chain of Craters Road is steam (and other, more lethal gases) rising in the far distance as the hot lava hits the cold ocean.

A good birding/hiking area in the park, and outside of the pay-to-enter section, is known as the Mauna Loa Strip Road. The entrance is off Route 11, the main highway, about 3.5 km (2 miles) west of the main park entrance. At the parking area at Kipuka Puaulu (also called Bird Park), about 2 km (1.2 miles) in, there is a very nice, 30-minute loop nature trail (Habitat Photo 3) at about 1160 m (3800 ft) elevation, where, with minimal effort, you can see Apapane, Common Amakihi, Hawaii Elepaio, Red-billed Leiothrix, cardinals, white-eyes, and pheasants, among others. It's a very quiet place and often deserted. Then, leaving your vehicle in the lot, you can walk (but not drive) more than 16 km (10 miles) farther along Mauna Loa Strip to Mauna Loa Lookout, at 2030 m (6660 ft) elevation. The deserted old road stretches through beautiful open and semi-open woodlands and shrublands. This is an absolutely gorgeous area. Recently I walked 2 hours up and 2 hours back on a cloudy cool weekday afternoon, saw no people, and had a great time.

Headquartered at Hawaii Volcanoes National Park is a full complement of government research biologists (the Pacific Island Ecosystems Research Center of the US Geological Survey's Biological Resources Division) working to conserve Hawaii's native plants and animals (for example, see Close-up, p. 158) *Warning*: Stay away from all volcanic eruptions and actively flowing lava; volcanic damage to rental cars is generally considered an act of God, not covered by most insurance policies. (Call 808-985-6000 or 808-967-7977 for updates on volcanic activity; www.nps.gov/havo.) (Area: 97,165 hectares, 240,000 acres. Elevation: sea level to 4169 m (13,677 ft) – Mauna Loa summit. Habitats: coastal zone, low, middle, and higher-elevation forests, rainforest, subalpine and alpine zones.)

Punaluu Beach County Park. Located along the main highway (Route 11) around the southern part of the island, south of the national park, this black-sand beach is popular as a nesting site with Green Sea Turtles. There are picnic and camping areas.

South Point. Widely touted as the southernmost point of the USA, the southern tip of the island is worth a short visit, even if only as a photo opportunity. Seabirds and shorebirds are sometimes in evidence. It's about 20 km (12 miles) south of the main highway along a narrow paved road through flat grazing land on old lava flows that slope down into the sea.

Manuka State Wayside Park. This small park (part of the much larger Manuka Natural Area Reserve) in the island's southwest corner is along the main highway (Route 11), so makes a good rest and picnic spot. A 4-km (2.5-mile) nature trail along an old lava flow passes through nice, shady forest with both native and non-native trees. A number of birds, including Common Amakihi, Hawaii Elepaio, cardinals, and pheasants are often seen, and skinks are common along the trail. Of interest is a side path to a "pit crater" (as it is labelled; an expert with me said it was actually part of a lava tube with a collapsed roof; see p. 27) – a large, deep hole in the ground with almost vertical rock walls and containing its own small green ecosystem with trees, ferns, and other plants. (Area, state wayside park: 5.5 hectares (13.4 acres) plus surrounding forest reserve. Elevation: about 550 m (1800 ft). Habitat: lowland moist forest.)

Polulu Valley Lookout. If you drive onto the Kohala Peninsula, on the island's northwest end, you will probably arrive eventually at the Polulu Valley Lookout. It's at the termination of Route 270. You get striking views of the beach far down below, cliffs over the ocean, and of the Polulu Valley. A steep trail leads down to the beach, about 90 m (300 ft) below. The Kohala Peninsula itself is a beautiful place, a different, isolated part of Hawaii. Route 250, through the center of the peninsula, is a narrow road that winds and climbs to a crest at 1086 m (3564 ft), and passes through cattle and horse ranches with pine-fringed pastures, some with stunning ocean views; if you had to be a horse or cow, this might be the place to do it.

Waipio Valley Lookout. West of the town of Honokaa on Route 240, north end of the island. From the end of the road you get a great panoramic view of the Waipio Valley and Waipio Bay (Habitat Photo 5). You can walk down a steep road to the beach, about 40 minutes, or pay for a ride in a shuttle van. From the beach area there is access to the Waimanu Trail, a strenuous up-and-down trail with many switchbacks, running about 14 km (8.5 miles) from Waipio to the Waimanu Valley.

Kalopa State Recreation Area. Located just 5 km (3 miles) off the main highway (Route 19), in the northern part of the island, near the town of Honokaa, this pleasant, lightly visited park has nice picnic and camping areas, as well as rental cabins (for reservations, 808–974–6200). There is a short nature trail with labelled trees and explanatory booklet, plus a more extensive trail system, along wilder, unkempt trails through towering forest – the cool, shady trails are great on hot, sunny days. There's also a small arboretum showing native plants. About 40 hectares (100 acres) of the park consists of native Hawaiian rainforest. (Area: 250 hectares, 615 acres. Elevation: 610 to 790 m (2000 to 2600 ft). Habitats: low and middle elevation wet forest.)

Akaka Falls State Park. This small, forested park is located just north of Hilo, about 5 km (3 miles) off the main highway, along route 220. You will spend probably no more than a half hour here, but it makes a nice stop. There is a 1 km (half-mile) paved loop trail through lush rainforest with heavily vined trees, and two waterfalls to gawk at, including the 135 m (442 ft) Akaka Falls. (Area: 26 hectares (65 acres). Elevation: about 365 m (1200 ft). Habitat: lowland wet forest.)

Waiakea and Loko Waka (Lokoaku) Ponds, Hilo. Waiakea Pond, in the center of the town of Hilo, is a favorite of birdwatchers for seeking water-associated birds such as marsh birds, ducks, and herons. The same types of birds can be found at Loko Waka (Lokoaku) Pond, located just behind the Hilo airport runways along coastal Route 19. The roadside pond is just opposite a beach-front park (James Kealoha County Park).

Kaloko-Honokohau National Historical Park. Located just north of Kailua-Kona on the leeward side of the Big Island, this park of 485 hectares (1200 acres) mostly preserves historical cultural sites. But a large pond there, Aimakapa Fish Pond, just inland from Honokohau Bay, is an important location for birdwatchers. The pond attracts many migrant marsh and water birds, and endangered Hawaiian Coots and Hawaiian Stilts can often be seen here. The pond is also the only breeding site in the islands for a newly established population of Pied-billed Grebes.

Puuhonua O Honaunau National Historical Park. This frequently visited park is about 30 km (18 miles) south of Kailua-Kona, on the Big Island's leeward coast. It is primarily a historical cultural site, but many consider the waters off the northern end of the park to be a great snorkeling spot.

Panaewa Rainforest County Zoo. Located just outside of Hilo on the way to Hawaii Volcanoes National Park, this small, pleasant, well kept, free zoo, is a good place to spend an hour or so. Most importantly, the zoo lets you get close views of some of the native Hawaiian animals you hear and read about but don't always

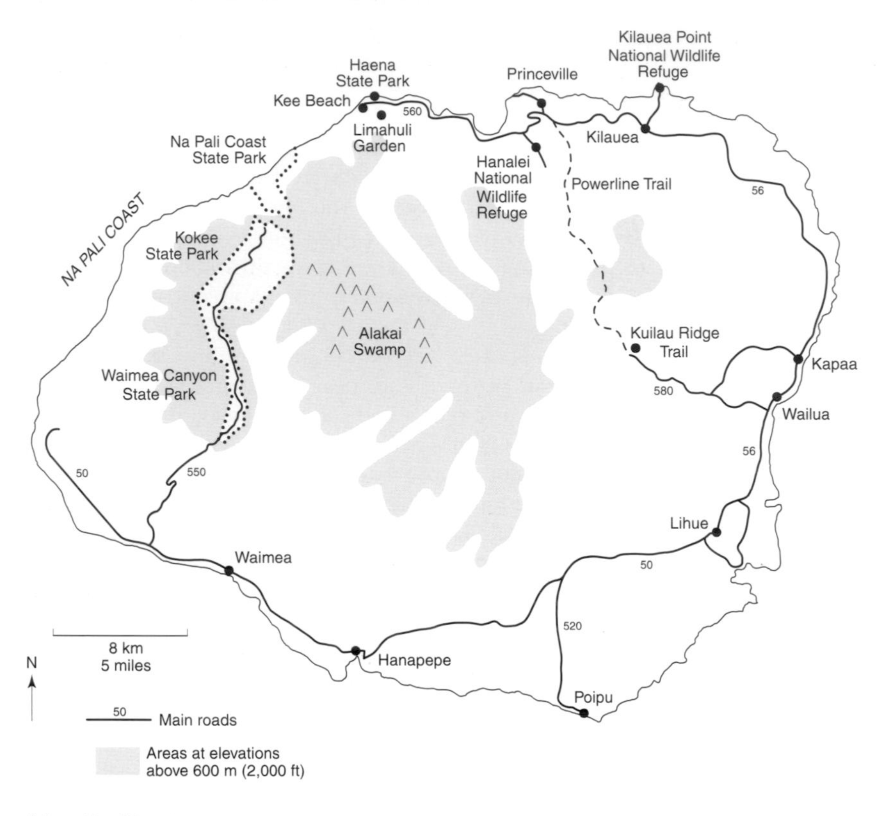

Map 3 Kauai.

see: Nene, Hawaiian Hawk, Short-eared (Hawaiian) Owl, and Hawaiian Duck. There are also some non-natives, looking slightly out-of-place (but generally well kept and happy-looking), such as squirrel monkeys, lemurs, a pygmy hippo, a tapir, vultures, and some reptiles and amphibians.

Kauai (Map 3)

Kokee State Park and Alakai Swamp Trail. If I were smart, I would tell you that this place is only so-so, skip it, so that there will be fewer people there the next time I visit. But I cannot help myself – you MUST see this place. You know, of course, about Beletsky's rating scale of scenic beauty and general eco-worthiness? (1 = generally OK, 2 = beautiful, 3 = gorgeous, 4 = magnificent, 5 = stupendous!) Well, this spot is a 5. There are absolutely gorgeous trails, some of them board-walked to keep you out of the sucking mud. What I would suggest is that you drive to the end of the main paved road through the park, which is at the Puu o Kila Lookout, park, then hike the Pihea and Alakai Swamp Trails. The habitat along the trails, and the views they provide, are almost too beautiful for words. These high elevation trails are also some of the last best places to see some of

Hawaii's native forest songbirds: Akekee, Kauai Amakihi, Anianiau, Apapane, Kauai Elepaio, Iiwi, Akikiki. In and around the muddy ponds of the Alakai Swamp are skinks, perhaps a frog, dragonflies, and some of the biggest, meanest-looking (but harmless) spiders I've ever seen. This park has one negative: it's one of the wettest spots on the planet – so prepare for rain and count yourself lucky if you hit a dry, sunny day.

The Pihea Trail (Habitat Photo 10) starts at the Puu o Kila Lookout, proceeds along a high ridge at about 1200 m (4000 ft), with spectacular views to the left (the Kalalau Valley, Na Pali Coast, and Pacific Ocean, with tiny insect-like helicopters giving coastal tours 1000 m (3300 ft) below) and right (the mountains and canyons of Kokee State Park, Waimea Canyon, and the Alakai Wilderness Preserve). The trail turns, descends a valley via boardwalks, and then intersects the Alakai Swamp Trail (Habitat Photo 7). If you have time, walk the entire Alakai trail – you won't be disappointed. (Note: The actual Alakai Swamp is a large, remote, mostly inaccessible area contained within the Alakai Wilderness Preserve, which adjoins Kokee. The Alakai Swamp Trail crosses through just the very northern edge of the Swamp.) If you turn left from Pihea onto Alakai, you move, mostly on boardwalks, through tropical forest that in many places provides good visibility – you can see all around because the treetops at this altitude are at only 3 to 5 m (10 to 16 ft) and sometimes at eye-level. This is a great area for forest birds. You see a number of them without much effort; they are along the trail, almost at eye-level, staring at you; all you have to do is raise your binoculars. The trail continues through a very open bog area (again with boardwalks saving you from bottomless mud) and eventually climbs to a magnificent high spot overlooking this true wilderness site.

There are many other trails in the area, probably more than 65 km (40 miles) of them if you include the ones in the adjacent Waimea Canyon State Park. Along the main road in Kokee you will find in one location a lodge (808–335–6061), restaurant, and museum (808–335–9975) where you can ask questions, obtain trail maps, etc. Some trails start from the museum area. There are also picnic and camping sites. Go early in the morning and you will have Kokee virtually to yourself. Take Route 50 west to Waimea, then proceed north on Waimea Canyon Drive, or continue to Kekaha and take Route 550 (Kokee Road) north. If you are not fond of exploring on your own, some local tour companies take groups into the area. (Area: 1760 hectares (4345 acres). Elevation: to 1270 m (4200 ft). Habitats: middle and higher elevation wet forest, wet shrubland, bogs, intermittent streams.)

Waimea Canyon State Park. Waimea Canyon (Habitat Photo 11) is a wide, deep, colorful, actually rather spectacular canyon in western Kauai, cut out by the Waimea River. It is sometimes referred to as "Hawaii's Grand Canyon" or "Grand Canyon of the Pacific" and the names fit. (I know, you're skeptical – how could they fit a huge canyon into a small island? Go. See for yourself.) The roadside scenic overlooks are worthwhile, as are trails in the park. One, for instance, the 4-km (2.5-mile) Kukui Trail, starts at the main highway and meanders down the steep canyon, switchback after switchback, descending 600 m (2000 ft) to the river below (do remember that if you walk easily downhill for an hour and a half, you will later have a difficult, uphill, 3-hour return hike). But there are stunning overlooks and sweeping valley views along the open trail, and birds (tropicbirds soaring overhead, francolins in the scrubby hillsides, Short-eared Owls hunting, Kauai Amakihi and Apapane in the trees) and feral goats to keep you company.

Clouds often temporarily obscure the views. You pass through this park on the way to Kokee State Park, which it adjoins. Take Route 50 west to Waimea, then proceed north on Waimea Canyon Drive, or continue to Kekaha and take Route 550 (Kokee Road) north. (Area: 755 hectares (1866 acres). Elevation: 100 to 950 m (330 to 3100 ft). Habitats: low and middle elevation dry shrubland and moist forest; cliffs, river, intermittent streams.)

Haena State Park. Haena State Park is a beach-front park at the western end of the main highway (Route 56) on Kauai's northern coast. There's the beautiful white-sand beach and ocean to enjoy, some lava-tube-formed (see p. 27) wet caves to explore, a few good coral reef dive sites just offshore, picnic and camping facilities, as well as views of the wild Na Pali Coast. But the main attraction here for many people is the trailhead for the Na Pali Coast (Kalalau) Trail (see below). (Area: 27 hectares (66 acres). Elevation: sea level. Habitats: beach, coastal forest, caves.)

Na Pali Coast State Park and Kalalau Trail. If you are someone who enjoys wild natural scenery (and I imagine that describes most people buying this book), then, to experience a great rush of intense pleasure, do the following: fly to Honolulu; hop a plane to Kauai; rent a car (a little sporty job will do, white perhaps); drive north from the airport on Route 56 (stopping at a market for trail essentials such as water, nuts, fruit, candy), then west, to the end of the road; park; get out; find the trailhead for the Kalalau Trail; hike for a few hours.

Much of Na Pali Coast State Park and adjoining protected areas consist of an inaccessible wilderness of mountains, cliffs, and steep-sided river valleys. But running here along the coast for about 18 km (11 miles), often at cliff-side, with stunning views of some of the Earth's most striking scenery (Habitat Photo 9), is an easily reached trail, probably Hawaii's most popular, most recommended, and most talked-about. The trail repeatedly climbs to cliff tops (Na Pali means "the cliffs") to traverse those sections of the coast that consist of sheer green cliffs falling down into the ocean, then drops down to low valleys to cross rivers and streams. The entire trail can be traversed in a single day, but it's a long, hard day usually in high heat, and recommended only for experienced hikers in good shape. Then you must camp at trail's end, Kalalau Beach, and hike back out the next day; or boat out if you made prior arrangements (possible only in summer). There are two other camping spots along the trail, so many people make it a three-day hike, in and out, with two nights camping. Seabirds are probably the most commonly seen wildlife along the trail, and wild goats are often spotted.

Most users of the trail day-hike only the first 3.2 km (2 miles), and for most people this will be sufficient. You will see some awesome views, have some small adventures, and hopefully return to your car safe and sound. For those who appreciate being on a cliff's edge, high above pounding surf, a few centimeters from certain death, well, the day-hike includes some of that. The trailhead is at the western end of Route 56, by Kee Beach (the beach at the end of the road) at Haena State Park. The first 3.2 km (2 miles) take you to Hanakapiai Beach where you can sit on shady beach boulders under trees or on the sand, rest, and contemplate some really rough, crashing surf (swimming here, as at the other beaches along the trail, is considered suicidal in some circles – the currents and surf are very strong and dangerous; we lose the occasional tourist here). Even if you only hike the first kilometer (half mile) or so, you will get great coastal views, great photos. If you get to Hanakapiai Beach, you can then opt to hike another 3 km (2 miles) or so up the Hanakapiai Valley, along a slightly more difficult trail, to a great waterfall (Hanakapiai Falls). *Warning*: The Kalalau Trail and peripheral trails can be dangerous. Be very careful with your footing. Even hiking only to Hanakapiai Beach, you will have to cross a wide stream either by walking on spaced rocks in the stream or by wading across. If you continue to Hanakapiai Falls, you will need to do this again; and if you hike the entire trail, you will do it frequently. The rocks can be very slippery – I saw several people fall into streams recently while trying to cross (I have personal experience here, also). There are trail areas where you must cross patches of muddy or slippery wet rocks – in places where a slip can cause you great damage or worse. You will be tempted to day-hike in sneakers or other light footwear, but light hiking boots are best or, minimally, some old high-top sneakers, so you have some ankle support. If you're injured, you had better have traveller's checks, a lot of them, because the rescue helicopter doesn't take Visa or MasterCard. There are also dangers from flash floods, during which the small, meek streams swell into raging rivers. I strongly suggest starting a hike early in the morning, because the parking area and trail sometimes fill up by late morning. Permits, obtainable at the state parks office at the state office building in Lihue, are required to hike along the coast past Hanakapiai Beach and for camping. (Area: 2500 hectares (6175 acres). Elevation: sea level to about 1200 m (4000 ft). Habitats: shoreline, sandy beaches, estuaries, lowland wet shrubland, middle elevation moist forest, cliffs, streams.)

Kilauea Point National Wildlife Refuge. Located about 3 km (2 miles) north of the town of Kilauea, just off the main highway around Kauai, Kilauea Point is the northernmost spit of land in the main Hawaiian Islands. This site is mainly a place to watch soaring seabirds around the lighthouse that occupies the point and nesting seabirds on adjacent cliffs and vegetation. The refuge was established in 1985 to protect and provide habitat for seabirds and Nene (a few Hawaiian Goose were transplanted to Kauai – to Kilauea Point – during the early 1990s, and their number is growing), to protect native coastal plants, and to act as an interpretive center for Hawaiian wildlife and natural history. Birds commonly seen here, depending on time of year, include albatrosses, boobies, tropicbirds, frigatebirds, and shearwaters; some nest at the refuge. Dolphins and whales, usually Humpbacks, are often seen from the point, far out in the coastal waters, and Monk Seals occasionally use the shoreline here to haul out and take the sun.

The refuge has very limited public access. You walk from your car a short distance on pavement to the actual Kilauea Point, on which sits a lighthouse. Soaring seabirds are visible all around – bring binoculars. You take in the ocean and cliff views, the pounding surf, birdwatch a bit, visit the gift shop, and it's time to go. There's nowhere to walk, nowhere to get away from people. But volunteers do give walking tours of closed parts of the refuge (the Crater Hill Trail). Usually you must register in advance (at the refuge or call 808–828–0168) for the 1.5–hour tour, which winds uphill through wooded and open areas to a very high bluff overlooking the sea. A guide provides natural history and cultural information. On the tour recently I saw nesting Red-footed Boobies, a Nene parent waddling around a meadow area with kids in tow, and a Short-eared Owl. Because it's near the main highway and offers a spectacular ocean view (and asks only a small donation), Kilauea Point has become part of the regular rental-car tourist circuit.

It does not open until 10:00 a.m. (Area: 82 hectares, 203 acres, Elevation: sea level to 173 m, 568 ft.)

Hanalei National Wildlife Refuge. A short distance west of Kilauea Point NWR, Hanalei NWR is a nice spot to spend an hour or so birdwatching and walking along a new trail. The refuge was established in 1972 to protect and provide habitat for native Hawaiian waterbirds, including Hawaiian Stilt, Hawaiian Duck, Hawaiian (Common) Moorhen, and Hawaiian Coot, all of which are threatened or endangered and can be seen here with little effort. Also in the area and commonly seen are Cattle Egrets, cardinals, frogs, and insects such as dragonflies and butterflies. About 50 bird species have been spotted in the refuge, including occasional migrants. The refuge consists of some of the flatlands of the Hanalei River Valley, some of the surrounding forested hills, and adjoining irrigated agricultural lands. Look for birds in the ponds by the parking area (walking around the ponds may or may not be legal) and in the wet agricultural impoundments along the road into the refuge. Volunteers recently built a short trail, which starts across the road from the parking area and winds up through a wooded section to a river valley overlook. If driving from Kilauea Point NWR, Hanalei NWR is reached by making the left turn (to Ohiki Road) immediately after passing over the large Hanalei bridge, before the town of Hanalei, and then driving a bit down the road to the small gravel parking area on the left. You can ask about Hanalei at Kilauea Point NWR (a much more heavily visited site) or call the US Fish & Wildlife Service, 808–828–1413. (Area: 371 hectares, 917 acres. Elevation: 6 to 12 m, 20 to 40 ft. Habitats: river, ponds, wet pasture.)

Kuilau Ridge Trail, Wailua. This wide, well maintained 3.5-km (2-mile) trail begins in a lower-elevation forested area but quickly ascends to open scenery as you hike along mountain ridges. It's a good, easy-to-access trail with spectacular views of mountains, valleys, and the coast. It's particularly nice in late afternoon: after a hard day lying around the pool or the beach, take a slow hike up the trail starting at 3:00 or 3:30, then start down at 5:00 or so. You will get good looks at the beautiful White-rumped Shama, a common trailside bird. To get to the trailhead, drive west on Route 580 from Wailua, about 11 km (7 miles); the trailhead is on the right just before the paved road turns to dirt.

Powerline Trail. This 20-km (13-mile) dirt track, marked on many maps, rough in spots, connects Princeville to Wailua. It's muddy when wet and incredibly dusty when dry. But it winds past some nice mountain scenery and, for those tired of staring at ocean views, is appropriate for a long inland hike or a morning bird walk.

Limahuli Garden. The Limahuli Garden is part of the National Tropical Botanical Garden, whose main purpose is the study and conservation of tropical plants; the NTBG has other facilities on Maui, the Big Island, and in Florida. The Limahuli was started in 1967 by a private citizen and donated in sections to the NTBG. The garden and adjacent Limahuli Preserve total about 400 hectares (1000 acres). Located very near Haena State Park and the trailhead for the Na Pali Coast (Kalalau) Trail, Limahuli is a splendid place with a beautiful setting, with both mountain and ocean views. The garden sports a loop trail that takes you (by yourself or with guide) past 30 or so botanical sites, and has huge, borrowable anti-sun/anti-rain umbrellas. Several parkland birds are common, such as plovers, shama, and mynas. If you begin a day-hike on the Na Pali Coast Trail, return to your car by 1:00 or 1:30 in the afternoon, rest and cool off for a half-hour on the nearby Haena State Park beach, then you can nicely finish your outing with an

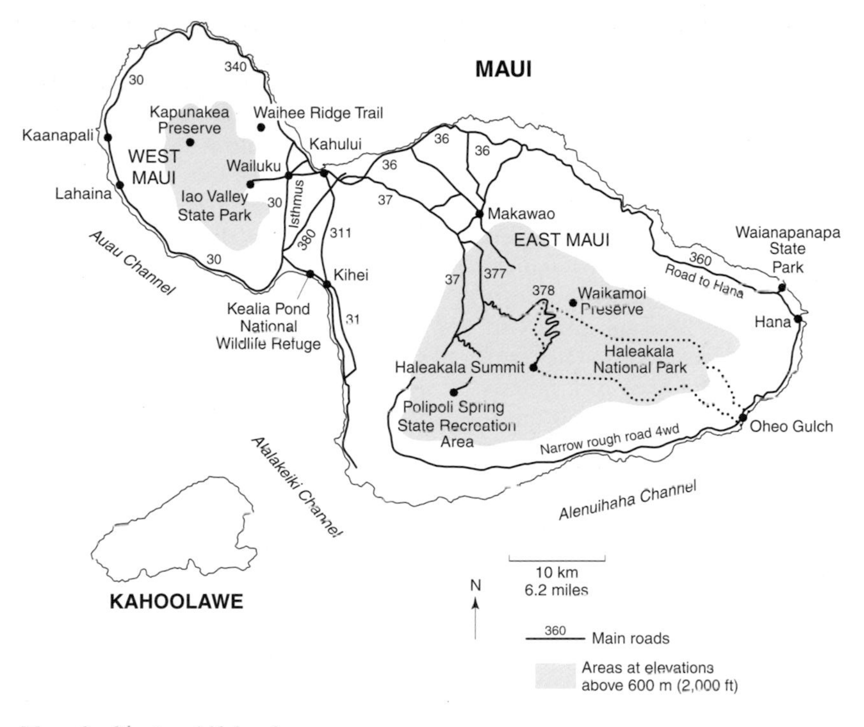

Map 4 Maui and Kahoolawe.

hour or two at the Limahuli Garden. There is a stiff admission fee, which sharply reduces tourist numbers; without the high price, however, the beautiful Garden with its tiny parking area would be over-run.

Maui (Map 4)

Haleakala National Park. This park occupies a big, beautiful chunk of East Maui and is one of Maui's, and the state's, most visited sites. The park includes a higher-elevation car-accessible summit area, where most people visit; about 44 km (27 miles) of strenuous high elevation trails along volcanic slopes and plateaus that begin in the summit area; and a large, closed-to-the-public tropical forest wilderness area (the Kipahulu Valley Biological Reserve) that runs from the higher reaches of Haleakala downslope to the sea. The park is reached by climbing Route 378, Haleakala Crater Road, with entry to the park at about 2130 m (7000 ft) elevation. You can drive slowly all the way to the Haleakala summit, where there is a glassed-in observation deck (be careful of cold and very high winds here above 3050 m, 10,000 ft). Trails that leave from the summit area cross large volcanic formations and fields of short-stature, scrubby high-elevation vegetation. The park contains some of the best-preserved native ecosystems in the islands – with both native plants and animals (although many non-native plants and some non-native animals occur here, too). The Haleakala Silversword (Plate 2), a highly distinctive, well known endangered plant that lives only above 2000 m (6500 ft), can be found growing around the visitor's center and along trails. About 250 species of flowering plants indigenous to Hawaii occur at Haleakala. A small population of the endangered Nene occurs here and the geese are often seen by visitors. The wooded Hosmer Grove area, near the park's entrance, with picnic and camping sites and a nature trail, is an excellent spot to see some of Hawaii's native forest birds – abundant Apapane and Common Amakihi, and less-common Iiwi, and Maui Alauahio – and the Short-eared Owl. Chukar and Ring-necked Pheasant are also common. You can easily visit this park yourself, park your car, and hike. Check at the park visitor center or call (808–572–9306; www.nps.gov/hale) for scheduled hikes with park rangers. Some guided hikes begin in the park but proceed to the adjacent Nature Conservancy Waikamoi Preserve (see below). *Warning*: It can be quite cool or even cold in this high-elevation park. Bring warm clothes and sun-, wind-, and raingear, particularly if you plan on hiking the highly exposed trails. (Area: 114 sq km (44 sq miles). Elevation: sea level to 3055 m (10,023 ft). Habitats: alpine zone, subalpine shrubland, dry forest, rainforest.)

Waikamoi Preserve. This nature preserve adjoining Haleakala National Park is actually owned by a private concern but has been managed for conservation purposes since 1983 by The Nature Conservancy of Hawaii (TNC). The reserve, together with adjacent state lands and adjoining Haleakala National Park, represents one of Hawaii's largest chunks of intact forest and is a major watershed area, actually the largest single source of surface water in the state. There are stunning rainforest and cloud forest habitats to see here, and it's a great place to spot some of Hawaii's native forest birds, which mostly survive only at elevations above 1200 m (4000 ft). Eleven or so native bird species occur in the reserve, six of them officially considered endangered. Apapanes, Amakihis, and Iiwis are all fairly common, and people hiking here with binoculars will probably see them. Maui Alauahio and, less often, Maui Parrotbills and Akohekohe are also spotted. The reserve contains 22 species of rare plants, six of which are endemic to East Maui. Access is limited to regular group hikes conducted by TNC (808–572–7849) and by rangers from Haleakala National Park (808–572–9306; www.nps.gov/hale). I went on an excellent 3-hour, low-strain hike guided by a park ranger who gave information on native birds, plants, and Hawaiian culture. We walked along a fairly muddy trail through mixed stands of native and introduced trees (Habitat Photo 14), saw some birds, and stopped at striking overlooks into areas of native Hawaiian cloud forest (Habitat Photo 12). If you are on Maui, this is a worthwhile destination. (Area: 2120 hectares (5230 acres). Elevation: 1350 to 2600 m (4400 to 8500 ft). Habitats: montane wet forest, subalpine dry forest, wet and dry shrublands, vegetated lava flows.)

Kapunakea Preserve. Kapunakea is a mostly high-elevation nature preserve located in the mountains of West Maui. Like the Waikamoi Preserve, the land is actually owned by a private concern but is managed for conservation purposes (since 1992) by The Nature Conservancy of Hawaii (TNC). It protects native plant communities (including 24 rare plant species, five of which are endangered) as well as native wildlife. Native forest songbirds such as Apapane, Amakihi, and Iiwi are common here, as is the Short-eared Owl. Access is limited to regular group hikes conducted by TNC (808–572–7849). (Area: 385 hectares (1264 acres). Elevation: 485 to 1645 m (1600 to 5400 ft). Habitats: lowland rainforest, lowland wet shrubland, highland rainforest, high elevation bogs.)

Road to Hana and Beyond. The Road to Hana is a popular tourist drive along the main drag that winds along East Maui's north coast (Hana being a small town on Maui's east coast with a beach and some cultural attractions; it's mostly just a rural town cited in travel literature to give tourists a destination for their scenic drives). It takes 3+ hours to drive from Kahului to Hana and beyond, even though it is a distance of only 90 km (55 miles) or so. The road is narrow and winding with hairpin turns and many single-lane bridges. You must drive much of the way at an incredibly slow 15 to 30 kph (10 to 20 mph). (I would advise going early in the morning to avoid some of the tourist traffic, but I would also advise waiting a bit after breakfast to start – all the turns can be quite stomach-churning.) There are several small parks and waterfalls along the roadway with tiny parking areas, picnic tables, and short trails through low-elevation coastal forest, for example, Puohokamoa Falls. Quite nice, just north of Hana, is the coastal Waianapanapa State Park. The site, with picnic and camping facilities, provides access to the coast, to a black-sand beach, and to a coastal trail. Noddies and other seabirds can be seen on the black rocks and boulders that stick out of the surf (Habitat Photo 13). Driving through Hana and continuing south for about 15 km (10 miles) you arrive at the southern end of Haleakala National Park (Oheo Gulch). You can park here and take short paths down to the sea and a 3.2-km (2-mile) trail (Habitat Photo 15) up to Waimoku Falls – a fairly spectacular falls that drops 120 m (400 ft) over a vertical rock face. Part of the trail is through towering, dense, bamboo forest, so dense that it's eerily dark on the trail even at mid-day.

Road Around West Maui. This is a scenic drive that offers some spectacular ocean views. The road follows the coastline and although it's now all paved, it's still dangerous in sections. The roadway is extremely narrow in places, essentially a one-lane road with two-way traffic, with many blind curves. Good luck. Most of inland West Maui consists of the West Maui Mountains and is largely inaccessible. There are few places along the rough east coast of West Maui to pull off the road and explore. One is the Kahakuloa Game Management Area. Its gate seems to be open to 4-wheel-drive vehicles on weekends; on weekdays the place appears deserted and you can climb the fence, walk the jeep trails, and birdwatch. If you hunt, there's a two pig-per-person limit (what's that come to – about 5 kg (11 lb) of ribs?).

Waihee Ridge Trail. This is a very nice, easy-to-reach, lightly used trail that provides fantastic views. The trail climbs, mostly along narrow, high ridge lines, from about 300 m (1000 ft) to a peak at about 760 m (2500 ft). You won't see much wildlife here, but the views make the hike worthwhile. From the top you get 360–degree mountain-vista views, down to the ocean in some directions. The trailhead is in a pasture area, good for seeing chicken-like birds such as pheasant. Leave about 3 hours for the hike (1.5 hours up, a half-hour at the top; an hour down). The trailhead is near a Boy Scout camp. Reach it by driving west from Kahului and Wailuku on Route 340. Make the sharp left just before the 7-mile marker and proceed up the road about 2 km (a mile).

Polipoli Spring State Recreation Area. The state recreation area here itself is small (4 hectares, 10 acres), but it is part of, and its trails lead into, a huge state reserve, the Kula Forest Reserve, that occupies some of Haleakala's western slopes. There are picnic and camping sites, and access to a network of excellent hiking trails. Polipoli is located off Route 377, about 16 km (10 miles), up Waipoli Road. The road is very narrow and steep; its pavement ends after about 10 km (6 miles) and the remainder is rough dirt. A standard sedan can often make the drive on the dirt road, but 4-wheel-drive is sometimes needed. If you do not want to chance the dirt road, you can park near where the pavement ends and walk farther along the road. In about 15 minutes you will come to the trailhead for the Boundary Trail, which provides access to the trail system; Polipoli Spring State Recreation Area is about a 10-km (6-mile) walk by these trails. Entering the Boundary Trail you descend into beautiful pine groves and other wooded areas, and move through open shrubby habitats. Birds such as Apapane, Amakihi, Maui Alauahio, Japanese White-eye, pheasants, and Short-eared Owl are commonly seen. (Elevation: to 1950 m (6400 ft). Habitats: higher elevation dry and moist shrubland and forest.)

Iao Valley State Park. This is a small park located just west of Kahului. It has paved walkways that descend to a small river (Iao Stream). More of a cultural and geological site. Tour buses stop here, so go early or late, before or after the buses. White-tailed Tropicbirds are seen here, as is the occasional Common Amakihi. (Area: 2.5 hectares, 6 acres.)

Kanaha Pond State Wildlife Sanctuary. This pond is in the middle of Kahului, the town harboring Maui's airport. It is easily reached, the entrance being right off a main road (Route 396, Haleakala Highway). There's a small shelter near the parking area, from which you can usually spot Hawaiian Stilt (an endangered species, whose habitat the pond is meant to protect), Hawaiian Coot (also endangered), Cattle Egret, Black-crowned Night-heron, shorebirds, and several species of ducks. You can view other sections of the pond from the shoulder of the main highway (Route 36).

Kealia Pond National Wildlife Refuge. Located on Maui's central isthmus, right along a main highway (Route 310), this natural basin shallow pond provides habitat for water and marsh birds. It is closed to the public during spring and summer, but there are some roadside pullouts where you can park and have views into the refuge. Hawaiian Stilt, Hawaiian Coot, Hawaiian Duck, Cattle Egret, Black-crowned Night-heron, and shorebirds, are regularly seen; about 30 bird species occur here. Birders consider this one of the best places in Maui to look for migratory ducks. Hawksbill Turtles nest on the ocean beach adjacent to the pond – probably the only such site on Maui. (Area: 280 hectares, 700 acres.)

Molokai (Map 5)

Kamakou Preserve and Waikolu Lookout. This is a great place! A beautiful, cool, lightly visited forested site perfect for hiking, birding, and simply taking in the breathtaking views (Habitat Photo 16). The opposite of, or antidote for, Waikiki on Oahu, is the Kamakou Preserve on Molokai. You won't forget this one any time soon. The preserve was established in 1982 on private land, when conservation rights were given to The Nature Conservancy of Hawaii (TNC), whose main interest here was to protect habitat for native forest birds. The preserve is

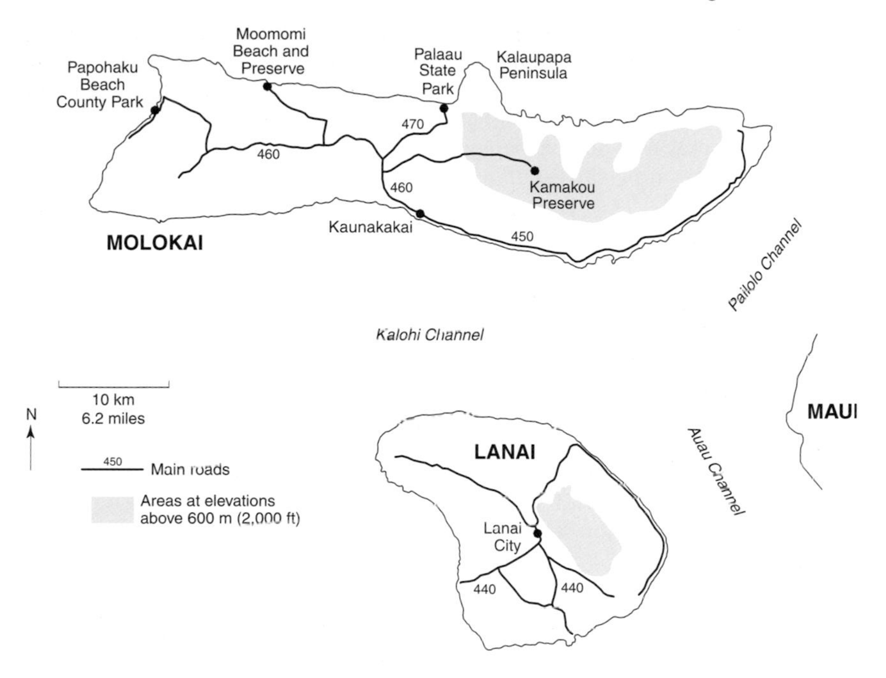

Map 5 Molokai and Lanai.

located at high elevation in the mountainous center of the island, so it's usually cool, even in full sun; parts are often misty as frequent clouds roll through. Together with adjacent protected lands, the reserve is part of an 8900-hectare (22,000-acre) parcel of mostly native habitats that range from sea level to 1500 m (4970 ft). Thirty-three rare plants occur here, including 17 considered to be endangered. Kamakou is located about 16 km (10 miles) from the main paved highway (Route 460) along a dirt road, almost all uphill, through dusty or muddy red soil. If you make it there, you will find wonderful narrow hiking trails (some quite overgrown) and many miles of jeep trails. Common birds are Apapane, Amakihi, Japanese White-eye, Short-eared Owl, Pacific Golden-Plover, and francolins.

Just as you get near the preserve, there is a map of trails posted alongside the road near a camphouse that TNC uses as a headquarters there. A bit farther on is the Waikolu Lookout, the start of the preserve. Here you find parking and picnic areas, along with great views overlooking lush green mountains and, drifting in and out through clouds, startling views down to the Pacific. Trails begin farther down the main road. You can park at Waikolu and walk to trailheads (thus seeing more wildlife), or try driving (parking places along the road, however, are very limited). It's about an hour's walk (but it's a very pleasant hour) to what is probably the best trail, the Pepeopae – just keep walking along the main road and you will eventually stumble across it. The trail, some of which goes through wet, marshy habitat, is nicely "board-walked" – lined with narrow wood boards, most of which have anti-skid surfaces. The trail winds through dense low-stature forest (Habitat Photo 17) with a canopy at 4 to 5 m (13 to 16 ft); trees are covered with mosses and other epiphytes. The trail breaks out of the forest into a stunning, open, delicate habitat, the "Pepeopae Bog," then proceeds to a sheer mountainside drop-off. People somewhat uncomfortable about great heights (like me) will not be happy here – it's like being up in an airplane, except there's no airplane. The trail is at times only a meter or two (a few feet) from the edge.

You can visit Kamakou either by yourself or go on one of TNC's periodic guided hikes. Because groups are small, you need to register for the hikes many weeks in advance. Contact TNC on Molokai (808–553–5236) for information about conditions of the road to and through the preserve before you go, even if you are going solo (the road is sometimes impassable). *Warning*: As far as I can see, the most dangerous part of visiting Kamakou is the risk of having a head-on collision on the single-lane road around a blind turn. Also, some of the side-road jeep trails in and around the preserve are very rough; even with your 4-wheel-drive, you could easily run into trouble. (Area: 1123 hectares (2774 acres). Elevation: 730 to 1370 m (2400 to 4500 ft). Habitats: grassland meadows, pine forests, mid-elevation wet forest and shrubland; highland wet forest, shrubland, and bogs.)

Moomomi Beach and TNC Preserve. This is a beautiful spot to visit and take a walk along the beach and upland, along sandy paths between delicately vegetated sand dunes. The preserve, established in 1988 to protect the dunes and highly specialized plants that grow on them, as well as to provide protected beaches for Green Sea Turtle nesting, harbors six species of highly endangered plants (being perhaps the only place they still occur in the wild). This is one of the last spots where native Hawaiian coastal vegetation, which was once plentiful but was largely lost to coastal industry and development, still exists as it must have been before heavy human settlement of the islands. The preserve, located on Molokai's north shore, is reached via a 4-km (2.5-mile) drive on a red dirt road that ends at a small parking area right above the ocean (at Moomomi Bay). A standard sedan can usually make the drive. Moomomi Beach (actually at Kawaaloa Bay) is about a one-km (half-mile) walk west, along the shore, on a coastal trail that runs right above the shoreline. The actual white-sand beach is nice but, as with many beaches worldwide these days, there can be as much colored plastic along the high tide line as you find in the soft-drink section of your local supermarket. The preserve includes the beach and dunes area upland from it. Contact The Nature Conservancy on Molokai (808–553–5236) far in advance if you would like to try to obtain space on one of their periodic tours of the preserve. You can also ask if it is possible to drive into the preserve, closer to Moomomi Beach; however, to do this, you must obtain a gate key from TNC, and have a 4-wheel-drive vehicle. *Warning*: The dune vegetation is extremely delicate and fragile, so stay on established roads and paths, or walk along the beach. This is an exposed coastal site with notoriously high winds that often hurl sand and sea spray at you with a fair amount of force. There's no shelter here – you will be fully exposed to wind, rain, and sun. (Area: 373 hectares (921 acres). Elevation: sea level to about 30 m (100 ft). Habitats: beach, vegetated sand dunes, petrified dunes.)

Papohaku Beach County Park and Papohaku Beach. Papohaku Beach is a very nice, very long (about 4 km (2.5 miles)), broad beach located at Molokai's western end. The beach is great for a long seaside walk or sun-fest, but less good for swimming; the surf is often dangerous. It's also frequently very windy. But the small county park that provides access to the beach has nice picnic and camping areas, and the trees and open grassy areas there provide good habitat for such birds as plovers, francolins, doves, cardinals, and finches. (Elevation: sea level. Habitats: beach, dunes, parkland, open grassy areas.)

Palaau State Park. This wooded park, on Molokai's northern coast, is located on a high bluff overlooking the Kalaupapa Peninsula (site of one of Molokai's most famous cultural tourist attractions, a residential medical community for people with leprosy, in existence for more than 100 years), with beautiful views down to the ocean and the peninsula. There are picnic and camping areas, and a short, forested trail. (Area: 95 hectares (234 acres). Elevation: about 460 m (1500 ft).)

Oahu (Map 6)

Kaena Point State Park and Natural Area Reserve. Located at the westernmost tip of Oahu, Kaena Point (Habitat Photo 20) is a scenic, open coastal area of sand dunes, dune vegetation, and seabirds. It is reached via a 90-minute hike along a trail/dirt road that hugs the coast on rocky outcroppings just above the crashing ocean surf. It's a nice hike but it's out in the open, so can be very hot. Once you arrive at the point, there are sandy trails between dunes to walk and from which to admire the plants and birds. The reserve is meant to protect the native coastal dune ecosystem, which is largely intact here but has been destroyed over most coastal areas of the main Hawaiian islands, and to provide protected habitat for nesting seabirds (Laysan Albatross) and perhaps Green Sea Turtles. The park is reached by driving to the end of Route 93, the main highway on the island's west coast, and parking where the road ends (leaving your car in the small lot by the beach bathrooms may be a good idea). Kaena Point also can be reached from the east. *Warning*: The dune vegetation is fragile – walk only on established pathways. (Area, state park: 315 hectares (780 acres). Habitats: shoreline, beach, dunes, tide-pools; dry shrubland.)

Honouliuli Preserve. This large Nature Conservancy of Hawaii (TNC) preserve, located along the southeast slopes of the Waianae Range, in the southwest part of the island, abutting the Schofield Barracks Military Reservation, protects some of the last remaining low- and middle-elevation native forest land on Oahu. Native forest birds such as Apapane, Oahu Amakihi, and Oahu Elepaio thrive here, as does the Short-eared Owl. The land was heavily grazed during the 1800s by goats and cattle. During the 1920s, the area was recognized as an important watershed region and received protected status. TNC has managed the area since 1990, when the preserve was established. More than 60 rare plant species occur in the preserve, two-thirds of which are considered endangered; at least four species are endemic to the preserve – they occur nowhere else (see p. 67). Endangered tree snails also occur here. The preserve, with several miles of trails, including ones along high ridges, is difficult to access because it is surrounded by private land. To visit, you must first get permission from The Nature Conservancy (808–537–4508), or go on one of their periodic guided hikes. The main reason for the controlled access is that there are rare and endangered plants near many of the trails. (Area: 1495 hectares (3690 acres). Elevation: 400 to 950 m (1300 to 3100 ft). Habitats: Low and middle elevation wet forest, shrubland, grassland.)

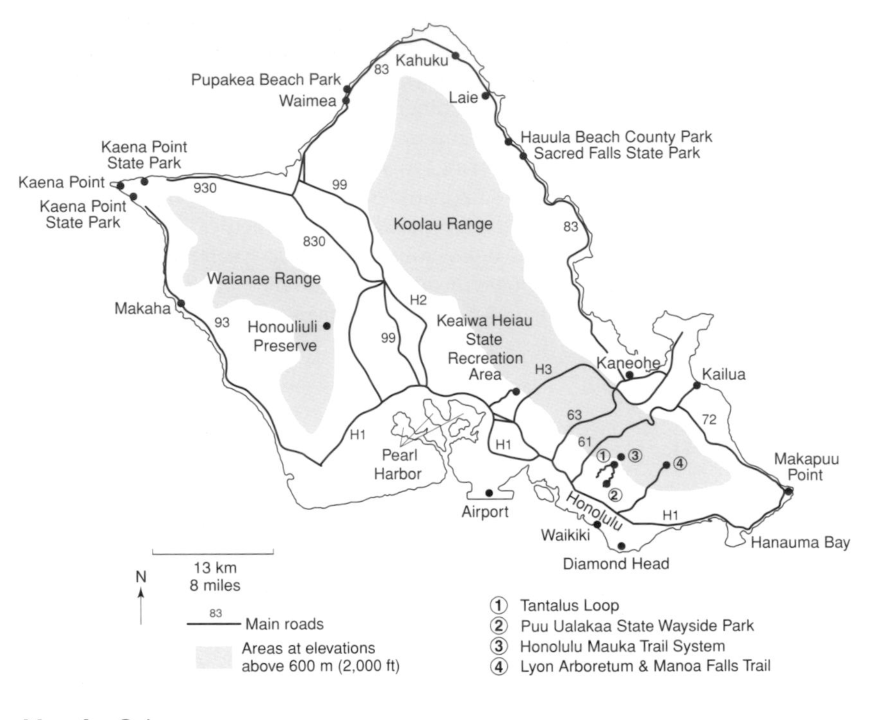

Map 6 Oahu.

Keaiwa Heiau State Recreation Area. This nice park, located only a half-hour from Waikiki or downtown Honolulu, provides trail access to mountain ridges and magnificent mountain views. Whether you're in Honolulu on business and have an important afternoon meeting or on vacation and have a hot beach date after lunch, you can reach this park by 08:00, hike for 2 or 3 hours, and be back in town by noon. The actual wooded park, with picnic and camping areas, is not that large, but it is surrounded by the huge Ewa Forest Reserve and backs onto the Koolau Mountain Range. The Aiea Loop Trail, about 7 km (4.5 miles) long, begins at one of the picnic areas, climbs through shady woods above the adjacent suburban neighborhoods, then moves along a ridge line, permitting good views all around. A little less than an hour along the main trail, a side trail, the Aiea Ridge Trail, leads uphill and then to the left. Its intersection is usually unmarked or marked only with a piece of flagging tape (the trail can be dangerous and the authorities presumably do not want children or inexperienced hikers venturing here accidentally). The Aiea Ridge Trail follows the high ridges of the Koolau Range – and does it quite impressively. The trail at times is extremely narrow, with the land falling off precipitously on both sides – it's not a particularly good place for people (like me) who do not appreciate heights. But the constant excellent views (Habitat Photo 21) of the green mountaintops moving in and out of the mist and clouds, and of the lushly green vegetated valleys below, make it worthwhile. This is a great place, not far from Honolulu, to see some of the native Hawaiian forest birds, such as Apapane and Oahu Amakihi. (Biology professors at the University of Hawaii bring their students here to give them their first glimpses of the state's native forest birds.) The park, located at the northern end of Aiea Heights Drive just north of the H-1 Freeway, opens at 07:00, so you can go early and birdwatch; weekday mornings the place can be almost deserted. (Area: 156 hectares, 385 acres). Elevation: about 275 to 500 m (900 to 1650 ft). Habitats: low and some middle elevation wet shrubland and moist forest.)

Sacred Falls State Park. This park, located along Route 83, the main highway along the eastern shore of the island, just south of Hauula, protects a large section of the Kaliuwaa Valley. There is a 3.5-km (2-mile) trail along an increasingly narrow canyon that terminates at the 25 m (80 ft) Sacred Falls. It's a scenic spot, but be careful of slippery rocks and stream crossings; flash floods during downpours occur in the canyon and are dangerous. Rock slides are also a hazard. (Area: 550 hectares (1375 acres). Elevation: sea level to about 460 m (1500 ft). Habitats: low elevation grassland, shrubland, forest.)

Hauula Loop Trail. Several nice trails climb into the Hauula Forest Reserve west of Hauula Beach County Park, just off Route 83, the main highway along the eastern shore of the island. The Hauula Loop Trail (Habitat Photo 18) is a lovely 4-km (2.5-mile) path through wooded and more open areas, with, as it ascends, great views of the Koolau Mountains and the Pacific. The trailhead for this and other trails is located half a kilometer, a quarter mile, west on Hauula Homestead Road (from its intersection with Route 83).

Tantalus Loop and Puu Ualakaa State Wayside Park. Tantalus Loop is a 13-km (8-mile) scenic drive described in many travel guides. The drive takes you on winding roads (Tantalus Drive and Round Top Drive) through a wooded suburban area at the northern fringe of Honolulu just a few minutes from Waikiki. The loop first climbs steeply (to about 500 m, 1600 ft) into and then descends out of the Makiki Valley of the Honolulu Watershed Forest Reserve, which occupies the slopes of the Koolau Mountains above Honolulu. Along the drive, with its hairpin turns and slow-moving tourist traffic, are high roadside pullouts that yield sweeping views of the city of Honolulu, Diamond Head, the airport area, and the Pacific. The pullouts are wedged between the fenced compounds of Hawaii's rich and famous, who apparently also appreciate beautiful views of the city. Puu Ualakaa State Wayside Park, near one end of the Tantalus Loop, sports some rolling parkland and splendid views of Diamond Head and Honolulu.

Honolulu Mauka Trail System. Just above Honolulu, literally 20 minutes from Waikiki's 12 billion tourists, is a network of many miles of wooded trails through quiet, wild areas. These trails, the Honolulu Mauka Trail System, are never crowded and often deserted; if you go in the early morning, say for birding, you will be all alone. The trails are associated with the Tantalus Loop (see above) in that the trailheads, with small parking areas, are along the Tantalus roadway. Look for the characteristic brown and yellow signs at the roadside that indicate trail access. Some of the trails are through the center of the Tantalus Loop (for example, the Makiki Valley Trail), whereas others start to the loop's north and then climb higher into the Makiki Valley (Manoa Cliff and Puu Ohia Trails, for instance). Some of the trails are partly along cliffsides, with steep drop-offs, and some terminate at high lookouts with gorgeous views of the green mountains and valleys of the Koolau Range (Habitat Photo 19). Bring mosquito repellent. Some travel guides include maps of the trail system, or contact the Hawaii Nature Center (808–955–0100) or the Hawaii Dept. of Land and Natural Resources/Na Ala Hele Trails Access Program (808–973–9782).

Manoa Falls Trail. The Manoa Falls Trail, which begins from the end of Manoa Road in Honolulu, makes for a nice, shady, 30–minute hike only about 20 minutes from Waikiki. The trail follows Waihee Stream into the Honolulu Watershed Forest Reserve to the 30-m (100-ft) drop of Manoa Falls. Birds such as shama, bulbuls, and leiothrix are commonly seen at close range. Bring mosquito repellent.

Lyon Arboretum. Near the trailhead for the Manoa Falls Trail, at the end of Manoa Road, you will find the entrance to the Univeristy of Hawaii's Lyon Arboretum. A great place to visit, only 20 minutes from Waikiki, nonetheless it appears to be a closely held secret; there are usually only a few tourists in evidence. This is a wilder arboretum than most people will be familiar with, but it's a good spot to spend an hour or two roaming shady pathways. The arboretum has a native Hawaiian plant section but also plants from around the world. About 5000 species are planted here. There is a herb and spice area, an ethnobotanical area, an economic plants section, etc. This is also a good place to see many of Oahu's introduced birds, although it opens at 09:00, fairly late for birding. Doves, bulbuls, shama, myna, and white-eye are seen regularly; and native Amakihi, occasionally. Guided tours are offered three times a month. Call 808–988–7378. Bring mosquito repellent. (Area: 78 hectares (194 acres). Elevation: 130 to 560 m (450 to 1850 ft.))

Honolulu Zoo. The Honolulu Zoo is a good little zoo in Waikiki, literally in the shadows of the resort hotels. It was started in 1938 as a bird exhibit and it still emphasizes birds. The zoo has a standard African savannah exhibit, perhaps because zoo-goers have come to expect giraffe and zebra, hippo and elephant, but there are also several parrot species from the Australian region and South America, as well as some of the native Hawaiian birds, including Nene. A few of the native forest birds – Apapane, Amakihi, Iiwi – are exhibited. Because they are so interesting and important from a conservation viewpoint, the zoo would like to show more of the forest birds, and plans to, but it turns out that it's an expensive proposition: enclosures for these birds must be totally sealed (to eliminate mosquitos, which would kill the birds by transferring avian malaria) and climate-controlled (the birds all live naturally at cooler, higher elevations, and would die in Waikiki's heat and humidity). The zoo plans to concentrate its collection in the future on Pacific island species and animals from the Australia/New Guinea region, species that most zoo-goers cannot see easily at other zoos.

National Wildlife Refuges and Offshore Seabird Sanctuaries. The US Fish and Wildlife Service maintains two pond and marsh-area refuges on Oahu, with the special purpose of protecting habitat for endangered Hawaiian waterbirds; both require special permission to access. James Campbell National Wildlife Refuge, northwest of the town of Kahuku in the northern corner of the island, is about 62 hectares (155 acres) of ponds and marshes where about 35 bird species have been seen. Pearl Harbor National Wildlife Refuge protects about 25 hectares (61 acres) of ponds in the Pearl Harbor military base area on the island's south shore; about 30 bird species have been noted there. You can ask about access at 808–637–6330/808–541–1201. The small islands off the eastern (windward) coast of Oahu are part of the Hawaii State Seabird Sanctuary. Many seabirds use the islands for nesting or roosting, including terns, boobies, noddies, shearwaters, and petrels, and migratory shorebirds use the islands' beaches to stock up on food. The islands are reachable only by boat, access is limited, and landing on two of them, Manana and Moku Manu, is prohibited.

Chapter 5

Environmental Threats and Conservation

Environmental Threats

Background

During the early 1990s, a committee of six senior biologists, operating under the auspices of the USA's National Academy of Sciences, issued a 136-page report detailing their investigation and assessment of the HAWAIIAN CROW (Plate 37) situation. Just what was, and still is, the Hawaiian Crow situation? Too many crows in Hawaii eating crops and garbage? Farmers irate and demanding federal action? No, the problem was just the opposite. The total global wild population of this separate, distinct species of crow, which occurs now only on the western slopes of the Mauna Loa volcano on Hawaii's Big Island, had fallen to only 12 individuals. The scientific committee of expert bird biologists had been convened to assess the situation and make recommendations about how best to prevent the almost-certain imminent extinction of the crow. I mention this for two main reasons. First, it's interesting and somehow comforting to note that persons in authority somewhere thought enough of a small bunch of crows to assemble a team of heavy-firepower scientists essentially to begin an effort to help the crows reproduce; that is, the aim of the panel's recommendations would be to increase the size of the crow population. Second, that the Hawaiian Crow is only one of a large number of threatened and endangered species in Hawaii and, although all do not rate their own panel of scientific consultants, there is increasing concern among ecologists, naturalists, environmentally aware Hawaiian citizens, and government officials, that unless quick actions are taken, large numbers of animal and plant extinctions in the state will occur in the near future. As an ecotraveller, you should have an understanding of why so many of Hawaii's native species are threatened and why so many have become extinct in the recent past.

Biologically, Hawaii is probably best known for its diversity of habitats, the high degree of endemism among its native plants and animals (see Close-up, p. 67), and for the fact that it is, in some ways, in the midst of an ecological collapse. The diverse habitats range from arid beach dune communities to wet mountainside cloud forests. Examples of the high level of endemism: about 90% of Hawaii's native flowering plants and between 95% and 99% of its native insects are endemic species – they occur nowhere else in the world. Among the native birds, almost all are endemic. But Hawaii is also now considered the extinction and endangered species capital of the world, with more threatened and endangered species than any other USA state or any comparable-sized region or island chain on Earth. Many animal and plant groups are threatened, but let me concentrate on birds; their biology is better known than that of say, the native plants or snails, they are the group in most immediate danger of total or near-total extinction, and, as a bird biologist, I know more about them.

The general numbers are these: of about 100+ native bird species that existed in Hawaii prior to human contact, about half became extinct between the time Polynesians first colonized the islands (perhaps 1600 years ago) and the arrival of Europeans (200 years ago); half the remaining species succumbed after the Europeans settled (21 of 59 historically known species and subspecies becoming extinct during the past 150 years, some of them just recently); and half (or more) of the surviving species are now endangered. Some entire groups are gone and others have been decimated. Five endemic members of Family Meliphagidae, the mostly-Australian *honeyeaters,* occurred in Hawaii – striking mid-sized to largish songbirds, mostly black with patches of yellow, with long tails and long downcurved bills for nectar-feeding at tree flowers. As of the mid-1980s, two species, the KAUAI OO and BISHOP'S OO, still lived; now all are probably gone. About 50 species of the most famous of the Hawaiian songbirds, the *honeycreepers* (endemic Family Drepanididae; p. 151) existed prior to human settlement of the islands. Today, only between 15 and 20 of the species still exist (the actual number depending on the classification scheme used to group them and on how many of a few species that recently were known to have been reduced to fewer than 20 individuals are now extinct).

What caused these drastic declines in Hawaii's bird biodiversity? As you might have guessed, if you didn't already know, people are responsible – in some direct and very indirect ways. Before you understand the reasons for the many species extinctions and for the continuing and immediate threats to Hawaii's habitats and remaining native animal and plant life, you must know something of island ecology – how species arrived in the first place, how they spread and diversified, and what happened when people discovered and settled the islands. To review briefly what was covered in Chapter 2 (pp. 9 to 12): isolated islands naturally have a relatively low *biodiversity* (the total number of different species that occur there). This is because only a few species, by chance, manage to cross thousands of kilometers of open ocean and land on the island's shores (see Close-up, p. 100).

These few *colonizing species*, finding few or no competitors for resources such as foods and nest sites, evolve rapidly to occupy the unused, or "unfilled," ecological niches; the result, technically called an *adaptive radiation*, is a diverse assemblage of closely related species that occupy an array of habitats but that often differ physically from one another only in rather small ways. For example, if each species in the group became highly specialized through evolution to pursue a particular food type or a specific foraging technique, then all might resemble one another in body shape and coloring, but have widely differing bills – such as we find among the Hawaiian honeycreepers (p. 151), perhaps the world's best bird example of such a radiation (50 or so species eventually arose from the arrival in Hawaii in the distant past of a single finch-like ancestor; a similar event occurred in the Galápagos Islands, where a single finch or finch-like colonizer diversified over time into the 14 or so species of the group now known as Darwin's Finches). Other adapative radiations of note in Hawaii occurred among its land snails, *Drosophila* pomace flies (p. 198), and plants such as silverswords (Plate 2) and lobeliads (including plants in genus *Cyanea* and genus *Clermontia*, Plate 3).

Causes of Extinctions

When early Polynesians settled in Hawaii, they immediately began altering the natural habitats, to the detriment of native birds. First and foremost, they began cutting lowland forests in drier coastal regions for their villages and farms. Eventually, as the human population expanded, most of this native habitat type disappeared, and the bird species that were adapted to these lowland dry forests died out. So outright habitat destruction undoubtedly contributed to the loss of Hawaii's native birds. But a more pervasively damaging force was also at work: the introduction by people, intentionally or not, of *alien* organisms – which we can define here as any plant, animal, or other organism that did not occur in Hawaii before people arrived.

Species and groups of species that evolved on highly isolated islands, like those of Hawaii, where there is little or no interaction between the island species and mainland organisms, are extremely vulnerable, physically and ecologically, when mainland animals, plants, and diseases finally arrive on the islands. Their vulnerability takes many forms and arises from several historical factors:

(1) *Absence of ground predators*. Mid-ocean islands, before their "contamination" by people, are usually free of large ground predators, such as carnivorous land mammals that might prey on adult birds or their eggs and young. The reason is that it is extremely unlikely, even over the span of millions of years, that a male and a female fox or cat or mongoose, or a small group of them, would arrive naturally on a highly isolated island and establish a successful resident population; it just does not happen. The "effect" on the islands' native birds is that they evolve in the absence of ground predators (but aerial predators may be present: large seabirds and the odd world-trotting raptor that might occasionally take some native birds, as well as native raptors such as the Hawaiian Hawk) and so "lose," or never develop, the appropriate anti-predation skills that might allow them to co-exist with these predators (as happens, of course, in continental habitats). They may not hide their nests, they may not attack mammals approaching their nests, they may not even try to escape when attacked themselves (there are, for instance, reports of

albatrosses incubating eggs not reacting as they are attacked and eaten by rats). This is the reason many seabirds such as albatrosses, which breed typically on remote oceanic islands, have no fear of people when they are nesting – they have no evolutionary experience with large land mammals and so do not regard them as dangerous. This is why the wildlife in the Galápagos Islands is famous for its "tameness," and why you often see photos of well-off, rather amazed tourists walking among unconcerned nesting penguins in Antarctica (another location with few or no land predators). In fact, this is why boobies (p. 114) are called boobies – people thought they were dumb birds because they did not react to people in their nesting colonies.

People have introduced a variety of mammal predators to Hawaii, starting probably 1000+ years ago when the PACIFIC RAT (Plate 47) was brought to the islands intentionally as a source of food or when it stowed away in early Polynesian boats and arrived accidentally (no one knows for sure). Other bird predators now common in the islands are two other rat species, BLACK and NORWAY RATS (Plate 47), the MONGOOSE (Plate 47), and feral house cats. The Black Rat, in particular, is a major menace to tree-nesting forest birds because it is a frequent tree-climber. Cat predation is now suspected of being a major cause of bird declines in Hawaii, but the actual amount of harm cats do is not known. Just a few examples of extinctions or near-extinctions caused by introduced predators: The HAWAIIAN RAIL, now extinct, probably succumbed to rat predation during the 1800s. Rats introduced to Midway Island during World War II led to the *extirpation* there of the LAYSAN RAIL and LAYSAN FINCH (meaning the elimination of those species there, but other populations of those species still exist elsewhere, on other islands). All of Hawaii's surviving native water birds (HAWAIIAN STILT, COMMON MOORHEN, HAWAIIAN COOT, HAWAIIAN DUCK; Plates 27, 28), in fact, are endangered because of the high level of predation on their nests and on adults by rats, cats, and mongooses. And the mongooses are considered to be a major factor in keeping NENE (HAWAIIAN GOOSE; Plate 27) populations so low – the mongooses eat eggs and young of the goose and also prey on adults. Luckily, mongooses are scarce around the higher-elevation natural nesting sites of Nene on Maui and the Big Island (see below); also, a new Nene population has been established artificially on Kauai – where there are no mongooses.

The effect of introduced mammalian predators on oceanic island bird species is summed up with this grim statistic: of the 130 or so birds known to have gone extinct during the past 400 years, 92% were island-dwelling species, and introduced predators accounted for at least 40% of these cases.

(2) *Absence of competitors*. Competition is one of the main ecological forces that acts on animal populations and species. Two or more different species may compete for the same resource, such as a type of food. The better competitors will win, that is, they will get more of the food and be able to breed successfully; their populations will be maintained or will grow. The "weaker" species, the poorer competitor, loses, obtains less of the food, breeds poorly or unsuccessfully; eventually the poor competitor dies out. Species of plants and animals we see today in mainland areas are all strong, successful competitors – the evolutionary survivors, if you will, of thousands of generations of competition between their ancestors and the ancestors of other, less successful, species. On isolated islands such as those of the Hawaiian chain, however,

relatively few species exist; habitats are less full of species (than in comparable mainland areas); there is, simply put, less competition. Continuing to think in terms of birds and of their food resources, the native Hawaiian birds were "accustomed" to being able to obtain their food without competition; they never evolved, never had reason to evolve, the behaviors (or anatomies or physiologies) that might allow them to be strong competitors – such as being super-efficient foragers or having the ability to aggressively defend feeding territories from other species seeking the same foods.

At last count, 47 different species of landbirds had been introduced to Hawaii's main islands during the past hundred years. Because these new birds came from mainland regions with high numbers of bird species, they are bound to be powerful competitors. We don't know as much about the effects of introduced-species competition on Hawaii's native birds as we do about the effects of introduced predators for the simple reason that predation is much easier to document (seeing predators near nests, finding destroyed nests, finding bunches of feathers where birds used to be, etc.). Competition, on the other hand, can be a subtle process, as subtle as one species being a more efficient locator of nectar-filled flowers than is another species. However, because some of the introduced birds eat the same foods as the natives, some competition is suspected, with, minimally, a weak negative effect on the natives. For example, the JAPANESE WHITE-EYE (Plate 39), a small alien, is now the most abundant land bird of Hawaii, present in a wide variety of habitats from sea level to treeline. White-eyes even penetrate into high-elevation native forests where they probably compete for flower nectar with threatened native honeycreepers (the white-eyes eat primarily bugs but also some fruit and flower nectar). Luckily, most of Hawaii's alien land birds are species of open habitats (they either came from such habitats in their native lands or, upon arrival in Hawaii, moved easily into these habitats because they were unfilled niches, habitats without competitors). In particular, the great majority of aliens do not occupy higher-elevation native forests; if they did, probably even more of the honeycreepers would now be extinct. The COMMON MYNA (Plate 38), now present in open and semi-open areas throughout Hawaii from sea level to about 2000 m (6500 ft), a hundred years ago was also very common in forest habitats. It is suspected of having been at that time a tough competitor for many native forest birds, perhaps to the point of having contributed to the declines of some of them. Also, introduced mammals may compete with native birds for food – rats are believed to have contributed to the decline of the almost-extinct POO-ULI by eating and so reducing the availability of the bird's arthropod and snail prey.

(3) *Absence of large grazers*. As is the case with large predatory beasts, large grazing mammals never make it on their own to isolated oceanic islands – pigs and cattle cannot cross seas themselves. But people brought their domesticated grazing mammals with them to Hawaii – cattle, goats, sheep, pigs (and deer) – and these domesticated animals and the feral herds they gave rise to, have severely damaged Hawaii's ecosystems. In fact, the experts believe that cattle and goats have probably caused more habitat destruction in the islands than any other non-human animals. Polynesians first brought pigs to Hawaii, but major ecosystem damage from grazers probably did not begin until the introduction of the larger European pigs in 1778, followed by sheep and cattle shortly before 1800.

These large feral mammals are particularly damaging to native habitats. They eat and so directly harm or kill plants, and are undoubtedly responsible for the elimination of some native plant species (but we have no way of knowing about these plants now). They also cause immense, far-reaching damage indirectly by actually altering habitats, by opening up dense forest habitats and allowing alien species to gain footholds – aliens that, once they are established, are powerfully competitive, fast-expanding and dominant members of plant communities. In other words, the grazing mammals allow "weed" species to gain access to native habitats from which they are then nearly impossible to remove. The weed species out-compete native plants for space and nutrients, causing their decline, and can negatively affect animals such as native birds because the native plants the birds depend on for food are reduced or eliminated.

Cattle were one of the worst offenders. At one point in Hawaiian history (including a period when there was a social prohibition on killing cattle), there were large herds of domestic and feral cattle on all the main islands, wandering through all types of habitats at all elevations. They caused large amounts of forest destruction and fragmentation and, with their grazing, probably caused some native plants to become extinct. Sheep on the Big Island greatly damaged extensive areas above 1000 m (3300 ft) with their grazing. They damage shrubs and trees (and so, for instance, greatly reduce the reproductive abilities of native Māmane trees; Plate 14), compact soil, and cause land erosion. Goats, now on all main islands but Lanai, damage drier open habitats at all elevations and eat rare native plants. Pigs, which damage a variety of habitat types, also occur on all main islands but Lanai.

How are native birds harmed by these grazing aliens? In a host of significant ways: the most obvious, as alluded to above, is that native birds often live and feed in native trees and the destruction of those trees by grazers such as cattle leads to bird declines. Cattle, goats, and possibly pigs, by their grazing and the habitat degradation they caused, are believed to be at least partly responsible for NENE (HAWAIIAN GOOSE; Plate 27) declines starting in the mid-1800s. The AKIAPOLAAU (Plate 43), an endangered Big Island honeycreeper, suffers from habitat loss and degradation owing to grazing by sheep and cattle (Akiapolaau also lost out to the lumber business; these birds feed on bugs on native Koa trees, Plate 6; but large Koa have been cut down extensively for their beautiful wood.) The PALILA (Plate 44), also an endangered Big Island honeycreeper, feeds mainly on the seedpods of one species of native tree, Māmane. Sheep and goats on the Big Island caused major damage to the Māmane habitats that the Palila depended on. The negative effect of the grazers on these habitats, and the link to the decline of the Palila, were so clear that, to protect the Palila (as mandated under provisions of the USA's Endangered Species Act), court decisions were obtained to allow the removal of sheep and goats from the Palila's last stronghold, the slopes of Mauna Kea (see Close-up, p. 158). Finally, feral pigs are a major and continuing problem in the islands. They eat rare native plants, especially those in wet forests. For instance, they stand accused of almost destroying or driving to extinction some species of the native group of lobeliad plants. Some nectar-eating native birds that fed at these plants are now probably extinct – such as BISHOP'S OO (a honeyeater) and the OU (a honeycreeper). The pigs root about in forests in the understory, eating and digging up roots and low-stature plants.

Unfortunately, some native birds, such as MAUI PARROTBILL (Plate 43) and POO-ULI, require a dense, complex understory in which to feed, and the pigs degrade the understory vegetation; both of these birds are now endangered – the Poo-uli about to go extinct. Another endangered honeycreeper, the AKOHEKOHE (Plate 44), which feeds mostly at ohia trees, also occasionally feeds on nectar produced by low-stature plants; so the pigs, by eating the plants, take away some of the birds' food. More about pigs below. Finally, let me point out that introduced organisms other than grazing mammals also damage Hawaiian habitats. Alien insects are suspected of having significant effects, too. For instance, the CHINESE ROSE BEETLE and BLACK TWIG BORER damage native trees.

(4) *Isolation from disease organisms.* Just as they were long isolated from predators and competitors, species that evolved on mid-ocean islands were also isolated from many *pathogens*, or disease-causing organisms. At first, this sounds like a good thing, and it is, until mainland pathogens are introduced to the islands. The problem then is that island species, such as Hawaii's native forest birds, having never been exposed to these pathogens, often have no natural immunity and are therefore extremely vulnerable. Animal immune systems work to kill or reduce the effects of pathogens that invade their bodies largely by having the ability to recognize the disease organisms as threats – either because the individual animal was previously exposed to the pathogen during its lifetime, survived, and developed a specific immune response to that pathogen, or some of its ancestors had natural recognition and immunity to it and passed them on genetically. If the pathogen is recognized as an invader in the body, immune system defenses can often kill it. But the immune systems of Hawaiian birds don't recognize many introduced pathogens, which therefore often are lethal to the birds.

A dark day for Hawaii's native birds occurred sometime around 1820, when the first disease-carrying mosquitos were introduced to the islands – they were probably stowaways on European ships. Mosquitos are *vectors* for many avian (and human) diseases, meaning they are transport mechanisms through which the actual pathogen gains access to the bird (or human) that will get the disease (called the *host* organism); the vector does not die of (or get sick from) the disease, the host does. Many pathogens, native and introduced, infect Hawaii's birds, including various viruses and bacteria. (Native pathogens that cause bird diseases probably first originated with migratory seabirds.) But the worst offender, carried frequently by the night-biting mosquito known to its friends as *Culex quinquefasciatus*, is thought to be a microscopically tiny one-celled parasitic organism, a protozoan, known as *Plasmodium relictum* – the source of *avian malaria*. Malaria is a parasitic disease of the blood and of tissues in the body that deal with blood: liver, spleen, bone marrow. It causes anemia, hemorrhaging, lesions and, eventually, death. The great majority of Hawaii's native forest birds that are exposed to avian malaria die of it. For instance, in an experiment, nine of ten IIWI (Plate 44) died of malaria soon after receiving *a single bite* of an infected mosquito. Both native and introduced birds serve as hosts for avian malaria in Hawaii, but the native birds are attacked more severely – with their long history of geographic and immunological isolation, they have less natural resistance.

Although it was probably around for quite some time, malaria was not identified as a bird disease in Hawaii until 1947; and it was not until 1968 that scientists realized that malaria was a major, if not paramount, factor in the elimination and reduction of many native bird species, and that, indeed, it was malaria that was limiting most species of native forest birds to high-elevation habitats. The *Culex* mosquito cannot reproduce in cold temperatures, so there are no mosquitos at high elevations, and the native birds at those altitudes are safe. But individual birds that fly down to middle and lower elevations are bitten and quickly succumb to malaria (or to other diseases, such as the viral disease known as *avian pox*). The maximum elevation in Hawaii at which the mosquitos can complete their life cycle in sufficiently warm temperatures is about 1200 m (4000 ft) – and this is why birdwatchers with local knowledge will often tell the uninitiated visitor to Hawaii "If you want to see native birds, you must go above 1200 m." (A few natives occur below 1200 m but have some immunity to malaria; see below.) Feral pigs have greatly exacerbated the malaria problem because they alter forest habitats, where the native birds live, to favor mosquito breeding. How? Simply put, their foraging behavior creates huge numbers of small "mosquito pools," into which female mosquitos lay their eggs and from which adult mosquitos eventually emerge. The pigs knock down treeferns and eat away parts of the thick trunks and branches, leaving hollowed out sections lying on the forest floor. Rainwater collects in the hollowed out sections, forming thousands of small mosquito breeding pools.

Other parasitic diseases and viral and bacterial diseases affect Hawaii's native birds, as do such external blood-sucking parasites as mites and ticks (both are spider relatives), but avian malaria is now considered to have caused devastating harm to native bird populations and to be the gravest current disease threat. In fact, the greatest fear Hawaiian bird conservationists have, aside from a successful invasion of the Brown Tree Snake (p. 11), which would likely exterminate most of the state's remaining native forest birds, is that a cold-adapted, malaria-carrying mosquito species is accidentally introduced. If that happened, the mosquitos could breed at high elevations and most of the surviving native birds, attacked in their last refuge, would soon be but a memory.

Conservation Programs

Hawaiian Eco-history

One of the first major acts in the conservation of unique native Hawaiian wildlife and habitats occurred in 1916 with the establishment, by an act of the USA Congress, of Hawaii National Park. The park (which in 1980 was recognized by UNESCO as an International Biosphere Reserve) includes both the Big Island's Hawaii Volcanoes National Park (HVNP) and Maui's Haleakala National Park. Although the park was initially established largely to preserve striking geological scenery, and although today most endangered species in Hawaii actually occur outside of the park's boundaries, the park is still very important for the habitats and wildlife it does protect. HVNP, for instance, preserves forever a huge chunk of land – nearly 10% of the Big Island, the largest area in the state in which native ecosystems are protected. An important characteristic of Haleakala National Park and (to a lesser degree) HVNP that may allow them to protect some kinds of wildlife is that they encompass the entire range of elevations at which Hawaiian animals occur, from sea level to mountaintops (Haleakala Summit on Maui, Mauna Loa Summit on the Big Island); wildlife, for example, that seasonally changes elevations, can be protected all year in the various habitats used. Although these large parks existed for decades and some scientists and naturalists realized that many native species were disappearing, it was not until the 1960s that researchers and National Park Service natural resource managers began to deduce the causes of the species declines – mainly the past and continuing habitat alteration state-wide, and the multitude of alien species introductions. With an initial understanding of the magnitude of the problems involved, an acceptance of the inevitable decline of Hawaii's native species may have at first prevailed. But during the early 1970s, experts began to meet to discuss the problems and to make plans to study them and formulate conservation actions.

To foster interaction between academic biological researchers and federal agency scientists and wildlife managers (US Fish and Wildlife Service, National Park Service), a Cooperative Park Studies Unit was established at the University of Hawaii/Manoa in 1973. In 1976, the first Hawaiian Natural Science Conference, held at HVNP, brought together resource managers, scientists, educators, and environmental administrators to discuss problems and plot conservation strategy. Because native birds were obviously a priority (they were disappearing fast!), one major undertaking, conducted from 1976 through 1983, was the Hawaiian Forest Bird Survey. Teams of experts at identifying birds censused birds in essentially all native forests above 1000 m (3300 ft) on Maui, Molokai, Lanai, the Big Island, and the Alakai Swamp region of Kauai. The idea was that before you could determine which areas most needed protection to save the forest birds, and before you decided on research and management strategies, first you had to know how many birds were left and where they were. The study resulted in a thick 1986 report (*Forest Bird Communities of the Hawaiian Islands: Their Dynamics, Ecology, and Conservation,* by J. M. Scott, *et al.*) that details for each species located, native and alien, its distribution according to elevation and habitats occupied and its population size. Many recent Hawaiian conservation and wildlife management projects aimed at native birds got their starting points, their background information and research, from the Forest Bird Survey. Finally, in 1992, with backing and money from the State of Hawaii and private foundations such as the John D. and Catherine T. MacArthur Foundation, the Bishop Museum in Honolulu began compiling a database listing and giving information on all species of plants and animals (not to mention fungi and some other living things) that occur in Hawaii. They are compiling information from scientific reports, from museum collections and records, and from past and new fieldwork. The idea behind the database (which can be accessed at www.bishop.hawaii.org) is the same as for the earlier forest bird survey: before you can devise conservation plans, you must know what species exist and where they are.

Conservation in Practice

Now that Hawaii's unique wildlife and habitats are widely appreciated, and the threats to them increasingly understood, many conservation projects have been established to try to preserve what's left and, if possible, restore some of what has been lost. In the limited space I have here, it's perhaps best to address some specific projects aimed at particular species (see also Close-up, p. 158). First, however, I should mention the efforts of one of Hawaii's foremost conservation organizations, The Nature Conservancy of Hawaii (TNC). Many people in the USA and other countries are familiar with The Nature Conservancy, an international non-profit organization that seeks to preserve wildlife and habitats primarily by purchasing tracts of land and then, through various arrangements, assuring that those lands are forever preserved in a wild state. But the local Hawaii affiliate of the larger organization faces a problem most other affiliates do not: buying land in Hawaii is difficult. The problem is that most of the state's land is owned by just a few large entities: the large corporations involved with sugarcane and pineapple plantations that in the past controlled much of Hawaii's economy (known sometimes as "the Big Five"), the Bishop Estate (created in the 1880s by the will of a Hawaiian princess and now Hawaii's largest private landowner), the State of Hawaii, and the USA federal government (the military alone controls about 107,000 hectares, 265,000 acres, of land in the state, including about a quarter of Oahu), among a very few others.

TNC's solution, in addition to buying land when possible (such as for Molokai's Moomomi Preserve), has been to pursue other kinds of land deals that allow for habitat preservation:

(1) They obtain from owners *conservation easements* (as gifts or for a fee) for parcels of ecologically important land – legal promises not to develop lands but to leave them wild and perhaps manage them with the aim of preserving current levels of biodiversity. Maui's Waikamoi and Kapunakea Preserves and Molokai's Kamakou Preserve were established under this kind of arrangement.
(2) They obtain *long-term leases* on tracts of land, such as the lease for the Honouliuli Preserve on Oahu.
(3) They arrange *management contracts* with owners, becoming in effect, the supervisors of tracts of lands, using their expertise for habitat and wildlife preservation.
(4) They *work with partners*, such as the University of Hawaii Arboretum, various botanical gardens, the State of Hawaii, and the US Fish and Wildlife Service, to preserve some lands and manage/improve habitats – for instance, by growing native plants to transfer to preserves. The state Natural Area Partnership (NAP) program, in particular, provides matching funds ($2 state to $1 private) to manage lands that have been designated for permanent conservation – such as Molokai's Kamakou Preserve.
(5) They *work with owners of large parcels of land adjacent to TNC preserves*, suggesting and teaching new conservation techniques. The result is larger contiguous chunks of land managed to conserve wildlife and habitats. For example, a large region of Maui is highly protected now because Haleakala National Park and TNC's adjacent Waikamoi Preserve have been in a sense "extended" by the cooperation of several adjacent landowners; in fact, by linking up protected lands owned by these different factions, TNC and others have succeeded in preserving an entire watershed area. In total, TNC has helped protect about 20,000 hectares (50,000 acres) of wild habitats on the six main islands and now manages about 8500 hectares (21,000 acres).

Another significant difference between the activities of TNC of Hawaii and other arms of that organization, is that because so many of Hawaii's native habitats have become terribly altered and degraded, TNC of Hawaii pursues more active management in the lands it owns or oversees. That is, elsewhere, TNC buys land and leaves it alone to stay wild; in Hawaii, the land needs some help to return to a state closer to what it was before people started altering habitats there. Examples: Fences are built in and around preserves to prevent pigs from entering and to reduce their movements among various sectors of a preserve. Very rare native plants are individually fenced to prevent their being eaten by pigs and goats. Some predatory mammals are killed, such as rats being eliminated with poisoned baits; the rats attack bird nests, but also eat threatened native land and tree snails. Invading (alien) plants are sometimes cut or otherwise eliminated or controlled, such as strawberry guava (Plate 10), Miconia (Plate 8), blackberry (Plate 11), and alien grasses; and native vegetation planted to replace the invaders.

Hawaiian Crow. What happened to the HAWAIIAN CROW after the 1992 National Academy of Sciences report mentioned in the first sentence of this chapter? Through some incredibly intense human efforts, the Hawaiian Crow still flies free over a small part of the Big Island. First, some background: several crow species probably ranged over the Hawaiian Islands in the past, but all but one was extinct by the time Europeans arrived 220 years ago. It ranged from about 300 m (1000 ft) to high elevations in the Big Island's dry and wet forests and parklands. By the 1890s, the Hawaiian Crow is known to have been declining in numbers; by the 1940s the crows no longer occurred at lower elevations and by the 1950s the species was reduced to small, isolated remnant populations. During the Hawaiian Forest Bird Survey, in 1978, the total wild population estimate was 76 individual birds. Through the 1960s, the crows were still commonly seen around ranches on the western (leeward, or kona) side of the Big Island, but by the 1970s, they were infrequently sighted. By the early 1990s, the last surviving group in the wild (perhaps three or four breeding pairs and a few other individuals), numbering perhaps 11 or 12 individuals in all, occurred on private property (owned by the McCandless Land and Cattle Company which, among other things, is in the Koa tree lumber business) on the western slopes of Mauna Loa in the South Kona District.

What pushed one very large black songbird to the brink of extinction? A combination of factors:

(1) *Habitat loss.* The crow's forest habitats were cleared for logging and for cattle and sheep grazing.
(2) *Food loss.* With habitat destruction came an associated reduction in the crow's food fruits, many of which grew in the forest understory.
(3) *Alien predators.* Introduced rats climbed trees and ate crow eggs and nestlings; introduced mongooses and feral cats killed adults and ate crow fledglings, which are helpless on the ground for a few days after popping out of nests. (A native predator, the Hawaiian Hawk, also occasionally kills crows, probably mainly fledglings.)
(4) *Exotic diseases.* The crows were susceptible to both avian malaria and avian pox, which killed or weakened many of them.
(5) *Shooting.* Crows were regarded as pests near farms and ranches, and by hunters, and were routinely shot.

As is often the case these days when people want to take action to try to save endangered species (and you don't get much more endangered than having a total world population reduced to 12 wild birds), controversial decisions needed to be made. The chief point of contention in this case was that the crow was listed as an endangered species according to the USA Endangered Species Act, and, according to the Act's rules, federal agencies were supposed to take action to try to preserve it; but the 12 crows lived on private land and the land's owners wanted the crows left alone (thinking, perhaps, that if the crows were simply left unmolested, the population would recover on its own; and perhaps thinking that if the USA government, in the guise of the Department of Interior's Fish and Wildlife Service, got involved, problems of one sort of another would inevitably ensue).

With the National Academy of Science report in 1992 advising that if active steps were not soon taken, the crow would most likely be extinct in the wild in 10 to 20 years, and with court cases brought by environmental organizations pending (to try to force the Fish and Wildlife Service to act to preserve the crows), all parties to the controversy worked out an agreement to try to save the Hawaiian Crow. One decision was to leave the wild birds in the wild (instead of capturing them all for a captive breeding program) but to take some of their eggs to protect them during the highly vulnerable egg, nestling, and fledgling stages (once the crows reach adulthood, their year-to-year survival in the wild is very high). Since the mid-1970s there had been a small captive flock of crows (about 10 individuals) held at a state-managed facility at Olinda on Maui, the Endangered Species Captive Propagation Facility. The idea had been to breed crows in the safety of captivity and release groups of them into the wild. But the captive group was highly inbred (all close relatives), so their nesting success was poor (of 33 fertile eggs produced between 1979 and 1991, only eight hatched live chicks and only six of those survived). The Peregrine Fund, a conservation organization with much experience raising birds in captivity to release them later into the wild, by 1996, was placed in charge of the Olinda facility (now called the Maui Bird Conservation Center), and a new one was built on the Big Island (Keauhou Bird Conservation Center; the Peregrine Fund also breeds other endangered species, including Palila and Puaiohi, in these facilities for eventual wild releases.) Clutches of eggs were taken from the wild crows (starting in 1993) and hatched in incubators (thus protecting them from nest predators). The adult nest-owning crows, when they discovered their eggs gone, laid new clutches of eggs (as will many bird species). Collected eggs are hatched and the young fed until they are of an age when the probability of predation by cats or mongooses is low. Then some are released to the wild, and some are brought into the captive breeding centers to increase those populations and bring more genetic diversity to young produced by the captive birds (that is, to reduce the degree of inbreeding, so more fertile eggs laid there will hatch and produce healthy chicks). Eventually, if enough birds are produced, perhaps other small wild populations, aside from the McCandless ranch population, can be started in the islands on protected public lands.

In addition to this captive propagation work, a host of agencies cooperate to try to save the crows (including the US Fish and Wildlife Service, the Biological Resources Division of the US Geological Survey, and the Hawaii Department of Land and Natural Resources). For instance, to decrease the threat of predation on wild crow fledglings, a comprehensive trapping program is pursued to try to eliminate cats, rats, and mongooses from the area around the wild crow population. Trees with crow nests are protected with metal barriers, or "predator guards," so predators on the ground cannot climb them. The result of all the efforts made on the crow's behalf? The news is not so good. As of mid-1999, the project has been only partly successful. The captive flock is breeding successfully, but the flock's offspring, released into the wild, have not fared well. Apparently the young crows, raised in captivity, lack the survival skills to deal in the wild with hawks, and several have been killed by the raptors. Some of the other released young crows have been recaptured to save them from the hawks. Meanwhile, the original wild flock is down to a single breeding pair and one other pair that has not yet bred.

The heroic effort made on the crow's behalf, which must cost quite a bit, raises an important and controversial question, that of resource/funds allocation in species conservation. Say an agency of the USA government has a million dollars to spend for conservation: is it better to use those funds for a massive effort to save a single species, one that has dwindled to just a few individuals left alive – such as the Hawaiian Crow? Or is it better to use the money to protect large tracts of wild or semi-wild land – entire ecosystems – thereby perhaps allowing some critically endangered species to die off, but perhaps saving, in the long run, many, perhaps hundreds (plants and animals) of others. Logically, the latter course is probably the better one to follow. But logic is usually only a small part of public policy decisions, and allowing a photogenic, charismatic species to go extinct, when something could be done to try to save it, is sometimes not an option. Such is the case of the NENE, Hawaii's state bird.

Nene. If you journeyed to Hawaii Volcanoes National Park during a typical winter, say in February, you might find some of the parking area entrances around Kilauea Crater blocked off with large brightly colored highway signs that, instead of saying "Closed For Construction" or "Closed – Volcanic Activity," actually say "Nene Territory – Closed For Nesting." Just what are Nene ("nay-nay") and why do we close parking lots to help their breeding efforts? The Nene, or Hawaiian Goose, an endangered species, is a close relative of the Canada Goose, which exists over much of North America in tremendous numbers. But it is a separate species, endemic to Hawaii, and with its idiosyncratic neck with furrowed rows of whitish/buffy feathers and fully terrestrial habits, it looks and behaves unlike its hugely abundant relative. Fossils tell us that Nene once roamed over many sites at middle and high elevations (and even some lower elevations) on Kauai, Lanai, Maui, Oahu, Molokai, and the Big Island. Population estimates are that, in the past, 25,000 or more Nene ranged over the islands. But various factors caused their populations to plunge, and by the 1890s, these geese were gone from all except the Big Island (and they may have survived there only because the parts of the island where they still roamed were fairly inaccessible or considered off-limits to hunters and other people). By the late 1940s there were only about 30 Nene left in the wild. What caused the decline? Because Nene evolved essentially in the absence of predators, they did not, and still do not, run away from people and predatory animals. Hunters easily killed thousands of them. Introduced predatory mammals killed many adults and ate their eggs and young. And habitat destruction limited the areas in which they could breed and forage well.

Although some captive breeding had been tried since the mid-1920s, in 1949 a large program was initiated, at Pohakuloa on the Big Island and in England at the Wildfowl Trust in Slimbridge. Between 1960 and 1995, more than 2150 captive-bred Nene were released, mostly at two sites: starting in 1962 on Maui at Haleakala National Park (thus re-establishing the bird on an island it had been absent from for 70+ years); and at Hawaii Volcanoes National Park on the Big

Island. Because the habitats to which the Nene are restricted in the national parks are of fairly low quality (nutritious food is apparently hard to come by for adults and young) and because parts of the parks still harbor cats and mongooses that prey on Nene, the populations are not self-sustaining, and new captive-raised birds are still added occasionally. In fact, many Nene that were released into high-elevation barren volcanic habitats during the earlier years of the release program died quickly; ones that survived were often those that left the release sites and set up shop in agricultural locations or areas with lawns and golf courses that provided plentiful and easy grass food. During the 1990s, in an effort to enlarge the number of wild populations, to try to create self-sustaining populations, and to provide a "safety-net" in case the high-elevation populations at the national parks suffered major setbacks, new populations were established: one on Kauai in low-elevation open grassland habitat on the north side of the island (the Kilauea Point area; this population was actually started accidentally when some Nene were unintentionally released on Kauai, the result of a 1983 hurricane), and one at Hakalau Forest National Wildlife Refuge on the Big Island in upper elevation grassland. On mongoose-free Kauai, at least, Nene are doing well. Still, probably less than a thousand of these geese exist in the wild. The general plan for them now includes enhancing the grassland areas in which they do well; maintaining predator control programs that keep adults and their eggs and young safe (trapping and poison-baiting rats, cats, and mongooses; fencing out feral pigs); maximizing the genetic diversity of present populations by translocating individuals among populations; and perhaps creating even more small populations, if suitable sites are located (such as mongoose-free Lanai).

The Future

To stabilize Hawaii's native biodiversity – to staunch the steady progression of plant and animal extinctions – and try to restore some of what has been lost, what are the main things that need doing? Again, let me focus on Hawaii's native forest birds, with the understanding that much of what needs to be done for their preservation would also protect other native species, plants and animals, as well.

The conservation of native forest birds now depends essentially on protection of habitats above 1200 to 1500 m (4000 to 5000 ft), where some of the birds still thrive but where none of the mosquitos that transmit avian malaria can live. The main conservation efforts should be to (1) protect the birds' habitats in preserves; (2) remove from the preserves predators such as cats, rats, and mongooses; and grazers and rooters such as pigs, goats, and sheep; (3) fence large areas to keep predators and grazers out of protected bird areas – as is being done by The Nature Conservancy of Hawaii in many of the preserves they manage, and by the National Park Service at Haleakala and Hawaii Volcanoes National Parks.

Because it is now generally believed that the introduction of alien species (mammals, birds, snakes) has caused more recent extinctions of island birds than any other factor, crucial conservation steps are reducing or even eliminating introduced species already in Hawaii and preventing new introductions. Preventing additional introductions is especially important: if the Mongoose becomes established on Kauai, populations of that island's remaining few native forest birds may plunge; if the Brown Tree Snake invades Hawaii, most of its native birds may face swift extinction; likewise, if new avian diseases reach the islands and become established, native birds could suffer terribly. Careful controls of bird imports and strict incoming animal quarantine laws seek to reduce the probability of new bird diseases reaching Hawaii, but each year's five or six million airline passenger arrivals, along with uncountable pieces of baggage and millions of tons of commercial and military cargo, make it difficult for such controls to be 100% effective.

It's now considered that diseases such as avian malaria are the most important factor preventing the recovery of several endangered forest bird species. What's to be done about this major problem? First, since the presence of feral pigs strongly facilitates malaria-mosquito breeding, pig controls are essential – fence them out of some areas, remove them completely from others. (Feral pig removal is a controversial issue in Hawaii because, although pigs could probably be eliminated or severely reduced in number over large areas, using trapping, poison baits, and other methods, many local residents hunt pigs for food or sport and very much want healthy pig populations roaming large parts of the islands.) Second, to eliminate many potential breeding sites for the mosquitos, forest restoration is necessary. Third, research should be pursued on the evolution of natural immunity to avian malaria and other diseases. A few birds have already developed some immunity: the OAHU AMAKIHI now thrives below 1300 m (4250 ft) elevation, where there are many disease-carrying mosquitos; and the APAPANE on the Big Island likewise seems to have adapted to avian malaria. Research might indicate ways to further or speed up the development of this natural immunity on other islands and perhaps in other species. For instance, scientists are considering translocating immune individuals from their home islands to islands where immunity has yet to develop, with the hope that the birds will breed in their new homes and pass on their immunity.

Other important conservation measures include trying to control some alien plants that increasingly choke out native vegetation, and changing some land-use policies. For instance, it would be helpful if there were economic incentives for local landowners to promote conservation, for example, by leaving intact any native forest habitat that occurs on their land. Currently, landowners obtain a lower tax rate from the state by clearing forests and introducing cattle to their land than they do by leaving forests intact. To build political support to pursue such policy changes, raising public awareness of the issues involved is essential.

It is very much up to local citizens to undertake and support these kinds of conservation efforts. But is there anything a visitor to Hawaii can do for conservation's sake, aside from making donations to non-profit conservation organizations and paying your federal taxes? Here are two main suggestions: First, obviously, never transport non-native animals or plants to Hawaii or between islands when you are in Hawaii. Even if you follow this advice consciously, you may still transport plant seeds from one island to another, or from lowlands to highlands on the same island, unintentionally – in the mud on your shoes or boots or on your clothes; clean up carefully after each hike. And second, in your travels, if you see new species that do not belong – any snake, for instance, or non-native birds or mammals not already established in the islands – report your sighting immediately to the Hawaii Department of Agriculture or appropriate other state or federal agencies; or if you see cats or pigs, etc., in fenced areas of nature preserves or national parks that should be free of these mammals, report the sighting to preserve or park managers.

Environmental Close-up 2 Endemism: Political System Advocating the End of the World, or Something More Interesting?

A few years ago, in New Zealand, I attended an international scientific conference on birds. The conference, a quadrennial event, moves from continent to continent, I suspect because this permits the participants – scientific researchers who usually double as birdwatchers – to see wild birds they have not seen before and cannot see back home. Indeed, overheard conversations at the conference centered on two topics: the awful cafeteria food at the host university, and seeing "*endemics.*" People would ask each other "Which endemics have you seen so far?" and "Where would I go to see this or that endemic?". What they were referring to were New Zealand species of birds that occur nowhere else on Earth; for many a birder, seeing such unique species is the paramount reason for visiting isolated spots of the world such as New Zealand or Hawaii. An organism is endemic to a place when it is found only in that place. But the size or type of place referred to is variable: a given species of frog, say, may be endemic to the Western Hemisphere, to a single continent such as South America, to a small mountainous region of Peru, or to a speck of an island off Peru's coast.

A species' history dictates its present distribution. When it's confined to a certain or small area, the reason is that (1) there are one or more barriers to further spread (an ocean, a mountain range, a thousand kilometers of tropical rainforest in the way), (2) the species evolved only recently and has not yet had time to spread, or (3) the species evolved long ago, spread long ago, and now has become extinct over much of its prior range. A history of isolation also matters: the longer a group of animals and plants are isolated from their close relatives, the more time they have to evolve by themselves and to change into new, different, and unique groups. The best examples are on islands. Some islands once were attached to mainland areas but continental drift and/or changing sea levels led to their isolation in the middle of the ocean; other islands arose wholly new via volcanic activity beneath the seas. Take the island of Madagascar, once attached to Africa and India. The organisms stranded on its shores when it became an island had probably 100 million years in isolation to develop into the highly endemic fauna and flora we see today. It's thought that about 80% of the island's plants and animals are endemic – half the bird species, about 800 butterflies, 8000 flowering plants, and essentially all the mammals and reptiles. Most of the species of lemurs of the world – small, primitive but cute primates – occur only on Madagascar, and an entire nature tourism industry has been built there around the idea of endemism: if you want to see wild lemurs, you must go there. Other examples of islands with high concentrations of endemic animals abound: Indonesia, where about 15% of the world's bird species occur, a quarter of them endemic; Papua New Guinea, where half the birds are endemic; the Philippines, where half the mammals are endemic.

The Hawaiian Islands arose initially as lifeless rock, the result of volcanic activity beneath the Pacific Ocean. Because they are so isolated – separated from any major patch of dry land by 4000 km (2500 miles) or so (and a few hundred trillion gallons) of occasionally churning saltwater – relatively few plant and animal species, which evolved in mainland areas, managed to immigrate naturally (how they did so – mostly by blowing or washing ashore – is addressed in a Close-up, p. 100). These few successful colonizers, finding essentially empty or only lightly filled habitats and ecological niches, quickly diversified evolutionarily (that is, they *speciated*, formed new species over evolutionary time) to fill the niches. This is itself a fascinating subject, one that many scientists study. Estimates, for example, are that Hawaii's 1100 or so species of native plants arose via evolutionary diversification from perhaps 270 colonizing species; 8000 insects arose from 250 or so colonizations (including 550+ species of the fruit fly genus, *Drosophila*, evolving from just a few – possibly only two – species colonizations); more than a thousand land snail species arose from 20 or so colonizations; and 80 land birds arose from between four and seven colonizations. The end result of a few million years of these processes? A large proportion of the native animals and plants in Hawaii today are endemic – globally unique (see table below). By some measures, Hawaii registers as number one in the world for the proportion of its biota that is endemic, even though the state encompasses less than two-tenths of 1% of the Earth's dry-land surface area. Depending on how the calculations are made, it is usually considered that almost 100% of Hawaii's native invertebrate animals are endemic, as are more than 95% of its native birds and 90% or so of its native flowering plants.

Why is a knowledge of endemism important for ecotravellers? Two main reasons, one practical, one environmental. First, like my friends the New Zealand birdwatchers, if there is a specific type of wildlife you'd like to see, you must first know where it occurs, then travel there. Madagascar for lemurs. Africa or Asia for elephants. Australia for koalas and wombats. New Zealand for Yellow-eyed Penguins. Hawaii for Drepanidid honeycreepers (p. 151) such as Amakihi, Apapane, and Iiwi. If you wanted to visit a region where you might encounter large varieties of strange, exotic wildlife, a region with a high degree of endemism would be just the ticket – such as some of the "hot spots" mentioned below. Second, species that are endemic to small areas often bear a special environmental vulnerability. Basically, when and if their numbers fall, these species or groups face a greater chance of extinction than others because they lack other places "to go,"

Hawaiian Endemism: Total Number of Species in Hawaii and Percent Endemic[1]

Group	Total number of species[2]	Number of species known to be endemic	% of species known to be endemic
Fungi and lichens	2023	240	12
Flowering plants	1894	850	45
Mollusks	1656	956	58
Insects	7906	5188	66
Fish	1195	139	12
Amphibians	5	0	0
Land reptiles	20	0	0
Land/sea mammals	44	1 (Monk Seal)	2

[1] Modified from L. G. Eldredge and S. E. Miller 1995.

[2] Number known as of the mid-1990s. Many species that are not endemic are alien, or introduced, species; also, a good many are of unknown origin, especially among such lesser-known groups as fungi, mollusks, and arthropods.

other populations in far-off places that might survive. Good examples are species that are endemic to islands. If a species of bird occurs only on a single island, all of its "eggs" are in one basket, so to speak: if there is a calamity there – a powerful hurricane, a volcanic eruption – the entire species could become extinct, because all individuals there die and there are no others elsewhere. This type of species extinction has apparently happened often on islands with birds over the past 400 years, as people colonized. People caused habitat destruction and brought animal predators that the native birds had no fear of or experience with. It's thought that about 108 bird species have become extinct in the last 400 years, 97 of them island endemics. (The problem persists: about 900 of the 9000 living bird species are island endemics, and so continually vulnerable; most of the remaining native Hawaiian forest birds, of course, all endemics, are now threatened or endangered.) Similarly, about 75% of mammals driven to extinction recently have been island dwellers.

Knowledge of the existence and distribution of endemic species is crucial for conservation of biodiversity. If we want to preserve biodiversity, then identifying areas with unique species (endemics) and areas with large numbers of unique species (*centers of endemism*), then targeting those areas for conservation attention, is a potentially profitable strategy. In other words, we don't have to make much of an effort to conserve species that are distributed worldwide or hemisphere-wide: their broad ranges often provide protection against quick extinction; but endemics, with their restricted distributions, are inherently more vulnerable and so deserving of immediate attention. A recent concept in conservation biology has been the idea of *"hot spots"* – relatively small areas of the world supporting very high numbers of endemic species, areas that should therefore receive priority conservation attention (meaning that time, effort, and funds allocated in these regions will result in greater conservation of biodiversity than efforts elsewhere). For instance, it's estimated that fully 20% of the Earth's endemic plants occur over just half a percent of the world's land area: preserve that half a percent and save 20% of endemic plants. Some reptile and amphibian hot-spots are Madagascar, certain regions of Colombia, and Atlantic coastal Brazil; chief mammal hot-spots are the Philippines, Madagascar, and northern Borneo. Hawaii, with its highly endemic native flora and fauna, can certainly be considered a "hot spot" of endemic biodiversity, a place where investments in conservation will hopefully pay big dividends in preserving large numbers of unique species.

Chapter 6

How to Use This Book: Ecology and Natural History

What is Natural History?

The purpose of this book is to provide ecotravellers with sufficient information to identify many common animal and plant species and to learn about them and the families to which they belong. Information on the lives of animals is generally known as *natural history*, which is usually defined as the study of animals' natural habits, including especially their ecology, distribution, classification, and behavior. This kind of information is of importance for a variety of reasons: researchers need to know natural history as background on the species they study, and wildlife managers and conservationists need natural history information because their decisions about managing animal populations must be partially based on it. More relevant for the ecotraveller, natural history is simply interesting. People who appreciate animals typically like to watch them, touch them when appropriate, and know as much about them as they can.

What is Ecology and What Are Ecological Interactions?

Ecology is the branch of the biological sciences that deals with the interactions between living things and their physical environment and with each other. *Animal ecology* is the study of the interactions of animals with each other, with plants, and with the physical environment. Broadly interpreted, these interactions take into account almost everything we find fascinating about animals –

what they eat, how they forage, how and when they breed, how they survive the rigors of extreme climates, why they are large or small, or dully or brightly colored, and many other facets of their lives.

An animal's life, in some ways, is the sum of its interactions with other animals – members of its own species and others – and with its environment. Of particular interest are the numerous and diverse ecological interactions that occur between different species. Most can be placed into one of several general categories, based on how two species affect each other when they interact; they can have positive, negative, or neutral (that is, no) effects on each other. The relationship terms below are used in the book to describe the natural history of various animals.

Competition is an ecological relationship in which neither of the interacting species benefit. Competition occurs when individuals of two species use the same resource – a certain type of food, nesting holes in trees, etc. – and that resource is in insufficient supply to meet all their needs. As a result, both species are less successful than they could be in the absence of the interaction (that is, if the other species were not present).

Predation is an ecological interaction in which one species, the *predator*, benefits, and the other species, the *prey*, is harmed. Most people think that a good example of predation would be a mountain lion eating a deer, and they are correct; but predation also includes interactions in which the predator eats only part of its prey and the prey individual often survives. Thus, deer eat tree leaves and branches, and so, in a way, they can be considered predators on plant prey.

Parasitism, like predation, is a relationship between two species in which one benefits and one is harmed. The difference is that in a predatory relationship, one animal kills and eats the other, but in a parasitic one, the parasite feeds slowly on the "host" species and usually does not kill it. There are internal parasites, like protozoans and many kinds of worms, and external parasites, such as leeches, ticks, and mites.

Mutualisms, which include some of the most compelling of ecological relationships, are interactions in which both participants benefit. Plants and their pollinators engage in mutualistic interactions. A bee species, for instance, obtains a food resource, nectar or pollen, from a plant's flower; the plant it visits benefits because it is able to complete its reproductive cycle when the bee transports pollen to another plant. A famous case of mutualism involves several species of acacia plants and the ants that live in them: the ants obtain food (the acacias produce nectar for them) and shelter from the acacias and in return, the ants defend the plants from plant-eating insects. Sometimes the species have interacted so long that they now cannot live without each other; theirs is an *obligate* mutualism. For instance, termites cannot by themselves digest wood. Rather, it is the single-celled animals, protozoans, that live in their gut that produce the digestive enzymes that digest wood. At this point in their evolutionary histories, neither the termites nor their internal helpers can live alone.

Commensalism is a relationship in which one species benefits but the other is not affected in any way. For example, *epiphytes*, such as orchids and bromeliads, that grow on tree trunks and branches obtain from trees some shelf space to grow on, but, as far as anyone knows, neither hurt nor help the trees. A classic example of a commensal animal is the remora (Plate 57b), a fish that attaches itself with a suction cup on its head to a shark, then feeds on scraps of food the shark leaves behind. Remora are commensals, not parasites – they neither harm nor help sharks, but they benefit greatly by associating with sharks. Cattle Egrets (Plate 27) are commensals – these birds follow cattle, eating insects and other small animals that flush from cover as the cattle move about their pastures; the cattle, as far as we know, couldn't care one way or the other (unless they are concerned about that certain loss of dignity that occurs when the egrets perch not only near them, but on them as well).

A term many people know that covers some of these ecological interactions is *symbiosis*, which means living together. Usually this term suggests that the two interacting species do not harm one another; therefore, mutualisms and commensalisms are the symbiotic relationships discussed here.

How to Use This Book

The information here on animals is divided into two sections: the *plates*, which include photographs and artists' color renderings of various species together with brief identifying and location information; and the *family profiles*, with natural history information on the families to which pictured animals belong. The best way to identify and learn about Hawaiian animals may be to scan the illustrations before a trip to become familiar with the kinds of animals you are likely to encounter. Then when you spot an animal, you may recognize its type or family, and can find the appropriate pictures and profiles quickly. Color photos of common plant species are shown in Plates 1 through 16, with accompanying captions; many of the plants shown are referred to in Chapter 3.

Information in the Family Profiles

Classification, Distribution, Morphology

The first paragraphs of each profile generally provide information on the family's classification (or *taxonomy*), geographic distribution, and *morphology* (shape, size, and coloring of the animals). Classification information is provided because it is how scientists separate animals into related groups and often it enhances our appreciation of animals to know these relationships. You may have been exposed to classification levels sometime during your education but if you are a bit rusty, a quick review may help: *Kingdom* Animalia: aside from plant information, all the species detailed in the book are members of the animal kingdom. *Phylum* Chordata, *Subphylum* Vertebrata: most of the species in the book, including fish, are vertebrates, animals with backbones (exceptions are the marine *invertebrate* animals discussed in Chapter 10 and the insects and other arthropods in Chapter 11). *Class*: the book covers several vertebrate classes such as Amphibia (amphibians), Reptilia (reptiles), Aves (birds), and Mammalia (mammals). *Order*: each class is divided into several orders, the animals in each order sharing many characteristics. For example, one of the mammal orders is Carnivora, the carnivores, which includes mammals with teeth specialized for meat-eating – dogs, cats, bears, raccoons, weasels. *Family*: families of animals are subdivisions of each order that contain closely related species that are very similar in form, ecology, and behavior. The family Canidae, for instance, contains all the dog-like mammals – coyote, wolf, fox, dog. Animal family names end in "-dae;" subfamilies, subdivisions of families, end in "-nae." *Genus*: further subdivisions; within each genus are grouped species that are very closely related – they are all considered to have evolved from a common ancestor. *Species*: the lowest classification level; all members of a species are similar enough to be able to breed and produce living fertile offspring.

Example:	Classification of the Laysan Albatross (Plate 20):
Kingdom:	Animalia, with more than a million species
Phylum:	Chordata, Subphylum Vertebrata, with about 40,000 species
Class:	Aves (Birds), with about 9000 species
Order:	Procellariiformes, with about 108 species; includes albatrosses, shearwaters, petrels, storm-petrels, and diving petrels
Family:	Diomedeidae, with 14 species; all the albatrosses
Genus:	*Phoebastria*, a genus of albatrosses, with 4 species
Species:	*Phoebastria immutabilis*, the Laysan Albatross

Some of the family profiles in the book actually cover animal orders; others describe families or subfamilies.

Species' distributions vary tremendously. Some species are found only in very limited areas, whereas others range over several continents. Distributions can be described in a number of ways. An animal or group can be said to be *Old World* or *New World*; the former refers to the regions of the globe that Europeans knew of before Columbus – Europe, Asia, Africa; and the latter refers to the Western Hemisphere – North, Central, and South America. Biogeographers – scientists who study the geographic distributions of living things – consider Hawaii to fall within the part of the world called the *Polynesian* sub-region of the *Australian* region (includes Australia, New Zealand, the New Guinea area, and oceanic islands of the Pacific). A Polynesian species is one that occurs naturally on one or more of the Pacific's oceanic islands. The terms *tropical, temperate,* and *arctic* refer to climate regions of the Earth; the boundaries of these zones are determined by lines of latitude (and ultimately, by the position of the sun with respect to the Earth's surface). The tropics, always warm, are the regions of the world that fall within the belt from 23.5 degrees North latitude (the Tropic of Cancer) to 23.5 degrees South latitude (the Tropic of Capricorn). The world's temperate zones, with more seasonal climates, extend from 23.5 degrees North and South latitude to the Arctic and Antarctic Circles, at 66.5 degrees North and South. Arctic regions, more or less always cold, extend from 66.5 degrees North and South to the poles. The main islands of Hawaii are located within the tropics; the chain of the Northwestern Hawaiian Islands falls mostly north of the Tropic of Cancer, and so within the north temperate zone (Map 7).

Several terms help define a species' distribution and describe how it attained its distribution:

Range. The particular geographic area occupied by a species.

Native or *Indigenous*. Occurring naturally in a particular place.

Introduced. Occurring in a particular place owing to peoples' intentional or unintentional assistance with transportation, usually from one continent to another, or from a continent to an island; the opposite of native. For instance, pheasants were initially brought to North America from Europe/Asia for hunting, Europeans brought rabbits and foxes to Australia for sport, and the British brought European Starlings and House Sparrows to North America. Other words

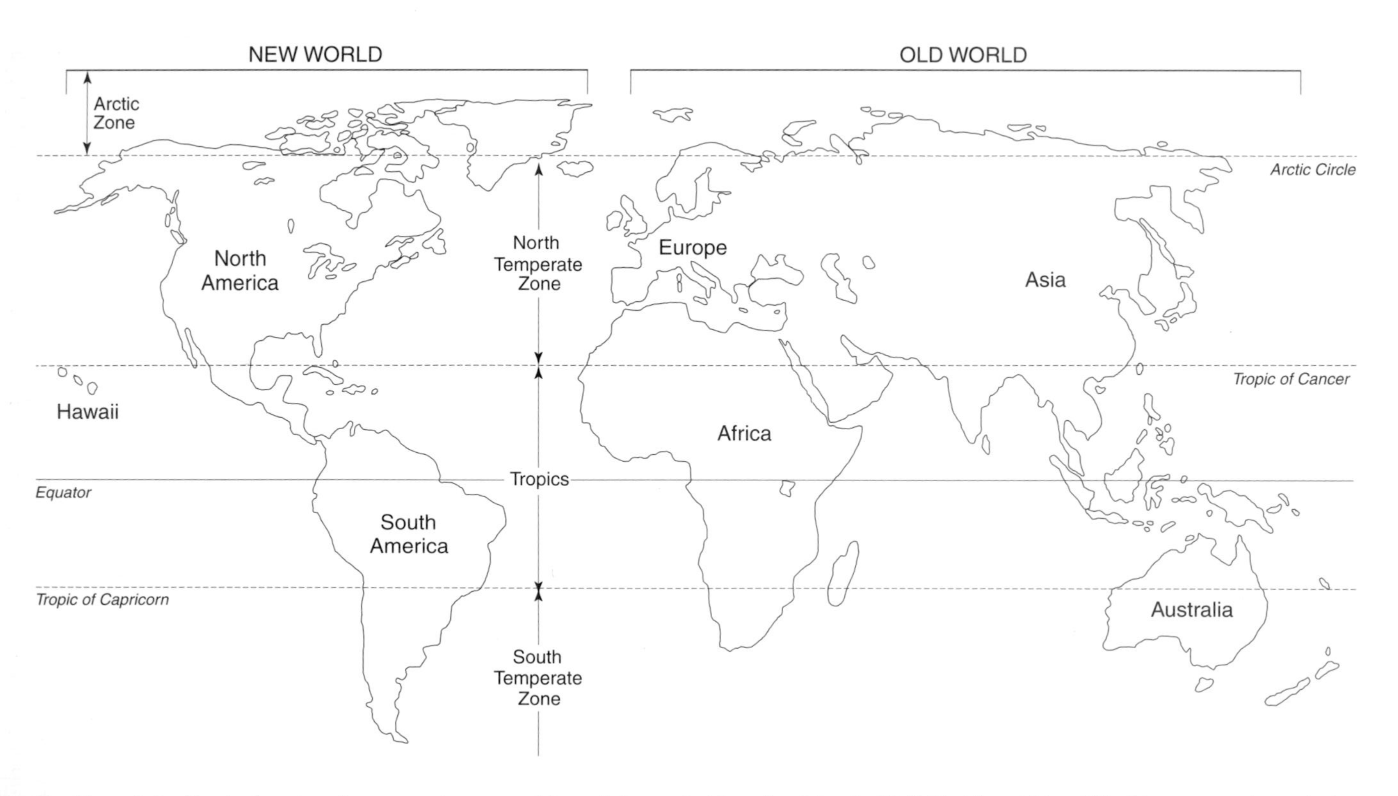

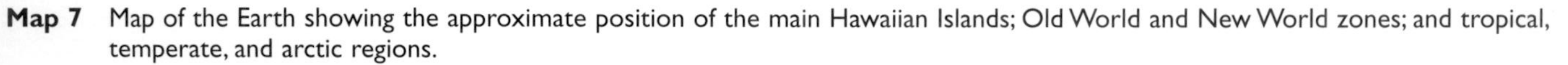

Map 7 Map of the Earth showing the approximate position of the main Hawaiian Islands; Old World and New World zones; and tropical, temperate, and arctic regions.

used somewhat interchangeably with "introduced" to describe non-native species spread by people are *alien, exotic,* and *invader.*

Endemic. A species, a genus, an entire family, etc., that is found in a particular place and nowhere else (see Close-up, p. 67). Galápagos finches are endemic to the Galápagos Islands; Hawaiian honeycreepers are endemic to Hawaii; nearly all the reptile and mammal species of Madagascar are endemics; all species are endemic to Earth (as far as we know).

Cosmopolitan. A species that is widely distributed throughout the world.

Some terms used especially for birds: A *resident* species stays in the same place (that is, Hawaii) all year and breeds there. A *migratory* species breeds in one location but spends the non-breeding portion of each year elsewhere. For instance, many shorebird species and ducks breed in North America but winter in Hawaii. *Visitors,* such as the PEREGRINE FALCON and OSPREY (Plate 31), are species that neither breed nor regularly spend long periods in a place, such as Hawaii – but they visit occasionally as they wander the world; some species are fairly regular visitors, while others are rare visitors. Seabirds can often be classed as *coastal* (spending all or much of their time along mainland or island coasts); *inshore* (occurring near the seashore, usually in or over water depths of 50 m (160 ft) or less; the terms "coastal" and "inshore" are sometimes used interchangeably); *offshore* (occurring at sea in or over water depths between 50 and 200 m (160 and 650 ft), usually over continental shelves; these species rarely come near shore, except to nest); or *pelagic* (open ocean species that do not come near land or even continental shelves, except to nest).

Ecology and Behavior

In these sections, I describe some of what is known about the basic activities pursued by each group. Much of the information relates to when and where animals are usually active, what they eat, and how they forage.

Activity Location – *Terrestrial* animals pursue life and food on the ground. *Arboreal* animals pursue life and food in trees or shrubs. *Cursorial* refers to animals that are adapted for running along the ground.

Activity Time – *Nocturnal* means active at night. *Diurnal* means active during the day. *Crepuscular* refers to animals that are active at dusk and/or dawn.

Food Preferences – Although animal species can usually be assigned to one of the feeding categories below, most eat more than one type of food. Most frugivorous birds, for instance, also nibble on the occasional insect, and carnivorous mammals occasionally eat plant materials.

Herbivores are predators that prey on plants.
Carnivores are predators that prey on animals.
Insectivores eat insects.
Granivores eat seeds.
Frugivores eat fruit.
Nectarivores eat nectar.
Piscivores eat fish.
Omnivores eat a variety of things.
Detritivores, such as vultures, eat dead stuff.

Ecological Interactions

These sections describe what I think are intriguing ecological relationships. Groups that are often the subject of ecological research are the ones for which such relationships are more likely to be known.

Breeding

In these sections, I present basics on each group's breeding particulars, including type of mating system, special breeding behaviors, durations of egg incubation or *gestation* (pregnancy), as well as information on nests, eggs, and young.

Mating Systems – A *monogamous* mating system is one in which one male and one female establish a pair-bond and contribute fairly evenly to each breeding effort. In polygamous systems, individuals of one of the sexes have more than one mate (that is, they have harems): in *polygynous* systems, one male mates with several females, and in *polyandrous* systems, one female mates with several males.

Condition of young at birth – *Altricial* young are born in a relatively undeveloped state, usually naked of fur or feathers, eyes closed, and unable to feed themselves, walk, or run from predators. *Precocial* young are born in a more developed state, eyes open, and soon able to walk and perhaps feed themselves.

Notes

These sections provide brief accounts of folklore associated with the profiled groups, and any other interesting bits of information about the profiled animals that do not fit elsewhere in the account.

Status

These sections comment on the conservation status of each group, including information on relative rarity or abundance, factors contributing to population declines, and special conservation measures that have been implemented. The definitions of the terms that I use to describe degrees of threat to various species are these: *Endangered* species are known to be in imminent danger of extinction throughout their range, and are highly unlikely to survive unless strong conservation measures are taken; populations of endangered species generally are very small, so they are rarely seen. *Threatened* species are known to be undergoing rapid declines in the sizes of their populations; unless conservation measures are enacted, and the causes of the population declines identified and halted, these species are likely to move to endangered status in the near future. *Vulnerable to threat*, or *Near-threatened*, are species that, owing to their habitat requirements or limited distributions, and based on known patterns of habitat destruction, are highly likely to be threatened in the near future. Several organizations publish lists of threatened and endangered species.

Where appropriate, I also include threat classifications from the Convention on International Trade in Endangered Species (CITES) and the United States Endangered Species Act (USA ESA). CITES is a global cooperative agreement to protect threatened species on a worldwide scale by regulating international trade in wild animals and plants among the 144 or so participating countries. Regulated species are listed in CITES Appendices, with trade in those species being strictly regulated by required licenses and documents. CITES Appendix I lists endangered species; all trade in them is prohibited. Appendix II lists threatened/vulnerable species, those that are not yet endangered but may soon be; trade in them is strictly regulated. Appendix III lists species that are protected by laws of individual countries that have signed the CITES agreements. The USA's Endangered Species Act works in a similar way – by listing endangered and threatened species, and, among other provisions, strictly regulating trade in those animals. The International Union for Conservation of Nature (IUCN) maintains a "Red List" of threatened and endangered species that often is more broad-based and inclusive than these other lists.

Information in the Color Plate Sections

Pictures. Among amphibians, reptiles, and mammals, males and females of a species usually look alike, although often there are size differences. For many species of birds, however, the sexes differ in color pattern and even anatomical features. If only one individual is pictured, you may assume that male and female of that species look exactly or almost alike; when there are major sex differences, both male and female are depicted.

Name. I provide the common English name for each profiled species, the scientific, or Latin, name, and Hawaiian name with Hawaiian language spelling, if known.

ID. Here I provide brief descriptive information that, together with the pictures, will enable you to identify most of the animals you see. The lengths of reptiles and amphibians given in this book are *snout–vent lengths* (SVLs), the distance from the tip of the snout to the vent, unless I mention that the tail is included. The *vent* is the opening on their bellies that lies approximately where the rear limbs join the body, and through which sex occurs and wastes exit. Therefore, long tails of lizards, for instance, are not included in the reported length measurements, and frogs' long legs are not included in theirs. For mammals, measurements I give are generally the lengths of the head and body, but do not include tails. For some mammals, I give approximate weights. Birds are measured from tip of bill to end of tail. For birds commonly seen flying, such as seabirds and hawks, I provide wingspan (wingtip to wingtip) measurements, if known. For most of the passerine birds (see p. 107), I use to describe their sizes the terms *large* (more than 30 cm, 12 in, long); *mid-sized* (between 15 or 18 cm, 6 or 7 in, and 30 cm, 12 in); *small* (11 to 15 cm, 4.5 to 6 in); and *very small* (less than 11 cm, 4.5 in). Lengths given for fish are *standard lengths*, the distance from the front of the mouth to the point where the tail appears to join the body; that is, tails are not included in the measurement.

Habitat/Islands. In these sections I give the habitat types in which each species occurs, symbols for the habitat types each species prefers, and the islands where each species may be found (BIG = the Big Island of Hawaii, KAU = Kauai, LAN = Lanai, MAUI = Maui, MOLO = Molokai, OAHU = Oahu, MID = Midway Atoll and nearby islands of the Northwestern chain). In general, when I say an animal is usually found at *low elevations*, I mean at elevations of less than 500 m (1600 ft); at *middle elevations*, between 500 and 1100 m (1600 and 3600 ft); and at *high elevations*, above 1100 m (3600 ft).

Explanation of habitat symbols:

= Lowland wet forest.

 = Lowland dry woodlands.

 = Middle and higher elevation forests, including cloud forests.

= High elevations, above treeline (usually at or above 3050 m, 10,000 ft); includes subalpine and alpine scrub areas.

= Forest edge/streamside. Some species typically are found along forest edges or near or along streams; these species prefer semi-open areas rather than dense, closed, interior parts of forests. Also included here: open woodlands, tree plantations, and shady gardens.

= Open habitats: pastureland; savannah (grasslands with scattered trees and shrubs); non-tree plantations; gardens without shade trees; roadside; old lava flows. Species found in these habitats prefer very open areas.

= Freshwater. For species typically found in or near ponds, reservoirs, streams, rivers, marshes, swamps, bogs.

= Saltwater/marine. For species usually found in or near the ocean or ocean beaches.

Origin. In these sections I tell if a species is native to Hawaii or was introduced; if the latter, I give the animal's native area, if known.

Example

Plate 31b

Hawaiian Hawk
Buteo solitarius
'Io

ID: Mid-sized bird-of-prey with yellow skin at base of bill, yellowish feet, and grayish tail with brown bars; dark form all brown; light form brown with whitish chest/belly often with dark streaks; light form immature bird is pale buffy/yellowish on head and below, with dark eyeline; to 46 cm (18 in).

HABITAT: Low, middle, and higher elevations; forest, woodlands, forest edges, and more open sites such as some agricultural areas, lava flows.

ISLANDS: BIG (reports of occasional visits to MAUI, OAHU)

ORIGIN: Hawaiian native; endemic.

Note: Threatened, USA ESA and CITES Appendix II listed.

Chapter 7

Amphibians and Reptiles

Amphibians and Reptiles of Hawaii

Of the small number of amphibian and reptile species in the Hawaiian Islands, few are abundant and widespread. Most of you will see very few of these animals during your trip, but just in case you do, they are treated here in good detail. None of the islands' terrestrial amphibians and reptiles is thought to be native to Hawaii; all were brought to the islands by people (some by Polynesians prior to European contact). Frogs, generally, were released into Hawaiian agricultural and settled areas in efforts to control pest insects. Reptiles such as geckos and anolis lizards were mostly introduced unintentionally – either as stowaways in cargo shipped to the islands or as escaped pets. In total, there are five well established amphibians and about 20 reptiles on the islands, plus six reptiles that occur in the surrounding oceans (one sea snake and five sea turtles).

Group Characteristics

Amphibians

AMPHIBIANS first arose during the mid-part of the Paleozoic Era, 400 million years ago, developing from fish ancestors that had lungs and thus could breathe air. The word *amphibian* refers to an organism that can live in two worlds, and that is as good a definition of the amphibians as any: most stay in or near the water, but many spend at least portions of their lives on land. In addition to lungs, amphibians generally have wet, thin skins that aid in gas exchange (breathing). Amphibians were also the first animals to develop legs for walking on land, the basic design of which has remained remarkably constant for all other land vertebrates. Many have webbed feet that aid locomotion in the water.

Approximately 4500 species of living amphibians have been described. (Owing to their all-terrestrial existence, almost all reptiles, birds, and mammals living on Earth have been identified but, most experts agree, many more amphibians, with their aquatic ways, remain to be discovered.) They are separated into three groups. The *salamanders* (Order Caudata, or "tailed" amphibians) comprise 450 species, the mysterious *caecilians* (sih-SIL-ians; Order Gymnophiona), large, worm-like amphibians, number about 160 species, and the *frogs* and *toads* (Order Anura, "without tails") make up the remainder, and the bulk, of the group; all five Hawaiian amphibians are *anurans*.

Most amphibians live in the water during part of their lives. Typically a juvenile stage is spent in the water and an adult stage on land. Because amphibians need to keep their skin wet, even when on land they are mostly found in moist habitats – in marshes and swamps, around the periphery of bodies of water, in wet forests. Adults of most species return to the water to lay eggs, which must stay wet to develop. Some amphibians – toads and some salamanders – are entirely terrestrial, laying eggs on land in moist places. That amphibians need to keep their skin moist with freshwater must be the reason that few of them arrive naturally on isolated oceanic islands – they cannot survive the long trips on, for example, rafts of vegetation, which many other types of animals apparently use to colonize isolated islands (see Close-up, p. 100). We can be fairly certain that no amphibians made it to the Hawaiian Islands naturally since the islands arose volcanically from the sea floor millions of years ago. Five species brought to the islands by people over the last hundred years have become firmly established. (Four are profiled here; the fifth, the CUBAN TREEFROG, a largish treefrog native to the Caribbean region, was introduced during the 1980s and is now found only over localized parts of Oahu.)

Most species of frogs and toads live in tropical or semi-tropical areas but some groups are abundant in temperate latitudes. Frogs can be either mostly aquatic or mostly terrestrial; some live primarily in vegetation on land. *Toads* constitute a group of frogs that have relatively heavy, dry skin that reduces water loss, permitting them to live on land. Most frogs and toads leave their eggs to develop on their own, but a few guard nests or egg masses, and some species actually carry their eggs on their backs or in skin pouches.

Frogs and toads are known for the vocal behavior of males, which during breeding periods call loudly from the edges of lakes and ponds or on land, attempting to attract females. Each species has a different type of call. Some species breed *explosively* in synchronous groups – on a single night thousands gather at forest ponds, where males call and compete for females and females choose from among available suitors. Many frogs, such as *bullfrogs*, fight fiercely for the best calling spots and over mates. Because frogs really have no weapons to fight with, such as teeth or sharp claws, size usually determines the winners. Some species of frogs and toads have developed *satellite* strategies to obtain matings. Instead of staking out a calling spot and vocalizing themselves (which is energy-draining and risky because calling attracts predators), satellite males remain silent but stay furtively near calling males, and attempt to intercept and mate with approaching females. Smaller males are more likely to employ such "sneaky-mater" tactics.

All amphibians as adults are predatory carnivores – as far as it is known, there is not a vegetarian among them (although one fruit-eating frog may exist in Brazil). Many animals eat amphibians, although it is not as easy as you might think. At first glance amphibians appear to be among the most defenseless of animals. Most are small, many are relatively slow, and their teeth and claws are not the types appropriate for aggressive defense. But a closer look reveals an array of ingenious defenses. Most, perhaps all, amphibians produce toxins in the skin, many of which are harmless to humans and so not very noticeable, but a few of which are quite poisonous and even lethal (the toxins produced by *poison-dart frogs*, for instance; p. 89). These toxins deter predators. Most amphibians are cryptically colored, often being amazingly difficult for people, and presumably predators, to detect in their natural settings. The jumping locomotion of frogs probably evolved as an anti-predator strategy – it is a much more efficient way of escaping quickly through leaves, dense grass, thickets, or shrubby areas than are walking or running. Some frogs hiss loudly and inflate their throat sacs when approached, which presumably makes predators think twice about attacking. Last, some frogs give loud screams when grabbed, which are startling to predators and so create opportunities for escape.

In general, amphibians are under less direct threat from people than are other vertebrate groups, there being little commercial exploitation of the group (aside from certain peoples' inexplicable taste for the limbs of frogs). However, amphibians are very sensitive to habitat destruction and, particularly, to pollution, because aquatic eggs and larvae are very susceptible to toxic substances. Many mainland amphibian populations have been noticeably declining in recent years, although the reasons are not entirely clear. Much research attention is currently focused on determining whether these reported population changes are real and significant and, if so, their causes. Not all ecologists agree yet on what is going on, but rapidly spreading water-borne diseases and parasites have been implicated as the lethal agents in some regions.

Reptiles

Many ecotravellers have ambivalent feelings about REPTILES. On one hand, almost all people react with surprise, fear, and rapid withdrawal when suddenly confronted with reptiles anywhere but the zoo (which is quite understandable, given the dangerous nature of some of their ilk). But on the other hand, these creatures are fascinating to look at and contemplate and many have highly intriguing lifestyles. Most reptiles are harmless to people and, if discovered going about their daily business, are worth a look. Unfortunately, to avoid predation, most reptiles are inconspicuous both in their behavior and color patterns, and often flee when alerted to people's presence; consequently most reptiles are never seen by people during a brief visit to a region. Overall, you should expect to see relatively few reptiles during your Hawaii trip because few species exist here and, of those that do, most are not widely distributed and/or they are difficult to spot. Still, it is a good idea to keep a careful watch for them and count yourself lucky for each one you see.

Reptiles have been around since the late portion of the Paleozoic Era, some 300 million years ago. Descendants of those first reptiles include about 7000 species that today inhabit most regions of the Earth. Chief reptile traits, aside from being scary-looking, are that (1) their skin is covered with tough scales, which cuts down significantly on water loss from their body surface. The development of this trait permitted animals for the first time to remain for extended periods on dry land, and most of today's reptiles are completely terrestrial (whereas amphibians, which lack a tough skin, need always remain in or near the water or moist places, lest they dry out). (2) Their heart is divided into more chambers that increase efficiency of circulation over that of the amphibians, allowing for a high blood presure and thus the sustained muscular activity required for land-living. (3) Some employ *oviparous* reproduction, placing their fertilized eggs in layers of tough membrane or in hard shells and then expelling the eggs to the external environment, where development of the embryos occurs; whereas others are *ovoviviparous* – eggs are not shell-encased or laid, but remain within the mother until "hatching," the young being born "live;" still others are *viviparous*, in which developing embryos are connected to mom via a type of placenta and derive nourishment from her until being born live. Most reptiles do not feed or protect their young, but desert their eggs after they are laid.

Reptile biologists usually recognize three major groups:

The *turtles* and *tortoises* (land turtles) constitute one reptile group, with about 260 species worldwide. Some turtles live wholly on land, the *sea turtles* live out their lives in the oceans (coming ashore only to lay eggs), but most turtles live in lakes and ponds. Although most eat plants, some are carnivorous. Turtles are easily distinguished by their unique body armor – tough plates that cover their back and belly, creating wrap-around shells into which head and limbs are retracted when danger looms. Three freshwater turtles and five sea turtles occur in and around Hawaii, but few are ever seen by short-term visitors.

The *crocodiles* and their relatives, large predatory carnivores that live along the shores of swamps, rivers, and estuaries, constitute a small second group of about 20 species. Last, and currently positioned as the world's dominant reptiles, the 3300 *lizard* species and 3500 *snakes* comprise a third group (lizards and snakes have very similar skeletal traits, indicating a close relationship).

Lizards, except for a few that are legless, walk on all four limbs. Most are ground-dwelling animals, but many also climb when the need arises; a fair number spend much of their lives in trees. Almost all are capable of moving quite rapidly. Most lizards are insectivores, but some, especially larger ones, eat plants, and several prey on amphibians, other lizards, mammals, birds, and even fish. Lizards are hugely successful and are often the most abundant vertebrate animals within an area. Ecologists suspect that they owe this ecological success primarily to their ruthlessly efficient predation on insects and other small animals and to low daily energy requirements.

Most Hawaiian lizards are insectivores, also opportunistically taking other small animals such as spiders and mites. Lizards employ two main foraging strategies. Some, such as many skinks, are *active searchers*. They move continually while looking for prey, for instance nosing about in the leaf litter on the forest floor. *Sit-and-wait* predators, highly camouflaged, remain motionless on the ground or on tree trunks or branches, waiting for prey to happen by. When they see a likely meal – a caterpillar, a beetle – they reach out to snatch it if it is close enough or dart out to chase it down.

Many lizards are territorial, defending territories from other members of their species with displays, such as bobbing up and down on their front legs and raising their head crests. Lizards are especially common in deserts and semi-deserts, but they are numerous in other habitats as well. They are active primarily during the day, except for many of the gecko species, which are nocturnal.

About 15 lizard species reside in Hawaii, but many occur over small, localized areas and others, such as the large GREEN IGUANA, are naturally sparse in the areas they do occupy, and so are rarely seen by visitors. The more common, more conspicuous, more frequently spotted species are profiled here. Lizards, with their tough, dry skins, are much more likely than amphibians such as frogs and toads to be able to make long ocean crossings from mainland areas and from other islands, "rafting" to oceanic islands (Close-up, p. 100). So it is possible that some of Hawaii's resident lizards arrived on the islands without people's help. It is known that four gecko species and three skinks occurred in the islands prior to European contact; but, given Hawaii's isolated, mid-ocean location, most biogeographers believe that these lizards arrived in ancient times with the Polynesian people, who probably spread them accidentally to many of the Pacific islands they colonized.

Snakes probably evolved from burrowing lizards, and all are limbless. Snakes are all carnivores, but their methods of capturing prey differ. Several groups of species have evolved glands that manufacture venom that is injected into prey through the teeth. The venom immobilizes and kills the prey, which is then swallowed whole. Other snakes pounce on and wrap themselves around their prey, constricting it until it suffocates. The majority of snakes are nonvenomous, seizing prey with their mouths and relying on their size and strong jaws to subdue it. Snakes generally rely on vision and smell to locate prey, although members of two families have thermal sensor organs on their heads that detect the heat of prey animals. As many snake species exist now as there are of all other reptiles combined. This success is thought to be attributable to their ability to devour prey that is larger than their heads (their jaw bones are highly mobile, separating partially and moving around prey as it is swallowed). This unique ability provides snakes with two great advantages over other animals: because they eat large items, they have been able to reduce the frequency with which they need to search for and capture prey; and owing to this, they can spend long periods hidden and secluded, safe from predators. Like lizards, snakes use either active searching or sit-and-wait foraging strategies.

Hawaii is essentially a "snakeless society," and the powers that be want to keep it that way (see p. 11). There are no native terrestrial snakes in the islands, and it is feared that the successful introduction of a largish snake, such as the BROWN TREE SNAKE, *Boiga irregularis*, would doom many native species, particularly birds. So it is not widely advertised that a single snake species *has already* successfully colonized the main islands: the ISLAND BLIND SNAKE,

Ramphotyphlops braminus. This small, worm-like, burrowing snake spends its life underground, sometimes moving into moist leaf litter in search of its food – insects and other small invertebrate animals. Native to southeastern Asia, this snake, usually only about 16 to 18 cm (6 to 7 in) long, has been shipped inadvertently to many parts of the world, packed in the soil that accompanies ornamental plants; it may have arrived in Hawaii during the 1930s. Unless you go digging about in gardens, you are unlikely to see one. Also, the yellow and black, venomous PELAGIC SEA SNAKE, *Pelamis platurus*, which occurs in warmer waters of the Pacific and Indian Oceans, is occasionally sighted in Hawaiian waters.

Seeing Amphibians and Reptiles in Hawaii

First, a warning: Skin secretions of some frogs, such as the GREEN AND BLACK POISON-DART FROG and GIANT TOAD (Plate 17), are toxic. The toxins can be dangerous if absorbed into your body through, for instance, a cut in your skin or through the mucous membranes of your eyes or nose. Although Hawaii's poison-dart frog is usually not harmful to humans, it may be wise, if it is encountered, to enjoy it visually and leave its handling to experts. If you do handle frogs, make sure to wash your hands thoroughly afterwards. With that warning in mind, the best way to see Hawaii's frogs and toad is to look for them in moist habitats – wet forests, near bodies of water, in small pools and puddles, along banks of streams, rivers, fish ponds, and reservoirs. The calls of male frogs can be of assistance in locating the little beasts. The poison-dart frog, the only species I can imagine anyone going out of their way to look for, is active during the day, especially on darker, overcast days; it is seen most often in its wet forest valley habitat during or after rains.

Most reptiles are difficult to observe. They spend much of their time concealed or still. Most do not vocalize like birds or frogs, so you cannot use sound to find them. Look for lizards during the daytime, when most, except for the geckos, are usually active. Weather is very important – lizards are often more active during sunny, warm times of the day. If you fail to see lizards along trails, on rocks, etc., you may look for some of them by carefully moving aside logs and rocks with a robust stick or with your boots; and you can carry out such actions secure in the knowledge that there are no dangerous snakes in Hawaii. Aside from seeing a few skinks along trails at such wilder areas as Kokee and Waimea Canyon State Parks on Kauai, the reptiles most often observed by ecotourists in Hawaii are probably sea turtles, seen by divers and snorkelers; the PACIFIC GREEN SEA TURTLE (Plate 51) accounts for 90% or more of such sightings.

Family Profiles

1. Toads

Scientists sometimes have trouble formally differentiating *toads* from frogs, but not so non-scientists, who usually know their toads: they are the frog-like animals with a rough, lumpy, wart-strewn appearance, not built for speed, that one finds on land. Actually, toads and frogs are both included in the amphibian group Anura; toads are a kind of frog. They have some special skeletal and reproductive traits that are used to set them apart (for instance, frogs lay eggs in jellied clusters, but toads lay them in jellied strings), but for our purposes, a modification of the common definition will do: toads are squat, short, terrestrial frogs with thick, relatively dry skins that prevent rapid water loss, short limbs, and glands that resemble warts spread over their bodies. A few families of frogs are called toads, but the predominant group, hugely successful, is the *true* toads, Family Bufonidae (*bufo* is Latin for "toad"), which spread naturally to all continents except Australia (and has been introduced by people to that continent and to many islands; all toads that occur in the USA and Canada are in this family). *Bufonids* usually have two prominent "warts" on each side of the neck or shoulder area, called *parotoids*. Often shades of olive or brown, toads vary in size, from small to quite large, with Hawaii's only toad, the GIANT TOAD (Plate 17), being anywhere from 9 to 19 cm (3.5 to 7.5 in) long, and weighing in at up to 250 g (a half pound). Worldwide there are perhaps 360 bufonid toad species.

Natural History

Ecology and Behavior

Although their relatively heavy, dry skin (as compared with that of other frogs) permits adult toads a permanently terrestrial existence, they experience some water loss through their skin, so unless they stay near water or in moist habitat (such as irrigated fields or sprinkled gardens), they dry out and die in just a few days. Although toads are freed from an aquatic life, water still governs their existence. Many are primarily nocturnal, avoiding the sun and its drying heat by sheltering during the day under leaf litter, logs, or rocks, coming out to forage only after sundown. Toad tadpoles are vegetarians, feeding on green algae and bacteria in their aquatic habitats, but adult toads are all carnivorous, foraging for arthropods, mostly insects, amid the leaf litter. As one researcher defines the toad diet, "if it's bite-sized and animate it is food, no matter how noxious, toxic, or biting/stinging." In Hawaii, beetles and other insects, spiders, earthworms, slugs, and snails are frequent prey for the GIANT TOAD; in other locations, they also take small vertebrates such as small frogs and lizards.

Slow-moving toads (with their short legs they are capable of covering only very short distances per hop) have two methods to escape being eaten. They can be extremely hard to detect in their habitats, concealing themselves with their *cryptic coloration* and habit of slipping into crevices, under leaves, or burying themselves in the earth. Also, apparently quite effectively, they exude noxious fluids from their skin glands – the warts are actually a defense mechanism. If grabbed, a viscous, white fluid oozes from the warts. The fluid is very irritating to mucous membranes, such as those found in a predator's mouth and nose. Toads also have muscle control over the poison glands and some can squeeze them to spray the poison more than 30 cm (1 ft). Most predators that pick up a toad probably do not do it twice; people's four-legged pets that put toads into their mouths have been killed by the poisons; dogs in Hawaii sometimes die from chewing on Giant Toads, which produce a strong toxin (bufotenine) in their parotoid glands. A few predators, however, such as raccoons and opossums, having learned their way around toad anatomy, avoid the warts on the back and legs by eating only the inside of the toad, entering through the mostly poison-gland-free belly.

Ecological Interactions

The GIANT TOAD, Hawaii's only toad, has somehow become semi-domesticated, and now, in many areas of the world (including the Philippines and Australia), is commonly found around human settlements at much higher densities than in wild areas. Adapting nicely to people's behavior, these toads will eat dog and cat food left outdoors for pets (which is quite a feat, given that most frogs will only eat live, moving prey). In some areas where these toads have been introduced, they are regarded as pests, because they are always around in the evening, occasionally poison curious pets, and even hop into homes and buildings; but other people, including many in Hawaii, appreciate them in their gardens for their insect-eating ways; they are especially fond of cockroaches.

Breeding

To breed, male GIANT TOADS at night slip into ponds, pools, or reservoirs, and call to attract females. After appropriate mating maneuvers of both sexes, sperm are released by a male in a cloud into the water, followed by a female releasing her eggs into the cloud. The eggs are laid within jellied strings, the jelly protecting the eggs physically and also, because it contains toxins, discouraging consumption by potential predators. A female may lay hundreds to thousands of eggs at a time, and may lay several times per year. Eggs hatch in only a few days, releasing young in the larval feeding stage known commonly as *tadpoles*. They feed, grow, and develop, transforming themselves into *toadlets* after about 4 weeks, which then swarm up the banks and disappear into their terrestrial existences. Toads generally reach sexual maturity in a year or two.

Notes

GIANT TOADS were introduced to the islands in 1932, when sugar growers imported them from Puerto Rico to Oahu, to try to control beetles and other insect pests in sugarcane fields. This toad is the most "successful" amphibian to be transported around the world by people, with thriving populations now in the continental USA (Florida, Louisiana), throughout the Caribbean, on many Pacific islands (Samoa, Fiji), Japan, Papua New Guinea, and Australia; it is very harmful to native species in most of these regions.

The claim that a person will contract warts by handling toads is not true. Human warts are caused by viruses, not amphibians. The glands on toads' skin that resemble warts release noxious fluids to discourage predators. Various poisons have been identified in these fluids that, among other effects, cause increased blood pressure, blood vessel constriction, increased power of heartbeat, heart muscle tissue destruction, and hallucinations. Because these fluids minimally are irritants, a smart precaution is to avoid handling toads or, after such handling, make sure to wash your hands. Caustic irritation will result if the fluids are transferred from hands to eyes, nose or mouth. Some reports have it that voodoo practitioners in Haiti use the skin secretions of toads in their zombie-making concoctions. Several hallucinogenic chemicals, which cause LSD-type effects when swallowed, have been isolated from the skin glands of Central and South American toads, for instance, from Giant Toads.

Status

A few bufonid toad species worldwide are known to be endangered, but for many species, there is inadequate information on the true health of their populations. In the USA, the YOSEMITE, HOUSTON, and AMARGOSA TOADS, are all endangered. Costa Rica's GOLDEN TOAD is either critically endangered or already extinct. Preservation of Golden Toads was one of the original reasons for the creation of cloud rainforest reserves in that Central American country, which, in turn, essentially led to the creation of a large ecotourism industry.

Profiles

Giant Toad, *Bufo marinus*, Plate 17a

2. True Frogs

Frogs of the Family Ranidae are known as *true frogs* not because they are better frogs than any other but because they are the most common frogs of Europe, where the early classification of animals took place. More than 650 species strong, the true frogs now have an almost worldwide distribution (the ones in Australia were introduced there by people). Most are in Africa; 250 species occur in the New World. True frogs include many of what most people regard as typical frogs – green ones that spend most of their lives in water. Among frogs familiar to many North Americans, BULLFROGS and LEOPARD FROGS are members of the Ranidae. Typically *ranids* are streamlined, slim-wasted frogs with long legs, webbed back feet, and thin, smooth skin. They are usually shades of green or mixed green and brown. Size varies extensively; the green BULLFROG (Plate 17), the species you are mostly likely to see, and a fairly large true frog, reaches lengths of 18 cm (7 in), excluding the long legs. The WRINKLED FROG (Plate 17), the other Hawaiian ranid, is smaller and gray or brown.

Natural History

Ecology and Behavior

These are aquatic frogs that are good swimmers and jumpers. With their webbed toes and long, muscular hind legs, they are built for speed on land or in water. This is fortunate for the frogs because most lack the poison glands in their skins that many other types of frogs use to deter predation. They have thin skin through which water evaporates rapidly and thus they tend to remain in or near the water, often spending much of their time around the margins of ponds or floating in shallow water. Except during breeding seasons, they are usually active only during the day. True frogs feed mainly on bugs and other small invertebrates such as snails, slugs, and worms, but also on fish and smaller frogs. In turn, they are common food items for an array of beasts, such as wading birds, fish, turtles, and small mammals. Because they are such tasty morsels to so many predators, these frogs are very alert to their surroundings, attempting escape at the slightest movement or noise; the splashing heard as you walk along a pond or stream is usually made by these frogs on the shore leaping into the water.

Ecological Interactions

The way that biologists view how animals distribute themselves within an area has been partially worked out by thinking about how true frogs might arrange themselves in a pond. Predation is a major worry for many animals, including aquatic frogs, and it doubtless influences many aspects of their lives. Often, there is safety in numbers, the rationale being that, although a group is larger than a single animal and therefore more easily located by a predator, the chances of any one animal in the group being taken is low, and so associating with the group, instead of striking out on one's own, is advantageous to an individual. But within the group, where should an individual frog, say one named Fred, position himself to minimize his chances of being eaten? Imagine a circular pond around which frogs tread in the shallow water near shore or sun themselves on the bank. If the predator is a snake or a bird that for lunch will take one frog, the best place for Fred is between two other frogs, so that his left and right are "protected," the neighbor frogs there to be eaten first. To gain such advantageous positioning, clearly Fred's best move would be to find two frogs near each other and move into the gap between them. If such a strategy exists, then the neighboring frogs should move in turn, also trying to fit into small gaps. Eventually, all the frogs in the pond, if playing this game, should end up in a small heap, the best protected frogs at its center. And indeed, as any small child who hunts frogs will tell you, they are often found in tight groups at the edges of lakes and ponds. This *selfish herd* explanation for the formation of frog and other animal aggregations was first proposed about 30 years ago.

Breeding

True frogs reproduce under what most would consider the standard amphibian plan. Eggs are released by the female into the water in ponds or streams, and are fertilized then by sperm released by the male. Eggs and tadpoles develop in the water. In some species, the jellied egg masses float, but in others, they are attached to the undersides of rocks in streams. Clutch size varies tremendously, but large females can release thousands of eggs; BULLFROG egg masses, in fact, can be as large as 30 cm (1 ft) in diameter. Sexual maturity occurs in from 1 to 4 years.

Notes

BULLFROGS were introduced from California to the Big Island during the late 1800s. The California population that provided the colonizing stock, in turn, was also introduced by people; Bullfrogs are native in North America only east of the Rocky Mountains. The WRINKLED FROG, native to eastern Asia, was introduced to Hawaii in the 1890s to control pest insect populations.

The true frogs have always provided a minor source of protein to people throughout the world. These frogs are often non-toxic, abundant, and large enough to make harvesting and preparation economically profitable. In much of Europe and parts of the USA, the long muscles of the rear legs, often dusted with flour and fried in butter, are eaten by otherwise civilized people. Bullfrogs, in fact, were apparently brought to Hawaii at least partly as a food source, as historical accounts exist from shortly after their introduction that record their legs being sold for consumption.

Status

Several species of true frogs are threatened throughout the world, and at least two, the ASIAN and INDIAN BULLFROGS, are CITES Appendix II listed. About three species of true frogs are threatened in the USA. As is the case for many of the Amphibia, relatively little is known about the sizes and health of the populations of many of these frogs; therefore, authoritative assurance that more are not endangered is impossible to provide.

Profiles

Bullfrog, *Rana catesbeiana*, Plate 17b
Wrinkled Frog, *Rana rugosa*, Plate 17c

3. Poison-dart Frogs

The *poison-dart frogs* of the Family Dendrobatidae, best-known for the deadly poisons some species produce in their skin glands, are a group of about 170 species that are native to tropical and subtropical Central America and northern South America. These treefrogs, often hugely colorful with striking large eyes, are perhaps the most dangerous of amphibians. Usually quite small, most being less than 5 cm (2 in) long and many about half that, these frogs take on some brilliant hues. Some for instance, are shiny black with bright red, orange, green, or blue markings. But about half the family's species have no bright colors. The GREEN AND BLACK POISON-DART FROG (Plate 17), a Hawaiian alien, although usually only about 2.5 cm (an inch) long, is one of the group's more striking members.

Natural History

Ecology and Behavior

These treefrogs are usually ant and mite specialists, although other small arthropods, mostly insects, are also eaten. They hunt actively during daylight hours in moist areas on the forest floor and low vegetation. One theory of the development of their lethal poisons is that, because they eat such small items – ants – they need to forage for many hours each day to meet nutritional needs. Therefore they are exposed to predators for extended periods each day, and so, in response to heavy predation, have evolved protective toxic skin secretions. The frogs apparently need ants in their diet to manufacture their alkaloid-based skin poisons; when fed other diets when in captivity, the frogs lose their toxicity. In many poison-dart species, males are territorial, defending small areas that contain essential breeding resources, such as bromeliad plants (see below).

Breeding

Poison-dart frogs show some of the most complex parental care behavior of any non-bird or non-mammal. Males call to attract females to their territories. When a female arrives, the male searches for a good egg-laying site – usually a leaf or a moist sheltered spot; the female follows him around. After some courtship rubbing and touching, the female lays eggs and the male fertilizes them. The male or female then stays or periodically visits the eggs to guard them. When hatching occurs, one of the parents carries the tadpoles on its back and places them into the pools of water that collect in the central parts, or *cisterns*, of bromeliad plants. In Hawaii, small forest pools and ponds are also used. Males will carry up to six tadpoles at a time, females usually just one or two. In these small, protected pools the tadpoles develop into frogs. Recently it has been discovered that the female's role in caring for her young is not finished after transporting the tadpoles; in at least some species, she also periodically returns to the pool and drops unfertilized eggs into it, feeding her young this nutritious resource.

Notes

A few score Panamanian GREEN AND BLACK POISON-DART FROGS were introduced to Oahu in 1932 as an experiment in mosquito control, and have been around ever since. A few have also been seen recently on the Big Island.

Although most species of poison-dart frogs are only mildly dangerous to people and can be safely handled with precautions, some can pack a wallop. The poisons in their skin secretions affect nerves and muscles, causing paralysis and, eventually, cessation of breathing. (Recently, the mixture of chemicals secreted from the skin glands of one South American species has been found to be an excellent pain-blocking drug, 200 times more effective than morphine!) Several groups of native American peoples, notably in Colombia, apply the skin secretions from some of these frogs onto their blowgun darts (hence, the name), using the poison's formidable toxicity to paralyze and kill the small animals they hunt. Even large ants and large spiders avoid poison-dart frogs: in experiments in which ants or spiders that normally eat small frogs were presented in cages with other small frogs or poison-dart frogs, they attacked and consumed the other frogs, but attacked and rejected the poison-darts. Not all poison-dart frogs avoid predation. Some, in their native lands, are taken by snakes immune to their poisons and some have been observed to be eaten by spiders, which avoid the skin by puncturing the body and sucking out the innards.

Status

As far as it is known, no species of poison-dart frog is currently endangered. As is the case for many of the Amphibia, relatively little is known about the sizes and health of the populations of many of these frogs; therefore, authoritative assurance that some are not threatened is impossible to provide. Because many species of these frogs have very restricted ranges in their home countries, and because they are popular with collectors, all poison-dart frogs are CITES Appendix II listed. The health of future populations of many of these frogs is clearly tied to preservation of their forest habitats.

Profiles

Green and Black Poison-dart Frog, *Dendrobates auratus*, Plate 17d

4. Chameleons

Chameleons are a distinctive group of lizards that resemble, truth be told, small, deformed dinosaurs. The family, Chamaeleonidae, an Old World group, is distributed mainly in Africa and Madagascar, but ranges east to India and north to Spain. There are about 135 species; one, JACKSON'S CHAMELEON (Plate 17), was recently introduced to Hawaii, and now occurs over many regions of Oahu as well as parts of other main islands. Chameleons, usually arboreal insect-eaters, range in size from tiny species only 5 cm (2 in) long, to large lizards up to 55 cm (21.5 in) long. They are known for their unusual looks (including some with horn-like head projections), color-changing ways, prehensile tails, and amazing, bullet-like projectile tongues.

Natural History

Ecology and Behavior

Chameleons eat insects and other small arthropods, and JACKSON'S CHAMELEON is no exception. They use their highly mobile eyes to spot a likely meal; the way their eyes can move around independently of each other permits excellent depth perception and aiming – necessary for what comes next. These lizards do not run down prey; rather, they stay some distance from their intended victims and shoot out their incredibly long tongues to grab them. The tongue, with a wet, somewhat sticky, muscular tip that holds the bug, can be as long as the lizard's entire head and body (it is curled up when not in use). Tongues shoot from a chameleon's mouth at up to 5 m (16 ft) per second, and can hit a bug in about 1/100 of a second. These lizards are usually highly arboreal; when on the ground they do not move well. When in trees and shrubs, their prehensile tail acts as a fifth limb, helping them to move efficiently around vegetation and reach at odd angles for prey. Males are strongly territorial, aggressively discouraging other males from entering or remaining in their territories. Male Jackson's have three "horns" on their head; aside from having a role in courtship (females may use them to recognize males as the correct species and sex to fool around with), they are used as weapons during male–male fights over territories. Chameleons tend not to be very social; they are often solitary animals, coming together in pairs only to mate.

Chameleons are among the best lizards at changing their colors in reponse to internal and perhaps external stimuli (see p. 92). One effect of their color changes is that, often, they take on the coloring of their background habitat, making them highly camouflaged. Jackson's Chameleons can appear green, yellowish, brown, gray, blackish, and even, at times, bluish. The camouflage allows them to sneak up on some of their prey and probably also renders them invisible to some predators. Another anti-being-eaten strategy is that, when cornered on a tree, a chameleon will drop to the ground and remain motionless.

Breeding

Most chameleons lay eggs, burying them in the ground or in moist, protected sites, but JACKSON'S CHAMELEON gives birth to live young after pregnancies of 5+ months. The female perches on a low branch and drops the young, between 10 and 40 of them, usually to the ground, where they break out of their protective membranous covering and, looking already much like mom, start roaming the habitat on their own and devouring small bugs. They reach adult size in about 8 months.

Notes

JACKSON'S CHAMELEONS were initially released on Oahu during the early 1970s by a petshop owner. They thrive on their own but are also kept as backyard pets and have been transported probably as pets to other main islands. Because they apparently eat mostly introduced insects and spiders, experts do not consider this lizard to be an ecological threat to the islands.

Status

Because of their status as prized reptilian pets and therefore, their interest to the international pet trade, all chameleons (genus *Chamaeleo*), are CITES Appendix II listed: commerce in them is highly regulated. Several species, particularly in South Africa, are rare or already endangered.

Profiles

Jackson's Chameleon, *Chamaeleo jacksonii*, Plate 17e

5. Anolis Lizards

Anolis lizards are a large group of small lizards. There are about 300 described species of anolis lizards, or *anoles*, distributed over Central and South America and the Caribbean, making the genus, *Anolis*, one of the largest among vertebrate animals. Anoles are members of a huge assemblage of lizards with an almost exclusively New World distribution, the Iguanidae. There are more than 700 *iguanid* lizard species (some scientists split the iguanids into several smaller families), including anoles, *iguanas, basilisks* (Jesus Christ lizards), and *spiny lizards* such as *horned lizards* and *fence lizards*. The group includes the spectacular, dinosaur-like, GREEN IGUANA, which is occasionally spotted high in trees in some of Oahu's wetter forests (an introduced alien to Hawaii, it occurs only on Oahu). Many in the family are brightly colored and have adornments such as crests, combs, helmets, spines, or throat fans. They range in size from tiny anolis lizards only a few centimeters in total length and a few grams in weight, to Green Iguanas, which, when their extremely long tails are counted, are up to 2 m (6.5 ft) long.

Anolis lizards are smallish brown, green, or gray lizards with long tails. They are commonly associated with human settlements – found in backyards, gardens, fences, agricultural sites; but many species are also common in natural areas. Males have throat fans, or *dewlaps*, usually red, yellow, or whitish, that are extended and shown to females in courtship and to other males in territorial contests. Three species occur in Hawaii; one, the KNIGHT ANOLE, a largish green species with a white bar at its shoulder, occurs only over very small parts of Oahu and so is not profiled here.

Natural History

Ecology and Behavior

Anoles are small, often arboreal lizards. Some are ground dwellers, but others spend most of their time on tree trunks or bush branches, perched head toward the ground, visually searching for insect prey. They are primarily *sit-and-wait* foragers: they sit still until they see a delicious morsel pass by on the ground or on the vegetation – say a beetle, an ant, a cockroach, or a spider – then dart out to grab it. One indication of how efficient these lizards are at hunting is that, in recent experiments in Puerto Rico, when anoles were removed from plots of forest, spiders in the plots, freed from anole predation, increased in numbers by factors of 10 to 30! Anoles themselves, small and presumably tasty, are frequent prey for many birds and, where present, snakes and larger lizards; the Hawaiian Mongoose eats them as well. Whereas geckos (p. 95), also usually associated with human habitations, are creatures of the night, anolis lizards are generally day-active.

Anoles are known especially for their territorial behavior. Males defend territories on which one to three females may live. In some species males with territories spend up to half of each day defending their territories from males seeking to establish new territories. The defender will roam his territory, perhaps 30 sq m (325 sq ft), occasionally giving territorial advertisements – repeatedly displaying his extended throat sac and performing *push-ups*, bobbing his head and body up and down. Trespassers that do not exit the territory are chased and even bitten.

Ecological Interactions

Via interactions between the external environment and their nervous and hormonal systems, anolis lizards, many other iguanids, and some other lizard groups, have the novel ability to change their body color. Such color changes presumably are adaptations that allow them to be more cryptic, to blend into their surroundings, and hence, to be less detectable to and safer from predators. Also, alterations in color through the day may aid in temperature regulation; lizards must obtain their body heat from the sun, and darker colors absorb more heat. Color changing is accomplished by moving pigment granules within individual skin cells either to a central clump (causing that color to diminish) or spreading them evenly about the cell (enhancing the color). It is now thought that the stimulus to change colors arises with the physiology of the animal rather than with the color of its surroundings. Some North American anoles, particularly CAROLINA ANOLES (the same species as Hawaii's GREEN ANOLE), owing to their color-changing ways, are hawked in pet stores as "chameleons" even though the real chameleons are strictly Old World lizards (p. 90).

Breeding

Female Anoles lay small clutches of eggs throughout the year; an individual female may produce eggs every few weeks. Hawaii's GREEN and BROWN ANOLES (Plate 18) produce very small clutches, usually one or two eggs at a time, which are placed on or slightly under the ground (the female digging a small hole and then covering her eggs with soil), or under leaf litter; hatching occurs in about a month.

Notes

All of Hawaii's anoles established themselves recently in the islands, the GREEN ANOLE during the 1950s and the BROWN ANOLE in the 1980s. The populations probably originated as escaped or released pets.

Status

Only a few anoles are known to be endangered. Among the many Caribbean species, for instance, Puerto Rico's CULEBRA ISLAND GIANT ANOLE is endangered (USA ESA listed). Like many small reptiles, not enough is known about many anole species to be able to say with any certainty that their populations are healthy and secure.

Profiles

Green Anole, *Anolis carolinensis*, Plate 18a
Brown Anole, *Anolis sagrei*, Plate 18b

6. Skinks

The *skinks* are a large family (Scincidae, with about 1100 species) of small and medium-sized lizards with a worldwide distribution. Over the warmer parts of the globe, they occur just about everywhere. Skinks are easily recognized because they look different from other lizards, being slim-bodied with relatively short limbs, and smooth, shiny, roundish scales that combine to produce a satiny look. Many skinks are in the 5 to 9 cm (2 to 4 in) long range, not including the tail, which can easily double an adult's total length. Skinks are usually common to forests at low to moderate elevations, and are especially prevalent in forest edge areas. Four species occur in Hawaii, but one is found only in a small region of Kauai, so is not profiled here.

Natural History

Ecology and Behavior

Many skinks are terrestrial lizards, particularly appreciative of moist ground habitats such as sites near streams, springs, and marshes. In fact, skinks generally are not seen unless searched for because they spend most of their time hidden under rocks, logs, vegetation, or wet leaf litter. A few species are arboreal, and some are burrowers. Skink locomotion is surprising; they use their limbs to walk but when the need arises for speed, they locomote mainly by making rapid wriggling movements with their bodies, snake-fashion, with little leg assistance. Through evolutionary change, in fact, some species have lost limbs entirely, all movement now being snake-fashion.

Skinks are day-active lizards, most activity on tropical islands being confined to the morning hours; they spend the heat of midday in sheltered, insulated hiding places, such as deep beneath the leaf litter. Some skinks are sit-and-wait foragers, whereas others seek their food actively. They consume many kinds of insects, which they grab, crush with their jaws or beat against the ground, then swallow whole. In Hawaii, main predators on skinks are probably birds, rats, and mongooses.

Ecological Interactions

Many lizards, including the skinks and geckos, have what many might regard as a self-defeating predator escape mechanism: they detach a large chunk of their bodies, leaving it behind for the predator to attack and eat while they make their escape. The process is known as *tail autotomy* – "self removal." Owing to some special anatomical features of the tail vertebrae, the tail is only tenuously attached to the rest of the body; when the animal is grasped forcefully by its tail, the tail breaks off easily. The shed tail then wriggles vigorously for a while, diverting a predator's attention for the instant it takes the skink or gecko to find shelter. A new tail grows within a few months to replace the lost one. Is autotomy successful as a lifesaving tactic? Most evolutionary biologists would argue that, of course it works, otherwise it could not have evolved to be part of lizards' present day defensive strategy. But we have hard evidence, too. For instance, some snakes that have been caught and dissected have been found to have in their stomachs nothing but skinks – not whole bodies, just tails! Also, a very common finding when a field biologist surveys any population of small lizards (catching as many as possible in a given area to count and examine them) is that a large proportion, often 50% or more, have regenerating tails; this indicates that tail autotomy is common and successful in preventing predation.

Breeding

Skinks are either egg-layers or live-bearers; METALLIC and SNAKE-EYED SKINK (Plate 18) females lay small clutches of eggs (usually one to five); the MOTH SKINK (Plate 18) gives birth to live young, usually two. Hawaiian skinks probably breed year-round, with individual females producing more than one clutch annually.

Notes

Two of Hawaii's skinks, MOTH and SNAKE-EYED SKINKS, preceded initial 18th century European contact with the islands, so they arrived either with early Polynesian settlers (who may have spread them to Pacific islands far and wide as stowaways in their small canoe-like boats), or on their own, most probably "rafting" in on floating oceanic debris (see Close-up, p. 100).

Status

The three skinks profiled here are all fairly common in Hawaii. A fourth Hawaiian species occurs over a very small region of Kauai. A fifth species, the AZURE-TAILED SKINK, a pretty striped skink with a bluish tail, common in the islands during the early 1900s, is now exceedingly rare or extinct in Hawaii (it is still common on many other Pacific islands); it probably succumbed to predation by the introduced mongoose and to competition with the METALLIC SKINK. Many skinks of Australia and New Zealand regularly make lists of vulnerable and threatened animals, and several Caribbean skinks are endangered. In the USA, Florida's SAND and BLUETAIL MOLE SKINKS (USA ESA listed) are considered vulnerable or threatened species.

Profiles

Metallic Skink, *Lampropholis delicata*, Plate 18c
Snake-eyed Skink, *Cryptoblepharus poecilopleurus*, Plate 18d
Moth Skink, *Lipinia noctua*, Plate 18e

7. Geckos

Geckos are most interesting organisms because, of their own volition, they have become "house lizards" – probably the only self-domesticated reptile. The family, Gekkonidae, is spread throughout tropical and subtropical areas the world over, 870 species strong. In many regions, geckos have invaded houses and buildings, becoming ubiquitous adornments of walls and ceilings. Ignored by residents, they move around dwellings chiefly at night, munching insects. To first-time visitors from northern climes, however, the way these harmless lizards always seem to position themselves on ceilings directly above one's sleeping area can be a bit disconcerting. Eight species occur in Hawaii, six with broad enough distributions to be profiled here.

Geckos are fairly small lizards, usually gray or brown, with large eyes. They have thin, soft skin covered often with small, granular scales, producing a slightly lumpy appearance, and big toes with well developed claws that allow them to cling to vertical surfaces and even upside-down on ceilings. The way geckos manage these feats has engendered over the years a fair amount of scientific detective work. Various forces have been implicated in explaining the geckos' anti-gravity performance, from the ability of their claws to dig into tiny irregularities on man-made surfaces, to their large toes acting as suction cups, to an adhesive quality of friction. The real explanation appears to lie in the series of miniscule hair-like structures on the bottom of the toes, which provide attachment to walls and ceilings by something akin to surface tension – the same property that allows some insects to walk on water.

Adult geckos mostly report in at only 5 to 10 cm (2 to 4 in) in length, tail excluded; tails can double the length. Because lizard tails frequently break off and regenerate (see p. 94), their length varies tremendously; gecko tails are particularly fragile. Lizards, therefore, are properly measured from the tip of their snouts to their *vent*, the urogenital opening on their bellies, usually located somewhere near to where their rear legs join their bodies. The geckos' 5 to 10 cm length, therefore, is their range of "SVLs," or snout–vent lengths.

Natural History

Ecology and Behavior

Although most lizards are active during the day and inactive at night, nearly all gecko species are nocturnal. In natural settings, they are primarily ground dwellers, but, as their behavior in buildings suggests, they are also excellent climbers; in fact, most of the Hawaiian geckos are considered at least partially arboreal. Geckos feed on arthropods, chiefly insects. It is their ravenous appetite for cockroaches and other insect undesirables that renders them welcome house guests in many parts of the world. They also eat ants, termites, flies, and mosquitos, among other insects. Geckos are *sit-and-wait* predators; instead of wasting energy actively searching for prey that is usually highly alert and able to flee, they sit still for long periods, waiting for unsuspecting insects to venture a bit too near, then lunge, grab, and swallow. Perhaps the only "negative" associated with house geckos is that, unlike the great majority of lizards, which keep quiet, geckos at night are quite the little chirpers and squeakers. They communicate with each other with loud calls – surprisingly loud for such small animals. Various species sound different; the word *gecko* approximates the sound of calls from some African and Asian species.

For escape from predators (which, in Hawaii, would include owls and several other larger birds, rats, pigs, and perhaps the mongoose), geckos rely chiefly on their *cryptic coloration* and their ability to flee rapidly. When cornered, geckos give threat displays; when seized, they give loud calls to distract predators, and bite. Should the gecko be seized by its tail, it breaks off easily, allowing the gecko time to escape, albeit tail-less; tails regenerate rapidly. When seized, geckos of the genus *Gehyra*, including Hawaii's STUMP-TOED GECKO (Plate 19), also lose large patches of their skin, which may also aid in their escaping an attack. Some geckos when seized also secrete thick, noxious fluids from their tails, which presumably discourages some predators. Almost all geckos can lighten or darken their skin coloring (see p. 92). The primary function of such color changing is often thought to be increased camouflage – to make themselves more difficult to see against various backgrounds.

Breeding

Geckos are egg-layers. Mating occurs after a round of courtship, which involves a male displaying to a female by waving his tail around, followed by some mutual nosing and nibbling. Clutches usually contain only a few eggs, but a female may lay several clutches per year. There is no parental care – after eggs are deposited, they and the tiny geckos that hatch from them are on their own.

Notes

Four of Hawaii's common geckos preceded initial 18th-century European contact with the islands, so they arrived either with early Polynesian settlers (who may have spread geckos to Pacific islands far and wide as stowaways in their small canoe-like boats), or on their own, most probably "rafting" in on floating oceanic debris (see Close-up, p. 100). Supporting the latter idea is that adult geckos, hearty lizards, hide well in trees under bark (so may have been able to survive for lengthy periods on floating, downed trees). Even if adults could not survive a long sea journey, their eggs may have; often deposited on trees under bark, they are surprisingly hard and tough and at least somewhat saltwater-tolerant. The very common HOUSE GECKO (Plate 19), now a permanent if alien resident over many parts of the world (in Asia, New Guinea, Australia, Mexico, Madagascar, eastern and southern Africa), probably originated in the Indian region of Asia. It arrived in Hawaii, undoubtedly as a stowaway in commercial or military cargo, in the 1940s, during or immediately after World War II. GOLD DUST DAY GECKOS (Plate 19), native to the Old World, were intentionally released by a private citizen on Oahu in 1974.

The world's smallest reptile, at 4 cm (1.5 in) long, is a gecko, the CARIBBEAN DWARF GECKO. As reptile biologists like to say, it is shorter than its name.

Status

More than 25 gecko species are listed by conservation organizations as rare, vulnerable to threat, or endangered, but they are almost all restricted to the Old World. Hawaiian gecko species are all thriving.

Profiles

Mourning Gecko, *Lepidodactylus lugubris*, Plate 19a
Stump-toed Gecko, *Gehyra mutilata*, Plate 19b
Tree Gecko, *Hemiphyllodactylus typus*, Plate 19c
Indo-Pacific Gecko, *Hemidactylus garnotii*, Plate 19d
House Gecko, *Hemidactylus frenatus*, Plate 19e
Gold Dust Day Gecko, *Phelsuma laticauda*, Plate 19f

8. Turtles

It is a shame that *turtles* in the wild are relatively rarely encountered reptiles (at least at close range) because they can be quite interesting to watch and they are generally innocuous and inoffensive. It is always a pleasant surprise stumbling across a turtle on land, perhaps laying eggs, or discovering a knot of them basking in the sunshine on rocks or logs in the middle of a pond. The 260 living turtle species are usually grouped into 12 families that can be divided into three types by their typical habitats. Two families comprise the *sea turtles*, ocean-going animals whose females come to shore only to lay eggs. The members of nine families, containing most of the species, live in freshwater habitats – lakes and ponds – except for the exclusively terrestrial *box turtles*. Finally, one family contains the *land tortoises*, which are completely terrestrial.

Three species of freshwater turtles occur over limted areas of Oahu, or on Oahu and Kauai, in streams, ponds, reservoirs, canals, and drainage ditches, but few visitors ever see them, so they are not profiled here. Of the three, one, the RED-EARED SLIDER, has become established just recently; the population originated with released pets. The other two, the WATTLE-NECKED SOFTSHELL and CHINESE SOFTSHELL TURTLES, have been in the islands since the early 1900s, brought here to be raised in fish ponds for food. Five sea turtles occur in Pacific waters off Hawaii's coasts, but only two, the GREEN SEA TURTLE (Plate 51) and HAWKSBILL SEA TURTLE (Plate 51), are commonly seen; in fact, more than 90% of sea turtle sightings in the region are of Greens – usually by divers and snorkelers.

Turtles all basically look alike: bodies encased in tough shells (made up of two layers – an inner layer of bone and an outer layer of scale-like plates); four limbs, sometimes modified into flippers; highly mobile necks; toothless jaws; and small tails. This body plan must be among nature's best, because it has survived unchanged for a long time; according to fossils, turtles have looked more or less the same for at least 200 million years. Enclosing the body in heavy armor above and below apparently was an early solution to the problems vertebrates faced when they first moved onto land. It provides both rigid support when outside of buoyant water and a high level of protection from drying out and from predators.

Turtles come, for the most part, in a variety of browns, blacks and greens, with olive-greens predominating. They range in size from tiny terrapins 11.5 cm (4.5 in) long to 250-kg (550-lb) GALAPAGOS TORTOISES and giant LEATHERBACK SEA TURTLES that are nearly 2 m (7 ft) long, 3.6 m (12 ft) across (flipper to flipper), and that weigh 550+ kg (1200+ lb). Leatherbacks, except for a few huge crocodile individuals, are the heaviest living reptiles. In many turtle species, females are larger than males.

Natural History

Ecology and Behavior

Sea turtles are large reptiles that live in the open oceans, with the result that, aside from their beach-nesting habits, relatively little is known of their behavior. Their front legs have been modified into oar-like flippers, which propel them through the water. Although they need air to breathe, they can remain submerged for long periods. At first, all sea turtles were assumed to have similar diets, probably sea plants. But some observations of natural feeding, as well as examinations of stomach contents, reveal a variety of specializations. GREEN SEA TURTLES eat bottom-dwelling sea grasses and algae, while HAWKSBILL SEA TURTLES eat bottom sponges and the occasional fish, crab, sea urchin, or jellyfish. LOGGERHEAD SEA TURTLES feed predominantly on mollusks (snails, etc.) and crustaceans (crabs, etc.), and LEATHERBACK SEA TURTLES eat mainly jellyfish; both species occasionally occur in Hawaiian waters. Green Sea Turtles, at times, apparently leave the ocean to bask in the sun on beaches of remote islands.

Breeding

Courtship in turtles can be quite complex. In some, the male swims backwards in front of the female, stroking her face with his clawed feet. In the tortoises, courtship seems to take the form of some between-the-sexes butting and nipping. All turtles lay their leathery eggs on land. The female digs a hole in the earth or sand, deposits eggs into the hole, then covers them over and departs. It is up to the hatchlings to dig their way out of the nest and navigate to the nearest water.

All sea turtle species breed in much the same way. Mature males and females appear in waters off nesting beaches during breeding periods (for example, GREEN SEA TURTLES in Hawaii, from May through about August). After mating, females alone come ashore on beaches, apparently the same ones on which they were born, to lay their eggs. Each female breeds probably every 2 to 4 years, laying from two to eight clutches of eggs in a season (each clutch being laid on a different day). All within about an hour, and usually at night, a female drags herself up the beach to a suitable spot above the high-tide line, digs a hole with her rear flippers (a half meter or more, 2 ft, deep), deposits about 100 golfball-sized eggs, covers them with sand, tamps the sand down, and heads back to the ocean. Sometimes females emerge from the sea alone, but often there are mass emergences, with hundreds of females nesting on a beach in a single night. Eggs incubate for about 2 months, then hatch simultaneously. The hatchlings dig themselves out of the sand and make a dash for the water (if tiny turtles can be said to be able to "dash"). Many terrestrial and ocean predators devour the hatchlings (although seabird predation is less intense in the Hawaiian Islands than in many other locations) and it is thought that only between 2% and 5% survive their first few days of life. The young float on rafts of sea vegetation during their first year, feeding and growing, until they reach a size when they can, with some safety, migrate long distances through the world's oceans. When sexually mature, in various species from 7 to 20+ years later, they undertake reverse migrations, returning to their birth sites to breed.

Major nesting beaches in the Hawaii area for Green Sea Turtles are located on small islets in the Northwestern Hawaiian Islands, which, starting west of Kauai, stretch northwestwards for about 1300 km (800 miles) – for example, on French Frigate Shoals, a set of about 10 tiny islands 800 km (500 miles) from Kauai, formerly a Coast Guard station, now part of the Hawaiian Islands National Wildlife Refuge; a few probably nest on secluded beaches on Molokai. HAWKSBILLS nest in small numbers on the Big Island and Molokai, and perhaps on Maui as well.

Ecological Interactions

There is an intriguing relationship between turtle reproduction and temperature that nicely illustrates the intimate and sometimes puzzling connections between animals and the physical environment. For many vertebrate animals, the sex of an individual is determined by the kinds of sex chromosomes it has. In people, if each cell has an X and a Y chromosome, the person is male, and if two Xs, female. In birds, it is the opposite. But in most turtles, it is not the chromosomes that matter, but the temperature at which an egg develops. The facts are these. In most turtles, eggs incubated at constant temperatures above 30 °C (86 °F) all develop as females, whereas those incubated at 24 to 28 °C (75 to 82 °F) become males. At 28 to 30 °C (82 to 86 °F), both males and females are produced. In some species, a second temperature threshold exists – eggs that develop below 24 °C (75 °F) again become females. (In some other reptiles – crocodiles and lizards – the situation reverses, with males developing at relatively high temperatures and females at low temperatures.) The exact way that temperature determines sex is not clear although it is suspected that temperature directly influences a turtle's developing brain. This method of sex determination is also mysterious for the basic reason that no one quite knows why it should exist; that is, is there some advantage of this system to the animals that we as yet fail to appreciate? Or is it simply a consequence of reptile structure and function, some fundamental constraint of their biology?

The discovery of this crucial relationship between temperature of egg development and sex determination has important consequences for reptile conservation efforts. For years conservationists concerned with saving sea turtles from the brink of extinction, for instance, would remove eggs from beach nests to protect them from predators, people, and bad weather. The eggs were placed in styrofoam boxes and incubated in shelters near the beach. Unwittingly, however, by having all the eggs develop under the same temperature conditions, the effect of these conservation efforts must have been to produce hatchlings that were almost all males. Artificially creating large numbers of male and few female sea turtles, of course, is a poor way to encourage reproduction and population growth. Thus, more recent efforts, fortified with knowledge of turtle physiological ecology, vary the conditions under which removed eggs are incubated, assuring mixed-sex groups of hatchlings. In fact, a recent controversy in the conservation of these turtles concerns whether it is worthwhile to try to increase growth of their populations by varying egg temperatures to produce large numbers of females.

Notes

GREEN SEA TURTLES, more common in past centuries, were probably an important food source for the early Hawaiian settlers, the Polynesians. They would have eaten eggs, but also the adult turtles. Greens are still hunted for food in some parts of the world. HAWKSBILLS, less than palatable, are not eaten.

Status

The ecology and status of populations of most freshwater and land turtle species are still poorly known, making it difficult to determine whether population numbers are stable or declining. However, it is mainly sea turtles, rather than freshwater or terrestrial turtles, that are exploited by people, and therefore that are most threatened. Sea turtle eggs are harvested for food in many parts of the world, and adults are taken for meat (only some species) and for their skins. Many adults also die accidentally in fishing nets and collisions with boats. It is possible that the turtles, attracted to natural lights in the oceans (the luminescence given off by some sea creatures), at night approach illuminated fishing boats or the chemical light-emitting devices they place in the water, with tragic consequences. One of the sea turtles, the HAWKSBILL, is the chief provider of tortoiseshell, which is carved for decorative purposes. The Hawksbill is under international protection, but some are still hunted. All sea turtles are listed as endangered by CITES Appendix I.

Profiles

Pacific Green Sea Turtle, *Chelonia mydas agassizii*, Plate 51a
Pacific Hawksbill Sea Turtle, *Eretmochelys imbricata bissa*, Plate 51b

Environmental Close-up 3
Hawaii: The Snake-less Society, and How it Got That Way

There's a certain calmness and freedom, I've found, associated with hiking in Hawaii. In past times I worked outdoors in rattlesnake-infested areas of the mainland USA and, understandably, became very sensitive and alert to the ground when walking and hiking. The somewhat docile but always scary venomous vipers that I dealt with, along with other local snakes, never emerged from their over-winter holes in the rocks until about mid-April. So I hiked with a carefree attitude until that time, and with a more concerned, looking-at-my-feet-all-the-time attitude afterwards (I managed to escape many years of this work with actually planting my boot on a rattler only once; we both lived to tell the tale.) Hiking in Hawaii at any time is like March hiking where I used to work: snake-free. As I stated on p. 83, Hawaii has essentially no snakes and, because a likely colonizer is the Brown Tree Snake (p. 11), which is blamed for causing the extinction of many bird species on Guam after the reptile reached that island, Hawaii is ever on its guard to prevent snakes from reaching its shores. (Indeed, the prohibition on snakes extends to the Honolulu Zoo, which, when I visited, displayed only a single snake, a large Burmese Python, beside a sign stating that Hawaii law limits the zoo to possession of only two snakes, which must both be males and which must be of different species – no baby snakes will be conceived or born in Hawaii; it's always nice to see state recognition of the principles of basic reproductive biology.)

So land snakes don't occur naturally in Hawaii. They never evolved there, they never made it there from other places. But some kinds of organisms do occur naturally in Hawaii. For instance, all those *endemic* species referred to throughout this book, species that occur in Hawaii and nowhere else on Earth (Close-up, p. 67). These endemic plants and animals are interesting both because of their Hawaiian uniqueness and because they arose evolutionarily from a few *colonizing species* that crossed oceans to arrive in initially life-free volcanic Hawaii. Estimates are, for instance, that Hawaii's 1100 or so species of native plants arose via evolutionary diversification from perhaps 270 colonizing species; 8000 insects arose from perhaps 250 colonizations, and 80 land birds arose from between four and seven colonizations (this process of the formation of a large group of new species from a single colonizing species that adapts quickly through evolution to wide-open, unoccupied niches, is known as an *adaptive radiation* (explained on p. 54).

The intriguing point here is that, obviously, some kinds of organisms can naturally colonize remote islands such as Hawaii, but others, such as snakes, cannot. This difference leads us to the field of *biogeography*, the study of the distributions of living things and of how they attained their current distributions. So, how do organisms initially reach islands? First, let's look at how species get *anywhere*. (A) They evolve there. Kangaroos evolved in Australia and still ocurr in Australia. (B) They spread there. Let's say species X, a squirrel, evolved, or first became a separate species, in New York State, a million years ago. Squirrel X now occurs over all parts of the USA east of the Rocky Mountains. The reason for its current distribution is that, after first developing in New York, generation by generation, as youngsters spread out in search of new habitat to claim as breeding territories, the species slowly, incrementally, colonized a huge region. But when, say a half-million years ago, the species reached the Rocky Mountains, its slow spread was halted; the individual squirrels simply could not survive and breed at the high, cold altitudes of the high mountain passes, so never were able to make it to the western side, where they might have been able to colonize to the USA's West Coast. (Many birds, on the other hand, could fly over the mountains as they dispersed from where they were born, and, in similar scenarios, perhaps could have evolved on the continent's East Coast and eventually spread to occupy the entire continent.)

Professional biogeographers (all three of them) might say that squirrel X spread from New York to the base of the Rockies, during a half-million-year period, along a *corridor route*; that is, given that a species evolved at one end of the corridor, and barring unforeseen circumstances (say the glaciers of a new ice age intervening, or the continent splitting apart), the species, given enough time, will almost certainly spread to the other end of the corridor. Sometimes, though, there are obstacles that a species would have to overcome in order to spread, and the obstacle, owing to its nature, would allow some types of animals to pass, but not others; in other words, it would filter out certain types of animals. This is known as a *filter route*. For instance, some small, cute mammal species, like a prairie dog, may have evolved in temperate-zone, savannah-like, central North America and spread south toward South America. There are certainly some similar temperate-zone habitats in southern South America in which this mammal could thrive – if it could reach the region. But blocking the way is the narrow isthmus of Central America, "clogged" with wet, humid tropical forest, as well as northern South America, also awash with wet tropical forest. The little mammal, adapted as it is to live in seasonal, cooler, drier temperate habitats, probably cannot survive and breed in tropical rainforests, and so it could spread southward, but the southern extreme of its range is bound to be the northern extreme of the range of tropical rainforest – probably southern Mexico.

What about remote oceanic islands? Corridor and filter routes to colonization do not apply. Islands such as Hawaii are too far away from mainland areas, where potentially colonizing species live, isolated by thousands of kilometers of saltwater. So only a few organisms will colonize (thus explaining why islands always have fewer species than nearest mainland areas), and which species succeed in doing so will be governed not so much by biology, geography, or topography, but by *chance*. Colonizing species naturally reach remote islands only by what biogeographers call *lottery*, or *sweepstakes, routes*: a large ocean in the way is probably nature's most effective barrier to the spread of plant and animal species. Only a few types of organisms have the ability to cross oceans (and then, by chance, only a few species of each type will). How do these species cross oceans? (A) They keep dry and fly – birds and bats, arriving under their own power. (Some colonizers, such as snakes or mammals, could swim to previously unsettled islands, but only over short distances; this would not apply to the Hawaiian Islands, with no near sources of colonizing immigrants.) (B) They arrive *passively*, not under their own power, by *drifting* or *rafting* ashore: tiny plant seeds, tiny spiders and bugs may drift in on the wind; seeds and fruit that can survive saltwater immersion may float and drift through the ocean – coconuts, for instance, appear to be adapted for long-term ocean drifting, and this is the reason that coconut palms line beaches worldwide. (C) They come by "boat," definitely tourist class – rafting in on floating bits of vegetation, for instance, or on trees washed off mainland areas during storms. Eggs or even adults of many kinds of animals may reach remote islands this way – spiders, insects, lizards (but not amphibians, whose thin wet skins needed for breathing do not mix well with saltwater), even small rodents. (Remember, of course, for the colonization to succeed, more than a single mouse or lizard must arrive on an island; generally you must have at least two, a male and a female – or, I guess, a single pregnant female; and actually, according to genetic principles, if you want the new population to be successful in the long term in its new home, you probably want more than two individuals – to increase the genetic diversity of the population so it is not too in-bred.) Finally, because seabirds and ducks cover vast distances in their ocean-crossing feeding and migratory flights, some organisms could travel over thousands of kilometers of open ocean in mud on birds' feet (plant seeds; insect, fish, and amphibian eggs) or even in bird intestines, eventually to be excreted on isolated islands (plant seeds). Many native plants of Hawaii, for instance, may have first arrived in the islands as seeds in bird guts.

Interesting, you say, but sounds far-fetched. Does it really happen? First of all, remember that we are talking about a few chance events – some small seeds making it to Hawaii from 4000 km (2500 miles) away – occurring over millions of years. To paraphrase a famous scientist speaking on probability and the occurrence of extremely rare events: given enough time, the seemingly impossible event becomes possible, the possible becomes probable, and the probable becomes a virtual certainty. Second, we actually have evidence that these kinds of events happen (albeit on a smaller scale than crossing huge oceans). One of the best documented cases concerns the recolonization of Krakatau, a small Indonesian island that suffered a cataclysmic volcanic explosion in 1883 that wiped out all life on the island (and killed perhaps 40,000 people on neighboring islands via the tidal waves the explosion generated). People started visiting Krakatau soon after the explosion to monitor recolonization of the island (just as many scientists now work at Mt. St. Helens in Washington state, USA, to monitor recolonization of the mountain's slopes following the massive eruption of 1980). The nearest island to Krakatau, from which living organisms might come, is about 18 km (11 miles) away. In the table below are the number of species of various kinds of organisms found on Krakatau after the explosion. Note especially the

Number of Years After Eruption	Number and Kinds of Organisms Found on Krakatau
3 years	some blue-green algae, 11 fern species, 15 flowering plant species
6 years	spiders, beetles, flies, moths, 1 monitor lizard species
23 years	114 plant species
25 years	263 animal species, including 240 arthropods (insects, spiders, crustaceans), 4 snails, 2 reptiles, 16 birds
37 years	573 animal species, including a python, 3 mammals and 26 birds
45 years	3 bat species, rats, 5 lizards, 1 crocodile, 36 birds

large reptiles: a monitor lizard at 6 years, a python at 37, a crocodile at 45; these are all species known to swim in seawater. A more recent event (1998), noted in major newspapers, was the discovery of a bunch of live Green Iguanas – very large lizards – found floating on a raft of vegetation – downed trees and the like – just about to land on one of the Caribbean Sea islands – an island free of iguanas. Apparently what had happened was that, during a hurricane a few months before, the iguanas were thrown into the ocean along with the trees, and had floated on the sea's currents until almost washing up ashore on what would have been, had they not been intercepted by people, their new home; and had enough males and females arrived, a new, viable population may have been formed to colonize the island. (Storms, in fact, are suspected of playing major roles in such colonizations. All land birds that colonized Hawaii naturally are suspected of having been first blown off course by storms and, thus, having landed accidentally, probably in small flocks, on the mid-Pacific islands.)

Today, of course, many new species of plants and animals reach remote oceanic islands with people's intentional or unintentional help, that is, they are *introduced*. Introduced organisms, to separate them from native species, which evolved in the islands or reached them naturally on their own, are usually called "aliens" or "exotics" or, depending on their ecology, "invaders." Most of this book, it seems, details the biology of alien species – so many have been brought to Hawaii by people, or have stowed away on boats or airplanes and reached Hawaii's shores accidentally. As discussed in Chapter 5, the main concerns today of conservation biologists in Hawaii are to stem the tide of more species introductions to the islands, reduce the harmful effects on native species of the aliens already there, and to begin to restore to abundant numbers some of the threatened and endangered remaining native species.

An unfortunate side effect of the struggle between native and alien animal and plant species is that one group, the natives, comes to be seen as "good" and the other group, the aliens, is seen as "bad." It is not the aliens' fault, of course, that their ancestors were brought to Hawaii by people. But the damage they do to native ecosystems, which many people treasure, causes us to seek their containment and, if possible, their elimination. Philosophy begins to creep into such arguments. An Asian gecko species, kept as a pet, escapes captivity and establishes itself with healthy populations on several of Hawaii's main islands. It is an alien species, not native to Hawaii. Is that forever the case? A thousand years from now, is it *still* an alien? Or, given enough time, as the plants and other animals in the islands adapt to the gecko's ecological presence, does the species pass into "native-hood"? I don't know. And, of course, natural colonizations are not finished. The Pied-billed Grebe (Plate 28) apparently reached the Big Island by itself during the 1980s; assuming that the colonization is successful and permanent, is the grebe now a native Hawaiian?

Chapter 8

Birds

Introduction

Most of the vertebrate animals one sees on a visit to just about anywhere at or above the water's surface are birds, and Hawaii is no exception. Regardless of how the rest of a trip's wildlife viewing progresses, birds will be seen frequently and in large numbers. The reasons for this pattern are that birds are, as opposed to other terrestrial vertebrates, most often active during the day, visually conspicuous and, to put it nicely, usually far from quiet as they pursue their daily activities. But why are birds so much more conspicuous than other vertebrates? The reason goes to the essential nature of birds: they fly. The ability to fly is, so far, nature's premier anti-predator escape mechanism. Animals that can fly well are relatively less prone to predation than those which cannot, and so they can be both reasonably conspicuous in their behavior and also reasonably certain of daily survival. Birds can fly quickly from dangerous situations, and, if you will, remain above the fray. Most flightless land vertebrates, tied to moving in or over the ground or on plants, are easy prey unless they are quiet, concealed, and careful or, alternatively, very large or fierce; many smaller ones, in fact, have evolved special defense mechanisms, such as poisons or nocturnal behavior.

A fringe benefit of birds being the most frequently encountered kind of vertebrate wildlife is that, for an ecotraveller's intents and purposes, birds are innocuous. Contrast this with too-close encounters with potentially dangerous fish (sharks), amphibians (frogs with toxic skin secretions), reptiles (venomous snakes), and mammals (bears and big cats). Moreover, birds do not always depart with all due haste after being spotted, as is the wont of most other types of vertebrates. Again, their ability to fly and thus easily evade our grasp, permits many birds, when confronted with people, to behave leisurely and go about their business (albeit keeping one eye at all times on the strange-looking bipeds), allowing us extensive time to watch them. Not only are birds among the safest animals to observe and the most easily discovered and watched, but they are among the most beautiful. Experiences with Hawaii's birds will almost certainly provide some of any trip's finest, most memorable naturalistic moments.

General Characteristics and Classification of Birds

Birds are vertebrates that have feathers and can fly. They began evolving from reptiles during the Jurassic Period of the Mesozoic Era, perhaps 150 million years ago, and saw explosive development of new species occur during the last 50 million years or so. The development of flight is the key factor behind birds' evolution, their historical spread throughout the globe, and their current ecological success and arguable dominant position among the world's land animals. Flight, as mentioned above, is a fantastic predator evasion technique, it permits birds to move over long distances in search of particular foods or habitats, and its development opened up for vertebrate exploration and exploitation an entirely new and vast theater of operations – the atmosphere.

At first glance, birds appear to be highly variable beasts, ranging in size and form from 135 kg (300 lb) ostriches, to 4 kg (10 lb) eagles, to 3 g (a tenth of an ounce) hummingbirds. Actually, however, when compared with other types of vertebrates, birds are remarkably standardized physically. The reason is that, whereas mammals or reptiles can be quite diverse in form and still function as mammals or reptiles (think how different in form are lizards, snakes, and turtles), if birds are going to fly, they more or less must look like birds, and have the forms and physiologies that birds have. The most important traits for flying are: (1) feathers, which are unique to birds; (2) powerful wings, which are modified upper limbs; (3) hollow bones; (4) warm-bloodedness; and (5) efficient respiratory and circulatory systems. These characteristics combine to produce animals with two overarching traits – high power and low weight, which are the twin dictates that make for successful feathered flying machines. (Bats, the flying mammals, also follow these dictates.)

Bird classification is one of those areas of science that continually undergoes revision. Currently more than 9000 separate species are recognized. They are divided into 28 to 30 orders, depending on whose classification scheme one follows, perhaps 170 families, and about 2040 genera (plural of genus). For purposes here, we can divide birds into *passerines* and *nonpasserines*. Passerine birds (Order Passeriformes) are the perching birds, with feet specialized to grasp and to perch on tree branches. They are mostly the small land birds (or *songbirds*) with which we are most familiar – blackbirds, robins, wrens, finches, sparrows, etc – and the group includes more than 50% of all bird species. The remainder of the birds – seabirds and shorebirds, ducks and geese, hawks and owls, parrots and woodpeckers, and a host of others – are divided among the other 20+ orders.

Features of Hawaiian Birds

Types of Birds

We can group Hawaii's birds into four main types, listed below. If during your visit to the islands you keep your eyes open, you will see a lot of (1), some of (2), many of (3), and, unfortunately, relatively few of (4).

(1) *Native (resident and migratory) water-associated birds*. Water-associated birds include seabirds that you see along the islands' coasts and on sea-outings, birds such as *albatrosses, frigatebirds, tropicbirds, terns,* and *noddies*; as well as freshwater marsh and pond birds such as *herons, ducks,* and *geese*; and shorebirds such as *plovers* that inhabit coastal sites as well as inland grassy areas.

(2) *Introduced chicken-like land birds*. These are 10 or more species of *quail* and *pheasant, francolin* and *turkey*, some now quite common, introduced to the islands as game birds or ornamental birds.

(3) *Introduced passerine land birds.* About 30 species of small land birds, a few of them forest birds but many of them birds of open habitats, occupy the islands. Some originated as escaped or released pet cage birds; others were imported to be released intentionally.

(4) *Native passerine forest birds.* About 20 species of generally small forest birds occur in the islands, most at higher elevations in protected national wildlife refuges, national parks, and state forest reserves. These birds evolved in Hawaii, occur only in these islands (that is, they are *endemic* species) and, currently, the continued existences of most of them are highly threatened. With a little effort, some of these species are fairly easily seen (see below) but some are so rare now that the average ecotraveller has essentially no chance of an encounter. In the recent past there were a good deal more of these species, but they succumbed to a variety of ecological threats (Chapter 5) and are now extinct.

Of Seabirds and Land Birds, Natives and Aliens

Aside from the famous, attention-getting, and highly endangered NENE, or HAWAIIAN GOOSE, most of the other Hawaiian birds of note, in which ecotravellers are usually most interested, fall into two groups, seabirds and native landbirds. A good number of seabird species occur in and around the Hawaiian Islands, but many of them are rarely if ever spotted inland or directly along the coasts of the main islands; that is, they only occur at sea, and are therefore seen only during cruises. (Seabirds can be divided into *inshore species*, which often inhabit coastal areas; *offshore species*, which mostly stay near continents but away from coastlines – feeding on fish in the shallow waters of the continental shelves; and *pelagic species*, which frequent the open ocean except when nesting on oceanic islands.) Many of these birds come ashore only to nest, and then usually not on the main islands but on smaller islands, unoccupied by people. Hawaii has numerous representatives of the three main seabird orders: the Procellariiformes (p. 111), or tube-nosed seabirds, such as albatrosses and petrels; Pelecaniformes (p. 114), pelican relatives such as boobies, tropicbirds and frigatebirds; and Charadriiformes (p. 117), including the terns, gulls, and noddies. Because they are large, and exotic to most people from north temperate regions, birds such as albatrosses and tropicbirds are probably the seabirds most watched for by visitors to Hawaii from the mainland USA, Europe, and Japan.

There are two main things to know about Hawaii's land birds. First, almost all of them that you see routinely – around cities, towns, and resorts, and in lowland parks along the coasts – have been introduced to the islands during the past 150 years (many of them within the past 50 years). Experts estimate that more passerine birds have been intentionally introduced to Hawaii than to any comparable place on the planet. Many of these birds have securely established themselves on the islands, now permanently occupying many habitats that the native forest passerine birds do not, and perhaps even competing with native forest birds for some resources, further pushing some of the natives toward extinction. Hawaii's most common and most frequently seen land birds – such as the JAPANESE WHITE-EYE and COMMON MYNA – are aliens. Even if you have neither the time nor opportunity to search out native birds, most of the alien birds are from Asia and Africa and therefore are exotic to most ecotravellers – worth taking a look at. Some, such as the WHITE-RUMPED SHAMA, RED-BILLED LEIOTHRIX, and RED

AVADAVAT, are gorgeous little birds, and you won't see them again until your next trip to Asia.

Second, the native landbirds, of which there are now only a few species, occur only in restricted regions of the main islands (higher elevation forested sites, for the most part) and many of them are severely threatened or already endangered. They can be divided into two groups, the Hawaiian Honeycreepers (Family Drepanididae, p. 151; also called the Hawaiian Finches), shown on Plates 42, 43, and 44, and a small group of miscellaneous species, members of three bird families, shown in Plate 37.

Nectar-eaters: the Hawaiian Honeycreepers

The forest-dwelling *honeycreepers* are Hawaii's most famous birds, at least among biologists (although the publicity-stealing Nene, the Hawaiian Goose, may be better known among tourists). They are known for two aspects of their evolutionary history (their actual development and their feeding specializations) and most recently, for their conservation status. The honeycreepers arose, apparently, when a single species of a finch or finch-like bird – a flock of them containing, we must assume, both males and females – long ago (perhaps 3.5 million years ago) left the North American mainland and, perhaps blown off-course by a storm, eventually spotted Hawaii's volcanic islands jutting from the sea and settled there. Perhaps finding no other forest-dwelling songbirds on the islands – so a complete lack of competition for resources and a wide assortment of "unfilled" ecological niches – the one finch species, over evolutionary time, diverged, or "speciated," into many different species, each specialized for a particular ecological niche. This process of a single species evolving into a group of closely related but individually highly specialized species is known as *adaptive radiation*, and the Hawaiian Honeycreepers are regarded as perhaps the most amazing, extreme example among birds (surpassing in number of evolved species even the similarly famous Darwin's finches of Ecuador's Galápagos Islands).

The honeycreeper's finch-like ancestor probably fed on seeds and fruit. During the adaptive radiation, some species retained the ancestor's feeding habits – continuing to eat seeds and fruit, some turned to nectar (mostly gaining this resource from the red flowers of a single native tree, *Metrosideros polymorpha*, known as ohia lehua, or simply, ohia; p. 232), and a few even specialized on insects. The most obvious physical manifestation of these feeding specializations is the tremendous variety of bill shapes these birds developed to aid them in their foraging. Some, seed-eaters, developed shortish, strong, stout, almost parrot-like bills, like the PALILA (Plate 44); some, eating only insects, developed mid-sized or longish down-curved bills to probe and grab bugs and their larvae on and under tree bark, like the MAUI PARROTBILL and AKIAPOLAAU (Plate 43); and some, feeding mostly on nectar but also grabbing some insects, developed long or shortish, narrow, down-curved bills, like the IIWI and APAPANE (Plate 44).

Few of these birds are now common and widespread in Hawaii; most are threatened or endangered. Several of the honeycreepers died out over the past two or three decades, and at the time of this writing, three or four of them – such as the OAHU CREEPER, OU, and POO-ULI – had just become extinct or had dwindled to such low numbers that their extinction was all but assured.

Seeing Birds in Hawaii

Illustrated in the color plates are 136 birds – just about every species you can see on and immediately around the main Hawaiian Islands (several can only be seen on Midway Atoll). A few species that may still exist but are critically endangered and almost extinct are not pictured – such as the POO-ULI, a small, chunky, brownish endemic honeycreeper, first discovered on Maui only in 1973 and which, at the time of this writing, was reduced in numbers to probably no more than three individuals (a male and two females). Essentially no one but researchers sees these birds now, and no one may ever see them again.

The best way to spot many of Hawaii's birds is to follow four easy steps:

(1) Look for them at the correct time. Birds can be seen at any time of day, but they are often very active during early morning and late afternoon, and so can be best detected and seen during these times. Unlike many other birds, Hawaii's birds are usually not exceptionally vocal during early mornings – that is, there is no "dawn chorus."
(2) Be quiet as you walk along trails or roads, and stop periodically to look around carefully. Not all birds are noisy, and some, even brightly-colored ones, can be quite inconspicuous when they are directly above you, in a forest canopy.
(3) Bring binoculars on your trip. You would be surprised at the number of people who visit parks and reserves with the purpose of viewing wildlife but don't bother to bring binoculars. They need not be an expensive pair, but binoculars are essential to bird viewing.
(4) If you are a serious birdwatcher, before your trip try listening to store-bought tapes of bird vocalizations to learn the songs of Hawaii's birds.

All visitors to Hawaii spend at least part of their stay in coastal areas and these are good places to spot seabirds. If they are a priority for you, spending a day on a fishing boat or taking a cruise – perhaps one of the islands' many whale-watching cruises – are good ideas. Kauai's Kilauea Point National Wildlife Refuge is an easily reached site where you can see albatrosses, boobies, and shearwaters, among other birds (not to mention having a chance to see some Nene; there is a small population there; p. 65). Marsh and water birds are often seen easily at ponds and other freshwater sites on many of the islands. Introduced land birds will be seen almost everywhere – whether you want to see them or not. To see some of Hawaii's native forest birds, however, you will have to go out of your way. If you enjoy wildlife (and because you bought this book it's a good bet that you do), it would be a shame to leave the islands without seeing at least some of the spectacular little native forest birds. A very few species are fairly common and if you know where to go (I'm going to tell you now) and what to look for (this book will help), you can definitely see some of them.

The common, oft-seen species are: ELEPAIO, OMAO, APAPANE, AMAKIHI, and if you're lucky, IIWI. In most cases, you need to be at an elevation *above* 1000 to 1200 m (3300 to 4000 ft) before you can see these birds (reason on p. 59). The easiest place in the islands to see some of them is probably the Big Island's Hawaii Volcanoes National Park. Drive to the park, pay your entry fee, look at the map they give you, find some forested trails, park your car, hike, listen, and look up occasionally; try the Crater Rim Trail, the Thurston Lava Tube area, the Kipuka Puaulu area (p. 34), or even around the main visitor center. Kauai's Kokee State Park and the access it allows to part of the Alakai Swamp region, worth visiting for the scenery alone, is the final refuge for many of the native honeycreepers, and several of these birds (Apapane, Kauai Amakihi, Iiwi, Anianiau), as well as Elepaio, can be seen there. Finally, Maui's Haleakala National Park and the adjacent Nature Conservancy Waikamoi Preserve are popular spots where Apapane, Common Amakihi, and Iiwi are routinely spotted, as is the MAUI ALAUAHIO; try the Hosmer Grove trail near the park's entrance.

If Nene are your goal, there are three spots to seek them out: Hawaii Volcanoes National Park, Haleakala National Park, and Kilauea Point National Wildlife Refuge.

If you have trouble locating the birds you'd like to see, ask people – National Park Service employees, preserve personnel, tourguides, resort employees – about good places to see them. A very helpful guidebook that details the islands' best birding locations is *Enjoying Birds in Hawaii: A Birdfinding Guide to the Fiftieth State*, by H. Douglas Pratt.

Family Profiles

Along Hawaii's coasts, as along coasts almost everywhere, *seabirds*, many of them conspicuously large and abundant, reign as the dominant vertebrate animals of the land, air, and water's surface. Many seabirds in Hawaii commonly seen by visitors from northern temperate areas are very similar to species found back home, but some are members of groups restricted to the tropics or subtropics (or, in the case of albatrosses, to the southern hemisphere and northern Pacific region) and, hence, should be of interest to the ecotraveller. A few of these birds will be seen by almost everyone. (Intriguingly, gulls, so common along coasts of many mainland areas, are rare or absent from mid-ocean warm-water islands, such as Hawaii; see p. 117). As a group, seabirds are extremely successful animals, present often at breeding and roosting colonies in enormous numbers. Their success surely is owing to their exceptionally rich food resources – the fish and invertebrate animals (crabs, mollusks, insects, jellyfish) produced in the sea and on beaches and mudflats.

1. Seabirds I: Albatrosses and Other Tube-nosed Seabirds

The seabird order Procellariiformes includes the *albatrosses, shearwaters, petrels,* and *storm-petrels*. All are found only in marine (sea water) habitats, and they spend their entire lives at sea (a *pelagic* existence) except for short periods of nesting on islands. They are called "tube-nosed" seabirds because their nostrils emerge through tubes on the top or sides of their distinctly hooked upper bills. Like many seabirds, they have a large gland between and above their eyes that permits them to drink seawater; it filters salt from the water and concentrates it. This highly concentrated salt solution is excreted in drops from the base of the bill and the nostril tubes then direct the salt drops to the end of the bill where they can be easily discharged. Some of these tube-nosed seabirds, or *tubenoses*, will be seen by almost all travellers to Hawaii, especially those who take cruises. Family Diomedeidae contains the 14 albatross species, distributed over the world's southern oceans and the northern

Pacific; three occur in Hawaiian waters. Family Procellariidae, with an essentially worldwide oceanic distribution, has 70 species of petrels and shearwaters; about eight species are seen regularly around Hawaii. Family Hydrobatidae contains the 20 or so species of small petrels known as storm-petrels; few species of storm-petrels occur regularly in the Hawaiian region. The tubenoses include the largest and smallest seabirds (the largest of all being the WANDERING ALBATROSS, with a wingspan to 3.5 m, 11.5 ft). Hawaii's tubenoses range from the huge SHORT-TAILED ALBATROSS (Plate 20), at 91 cm (3 ft) long, to the small BULWER'S PETREL (Plate 22), at 25 cm (10 in), and BAND-RUMPED STORM-PETREL, at 23 cm (9 in). Albatrosses are large, heavy birds with very long wings and long, heavy, hooked bills. Shearwaters are small to mid-sized seabirds with slender, hooked bills. The petrels comprise a large group, but most of Hawaii's (genus *Pterodroma*) are medium-sized seabirds with long wings and shortish, stout, black bills. Tubenoses are often dark above and lighter below, although some are all dark.

Natural History

Ecology and Behavior

Albatrosses feed, either solitarily or in small groups, on fish, squid, and other invertebrates (crabs, krill) near the surface at night (and sometimes during the day). Larger species sit on the water and seize prey in their bills; smaller, more agile species can also seize prey from the surface while flying. Also, they are not above eating garbage thrown overboard from ships, as well as floating carrion such as dead whales and seals. They often go out to sea for several days to bring food back to nestlings. Albatrosses use a type of non-flapping flight, known as *dynamic soaring*, that takes advantage of strong winds blowing across the ocean's surface. Their efficient but peculiar soaring flight takes them in huge loops from high above the ocean surface where the wind is fastest, down toward the surface, where friction slows the wind, and then up into the faster wind again to give them lift for the next loop – a kind of roller coaster flight that requires virtually no wing flapping. Albatross wings are so long and narrow that these birds literally need wind to help them fly, and on absolutely calm days (which, luckily, are rare on the open ocean) must wait out the windless hours sitting on the sea's surface. Even taking flight for these long-winged birds is problematic: in very windy conditions, whether they are on land or the water, they need only spread their wings, and the wind moving over and under the wings provides sufficient lift to make them airborne. But in low winds, they face into the wind (like an airplane taking off) and make a take-off run; at island breeding colonies, often there are actual "runways," long, clear paths on the islands' windier sides, usually on slopes, along which the large birds make their downhill take-off runs.

Petrels and shearwaters are often excellent flyers, some using dynamic soaring like albatrosses, some alternating flapping flight with gliding. They feed at sea by day or by night, often in groups, on squid, fish, and crustaceans. Some (particularly storm-petrels) pluck prey from the surface of the sea using their wings to flutter and hover just above the water.

Tubenoses have perhaps the best developed sense of smell of any birds, and they use this ability to locate young and nest sites when returning from extended foraging trips and also probably to locate some types of food. Tubenoses produce a vile-smelling stomach oil that they regurgitate to feed to nestlings and to squirt at enemies. (The oil reputedly makes an excellent suntan lotion, but its smell, I'm sure, would wreak havoc on your social life.)

Breeding

Tubenoses breed usually in large, dense colonies, often but not always on small oceanic islands. Albatrosses are monogamous and most breed with the same mate for life. On their remote breeding islands they engage in elaborate courtship dances in which male and female face each other, flick their wings, bounce their heads up and down, and clack their bills together. This behavior probably strengthens the pair-bond and also coordinates hormone release and synchronizes mating readiness. Nests vary from a scrape on the bare ground to scrapes that are surrounded by vegetation, soil, and pebbles. The female lays one large egg. Male and female alternately incubate and brood the young, in shifts lasting several days to weeks. The other adult flies out to sea and searches for food. When it returns, it feeds the chick regurgitated fish and squid and stomach oil and takes over its turn at brooding again. When the chick's demand for food becomes overwhelming, both adults leave it alone for long periods as they search great distances over the ocean for enough food. At some points in its nestling period, a young albatross can actually weigh more than its parents. After about 7 months, the chick is large enough to fledge and fly. Albatrosses take 5 to 7 years to mature, staying at sea during this period before finally returning to their birth place to breed. Many individuals do not breed until they are 7 to 9 years old; some albatrosses in the wild live 40+ years.

Petrel and shearwater breeding is similar to that of albatrosses. Some nest in the open, with nests simply small depressions in the ground, but most are burrow or cavity nesters, nesting in a burrow they dig themselves or take over, or in a natural cavity, such as holes in cliffs. The burrow or cavity is re-used each year by the same pair. One large egg is incubated for 6 to 9 weeks, both parents incubating the egg and feeding and brooding the chick. These birds probably live an average of 15 to 20 years. Several species in Hawaii nest now only in the relatively predator-free precincts of high mountainous areas, such as in Maui's Haleakala Crater and high on the Big Island's Mauna Kea and Mauna Loa.

Notes

Most albatross species nest on only one or a few islands, and their survival can be precarious. In the case of the SHORT-TAILED ALBATROSS, the long wandering phase of the juveniles saved the species. Intense hunting, volcanic activity and a mid-20th Century world war virtually wiped out the nesting population of this species on its single breeding island southeast of Japan. But a reservoir of younger individuals was at sea, and by the time they returned to the nesting island, the war was finished and the island was a preserve. The population is growing slowly now, with the total world population perhaps less than a thousand.

Petrels get their name from the Greek word "petros," which refers to the biblical disciple Peter, who tried to walk on water, just like *storm-petrels*, a group of small petrels, appear to be doing when feeding (they have a fluttering flight low over the water's surface, and some species even patter their feet on the surface).

Sailors have long believed albatrosses contained the souls of lost comrades, so it was considered bad luck for one's ship and shipmates to kill one. Albatrosses are highly respected birds in many cultures, perhaps owing to their amazing flight capabilities, and killing them is frowned upon. Albatrosses are sometimes called "gooney birds" because of their awkwardness moving on land and the untidiness of their take-offs and landings (crashing into beach shrubbery is a common occurrence).

Status

Because of their often highly restricted nesting sites on small islands and vulnerability during the nesting period, many species of tubenoses are at risk. Albatrosses cannot become airborne readily from land and are easy victims for humans and introduced predators. Cats and rats, for instance, if carelessly released by people, can wreak havoc on an entire island's population of petrels. Albatrosses during the 1800s and early 1900s were widely killed for their feathers, entire breeding colonies destroyed. In fact, albatrosses were largely absent from the main Hawaiian Islands from the middle of the 19th to the middle of the 20th Centuries. Then, during the mid-1970s, LAYSAN ALBATROSSES began nesting at Kauai's Kilauea Point.

Two albatrosses are endangered (USA ESA listed), the AMSTERDAM and SHORT-TAILED ALBATROSSES. Both are restricted now to single breeding colonies. BLACK-FOOTED and LAYSAN ALBATROSSES (Plate 20), both of which breed in the Hawaiian region, often lose eggs and chicks to introduced predators. Pacific Rats (Plate 47) on some islands, for instance, eat eggs and chicks and, because adult albatrosses often don't react to their presence (having lost, or never having evolved, anti-predator responses), sometimes are able to eat adults also as they incubate eggs. (Fortunately, rats were recently eradicated from Midway Atoll, greatly increasing nesting success of albatrosses and BONIN PETRELS.) About 17 species of petrels and shearwaters are listed as threatened (IUCN Red List). The HAWAIIAN PETREL (Plate 22; formerly known as the Dark-rumped Petrel), which breeds only in Hawaii (in burrows in high mountainous areas of Maui's Haleakala National Park, and in small numbers on Kauai, the Big Island, and perhaps Molokai), is endangered. Its nests were regularly raided by rats, cats, and mongooses. Predator elimination programs at Haleakala have vastly improved this petrel's rate of breeding success. NEWELL'S SHEARWATER (Plate 21), which also breeds only in Hawaii in high mountainous regions, is threatened.

Profiles

Laysan Albatross, *Phoebastria immutabilis*, Plate 20a
Black-footed Albatross, *Phoebastria nigripes*, Plate 20b
Short-tailed Albatross, *Phoebastria albatrus*, Plate 20c

Wedge-tailed Shearwater, *Puffinus pacificus*, Plate 21a
Christmas Shearwater, *Puffinus nativitatis*, Plate 21b
Sooty Shearwater, *Puffinus griseus*, Plate 21c
Newell's Shearwater, *Puffinus newelli*, Plate 21d

Bulwer's Petrel, *Bulweria bulwerii*, Plate 22a
Hawaiian Petrel, *Pterodroma sandwichensis*, Plate 22b
Bonin Petrel, *Pterodroma hypoleuca*, Plate 22c
Black-winged Petrel, *Pterodroma nigripennis*, Plate 22d
Mottled Petrel, *Pterodroma inexpectata*, Plate 22e

2. Seabirds II: Pelican Allies (Boobies, Tropicbirds, Frigatebird)

The large seabirds treated in this section are members of the Order Pelecaniformes: *boobies*, of the Family Sulidae (nine species worldwide, with three occurring off Hawaiian shores); *tropicbirds*, Family Phaethontidae (three species, all with tropical distributions, two in Hawaii); and *frigatebirds*, of the Family Fregatidae (five species with mainly tropical distributions; only one occurs around Hawaii). (The order also includes the pelican family, Pelecanidae, but none of the eight pelican species occur in or around Hawaii.) Boobies are large seabirds known for their sprawling, densely packed breeding colonies, spots of bright coloring, and for plunging into the ocean from heights to pursue fish. They have tapered bodies, long pointed wings, long tails, long pointed bills, and often, brightly colored feet. Tropicbirds, considered among the most striking and attractive of the tropical seabirds, are mid-sized white or white-and-black birds with two very long, thin tail feathers, called *streamers*, that provide the birds an unmistakable flight silhouette. They have very short legs unsuited for walking on land (and so they spend most of their time in the air or out at sea), and webbed feet. The GREAT FRIGATEBIRD (Plate 20) is a very large soaring bird, mostly black, with huge pointed wings that span up to 2 m (6.5 ft) or more, and a long forked tail. Males have red throat pouches that they inflate, balloon-like, during courtship displays.

Natural History

Ecology and Behavior

Seabirds feed mainly on fish, and have developed a variety of ways to catch them. Boobies, which also eat squid, plunge-dive from the air (from heights of up to 15 m, 50 ft, or more) or surface-dive to catch fish underwater. Sometimes they dive quite deeply, and they often take fish unawares from below, as they rise toward the surface. Unlike most other birds, boobies do not have holes or nostrils at the base of the upper bill for breathing; the holes are closed over to keep seawater from rushing into their lungs as they plunge-dive. Tropicbirds are often seen flying alone or in pairs over the ocean or near shore, or inland, circling in valleys and canyons. They eat fish, squid, and crustaceans, which they obtain by flying high over the water, spotting food, hovering a bit, then plunging down into the water to catch the meal. They rarely feed within sight of land, preferring the open ocean. They travel far and wide on oceanic winds, and individuals banded in the Hawaiian region have been spotted at sea 8000 km (5000 miles) away. Frigatebirds feed on the wing, sometimes soaring effortlessly for hours at a time. They swoop low to catch flying fish that leap from the water (the fish leap when they are pursued by larger, predatory fish or dolphins), and also to pluck squid and jellyfish from the wavetops. They even drink by flying low over the water's surface and sticking their long bill into the water. Although their lives are tied to the sea, frigatebirds cannot swim and rarely, if ever, enter the water voluntarily; with their very long, narrow wings, they have difficulty lifting off from the water. To rest, they land on remote islands, itself a problematic act in the high winds that are common in these places.

Ecological Interactions

Frigatebirds, large and beautiful, are a treat to watch as they glide silently along coastal areas, but they have some highly questionable habits – in fact, patterns of behavior that among humans would be indictable offenses. Frigatebirds practice *kleptoparasitism*: they "parasitize" other seabirds, such as boobies, frequently chasing them in the air until they drop recently caught fish. The frigatebird then steals the fish, catching it in mid-air as it falls. Frigatebirds are also common predators on baby sea turtles (p. 98), scooping them from beaches as the reptiles make their post-hatching dashes to the ocean.

Breeding

Pelican allies usually breed in large colonies on small oceanic islands (where there are no mammal predators) or in isolated mainland areas that are relatively free of predators. Some breed on cliffs or ledges (tropicbirds, some boobies), some in trees or on tops of shrubs (frigatebirds, some boobies), and some on the bare ground (tropicbirds, most boobies, and frigatebirds if their preferred nesting sites are unavailable). Where people have introduced small mammals such as rats to islands, which feed on seabird eggs and nestlings, reproductive success is often dramatically reduced. Most species are monogamous, mated males and females sharing in nest-building, incubation, and feeding young. High year-to-year fidelity to mates, to breeding islands, and to particular nest sites is common. Booby females usually lay one or two eggs, which are incubated for about 45 days. Usually only a single chick survives to fledging age (one chick often pecks the other to death). Tropicbirds breed in nests hidden in rock cavities on steep cliffs, including inland sites such as on the cliffs surrounding Kilauea Crater at the Big Island's Hawaii Volcanoes National Park and in Waimea Canyon, Kauai. A single egg is incubated by both parents, in shifts of 2 to 5 days. The egg hatches in 40 to 45 days, after which the youngster is fed by the parents for 10 to 14 weeks, until fledging. GREAT FRIGATEBIRDS lay a single egg that is incubated for about 50 days; male and female spell each other during incubation, taking shifts of up to 12 days. Young remain in and around the nest, dependent on the parents, for up to 6 months or more. In most seabirds, young are fed when they push their bills into their parents' throats, in effect forcing the parent to regurgitate food stored in its *crop* – an enlargement of the top portion of the esophagus. Seabirds reach sexual maturity slowly (in 2 to 5 years in boobies; 5 years in tropicbirds; 7+ years in frigatebirds) and live long lives (frigatebirds and boobies live 20+ years in the wild).

Notes

Boobies are sometimes called *gannets*, particularly by Europeans. The term *booby* apparently arose because the nesting and roosting birds seemed so bold and fearless toward people, which was considered stupid. Actually, the fact that these birds bred on isolated islands and cliffs meant that they had few natural predators, so had never developed, or had lost, fear responses to large mammals, such as people. Frigatebirds are also known as *man-of-war* birds, both names referring to warships, and to the birds' kleptoparasitism; they also steal nesting materials from other birds, furthering the image of avian pirates. The long central tail feathers of tropicbirds are used as adornments by some South Pacific islanders, and were used as such by early Hawaiian settlers.

Status

Most of the seabirds that occur in Hawaiian waters, including boobies, frigatebirds, and tropicbirds, are quite abundant and not considered threatened. ABBOT'S BOOBY, now limited to a single, small breeding population on the Indian Ocean's Christmas Island, is endangered (CITES Appendix I and USA ESA listed). Two of the frigatebirds, the CHRISTMAS ISLAND FRIGATEBIRD (CITES Appendix I and USA ESA) and ASCENSION FRIGATEBIRD are endangered. None of the tropicbirds are threatened.

Profiles

Great Frigatebird, *Fregata minor*, Plate 20d
Masked Booby, *Sula dactylatra*, Plate 23a
Brown Booby, *Sula leucogaster*, Plate 23b
Red-footed Booby, *Sula sula*, Plate 23c

White-tailed Tropicbird, *Phaethon lepturus*, Plate 24a
Red-tailed Tropicbird, *Phaethon rubricauda*, Plate 24b

3. Seabirds III: Gulls and Terns

The most common and conspicuous birds over mainland seacoasts and nearshore and offshore islands almost anywhere are *gulls* and *terns*. These highly gregarious seabirds – they feed, roost, and breed in large groups – are also common in offshore waters and even, in certain species, inland. Family Laridae (about 100 species worldwide), which is allied with the shorebirds (p. 131) in Order Charadriiformes, includes the gulls, terns, *noddies*, and *jaegers* (often called *skuas*, especially by the British). The approximately 45 species of gulls are distributed worldwide, but they are mainly birds of cooler ocean waters, and even of inland continental areas. Few occur in the tropics or around isolated, oceanic islands. Furthermore, gulls are adapted to feed in the shallow waters of continental shelves, not the deep waters that surround mid-ocean islands. No gulls breed in Hawaii; the two profiled here are only occasional visitors. In total, about 10 species visit Hawaii, most of them being seen only rarely in or around the main islands. Gulls generally are large white and gray seabirds with fairly long, narrow wings, squarish tail, and sturdy, slightly hooked bill. Many have a blackish head, or *hood*, during breeding seasons. The 40 or so tern species are distributed throughout the world's oceans; about 10 species occur in Hawaii. Terns (and noddies, which are a type of tern) are often smaller and more delicate-looking than gulls. They have a slender, light build, long, pointed wings, a deeply forked tail, a slender, tapered bill, and webbed feet. They are often gray above and white below, with a blackish head during breeding. The seven species of jaegers, dark, gull-sized birds, known for their predatory habits, are mostly birds of the polar seas; one species, the POMARINE JAEGER (Plate 25), is a winter resident in Hawaiian waters.

Natural History

Ecology and Behavior

Gulls and terns feed on fish and other sealife that they snatch from shallow water, and on crabs and other invertebrates they find on mudflats and beaches, often in the intertidal zone. Also, they are not above visiting garbage dumps or following fishing boats to grab whatever goodies that fall or are thrown overboard. They also scavenge what they can, taking bird eggs and nests from seabird breeding colonies when parents are gone or inattentive. Many larger gulls also chase smaller gulls and terns in the air to steal food the smaller birds have caught – an act termed *kleptoparasitism*. Terns, which eat mainly fish, squid, and crustaceans, feed during the day but also sometimes at night (like the SOOTY TERN, Plates 25 and 26). Their main food-gathering technique is a bit messy: they spot prey near the water's surface while flying or hovering, then plunge-dive into the water to grab the prey, then rapidly take off again, the prey held tightly in the bill. Most terns seem to fly continuously, rarely setting down on the water to rest. They are often distinguished from other seabirds by their continually flapping flight. Jaegers have a reputation of being predatory birds: in addition to catching and eating fish, they also take small seabirds and, from the land, seabird nestlings, newly hatched sea turtles, and even small rodents such as lemmings. They are also well known for their kleptoparasitism.

Ecological Interactions

Many gulls have a commensal relationship with people. They make good feeding use of human-altered landscapes and human activities, such as garbage dumps, agricultural fields, and fishing boats, but neither help nor harm people. Human activities have reduced the populations of many other bird groups, but people's exploitation of marine and coastal areas has, in many cases, enhanced rather than hurt seabird populations. In fact, many gull species, worldwide, are almost certainly more numerous today than at any time in the past.

Breeding

These seabirds usually breed in large colonies, often in the tens of thousands, on small islands, where there are no mammal predators, or in isolated mainland areas that are relatively free of predators. Gulls and terns usually breed on flat, open ground near the water (but some breed on cliffs, or inland, in marsh areas; and noddies breed in trees, bushes, or cliffs). Most species are monogamous, mated males and females sharing in nest-building, incubation, and feeding young. Gulls and terns typically lay one to three eggs, and both sexes incubate for 21 to 30 days; young fledge after 28 to 35 days at the nest. Young are fed when they push their bill into their parent's throat, forcing the parent to regurgitate food stored in its *crop* – an enlargement of the top portion of the esophagus. These seabirds reach sexual maturity slowly, often not achieving their full adult, or breeding, plumage until they are 3 to 4 years old (terns) or 4 to 5 years old (gulls).

Notes

Most terns are migratory to varying degrees, breeding in one region but spending the non-breeding portions of the year elsewhere. The three tern and two noddy species profiled here all breed in the Hawaiian Islands, with the GRAY-BACKED and SOOTY TERNS (Plates 25 and 26) being migratory (but some individuals are seen offshore at any time of year); the other three species are *resident*, meaning they do not migrate and can be found in the region all year. The bird that undertakes the world's longest migration is a tern, the ARCTIC TERN. Individuals that breed north of the Arctic Circle migrate to spend their non-breeding season in Antarctic waters, a minimum flight distance away of 12,800 km (8000 miles). Since they cover this incredible distance twice a year, these terns are considered among nature's champion flyers.

When Polynesians first reached the main Hawaiian Islands, there would have been large coastal colonies of seabirds such as terns in many areas (as there are now on some of the small offshore islets around the main islands and on the islands of the Northwestern Hawaiian chain). These early settlers undoubtedly raided the colonies for food – tern eggs and nestlings would have been easy, abundant, and delicious prey. Needless to say, these large colonies no longer exist.

Because of their flying prowess, their often sustained flight, and some physical similarities to swallows (the pointed wings and forked tail), terns are sometimes called "sea swallows."

Status

Most gulls, terns and jaegers are abundant seabirds and, as noted above, some species have even been able to capitalize on people's activities and during the past few hundred years have succeeded in expanding their numbers and breeding areas. BLACK NODDIES and COMMON FAIRY-TERNS (Plates 24, 25 and 26), for instance, which nest in trees, started nesting in large numbers on several small islands in the Northwestern Hawaiian chain (such as Midway) only during the mid-20th Century, when military-associated landscape modification resulted in trees growing where none had been before. The LEAST TERN, a very small tern which ranges over parts of North, Central and South America, is considered endangered (USA ESA listed) over parts of its range, particularly in California (USA).

Profiles

Common Fairy-Tern, *Gygis alba*, Plates 24c and 26a
Ring-billed Gull, *Larus delawarensis*, Plate 24d
Laughing Gull, *Larus atricilla*, Plate 24e
Gray-backed Tern, *Sterna lunata*, Plates 25a and 26b
Sooty Tern, *Sterna fuscata*, Plates 25b and 26c
Brown Noddy, *Anous stolidus*, Plates 25c and 26d
Black Noddy, *Anous minutus*, Plates 25d and 26e
Pomarine Jaeger, *Stercorarius pomarinus*, Plate 25e

4. Marsh, Stream, and Water Birds

Along with ducks and geese (p. 122), Hawaii has a few other species of water birds associated with ponds, marshes, and wet fields, several of which are readily seen by visitors to the islands' freshwater sites and wetlands. Most of these birds are native to the islands, and all but one of the native ones are now endangered. *Herons* and *egrets* are beautiful medium to large-sized wading birds that enjoy broad distributions throughout the world's temperate and tropical regions. Herons and egrets, together with the similar but quite elusive wading birds called *bitterns*, constitute the heron family, Ardeidae, which includes about 58 species (it is included in the order Ciconiiformes, along with *ibises* and *storks*). Two species occur regularly in Hawaii, one native and one introduced. Herons frequent all sorts of aquatic habitats: along rivers and streams, in marshes and swamps, and along lake and ocean shorelines. They are, in general, highly successful birds. Why some in the family are called herons and some egrets, well, it's a mystery; but egrets are usually all white and tend to have longer *nuptial plumes* – special, long feathers – than the darker-colored herons.

Most herons and egrets are easy to identify. They are the tallish birds standing upright and still in shallow water or along the shore, staring intently into the water. They have slender bodies, long necks (often coiled when perched or still, producing a short-necked, hunched appearance), long, pointed bills, and long legs with long toes. Most are attired in soft shades of gray, brown, blue, or green, and black and white. From afar most are not striking, but close-up, many are exquisitely marked with small colored patches of facial skin or broad areas of spots or streaks. Some species, such as the BLACK-CROWNED NIGHT-HERON (Plate 27), during breeding seasons have a few very long feathers (nuptial plumes) trailing down their bodies from the head, neck, back, or chest. The sexes are generally alike in size and plumage, or nearly so.

Marsh and *stream* birds are small and medium-sized birds adapted to walk, feed, and breed in swamps, marshes, wet fields, and along streams. The chief characteristics permitting this lifestyle usually are long legs and very long toes that distribute the birds' weight, allowing them to walk among marsh plants and across floating vegetation without sinking. Some resemble ducks when swimming, but usually they do not have duck-like bills, and their feet are not webbed. Two representatives of one group, the *rails* (Family Rallidae), occur in Hawaii. The rails are a large, worldwide group of often secretive small and medium-sized marsh and dense vegetation birds, about 130 species strong, that includes the *rails, wood-rails, crakes, coots,* and *gallinules.*

A fifth frequently seen Hawaiian marsh and pond bird is the HAWAIIAN STILT (Plate 27). A member of the family Recurvirostridae, it is closely related to the shorebirds (included with gulls, terns, and shorebirds in Order Charadriiformes) and is detailed in that book section (p. 131).

Natural History

Ecology and Behavior

Herons and egrets walk about slowly and stealthily in shallow water and sometimes on land, searching for their prey, mostly small vertebrates, including fish, frogs, and the occasional turtle, and small invertebrates like crabs. On land, they take mostly insects, but also other invertebrates and vertebrates such as small rodents. CATTLE EGRETS (Plate 27) have made a specialty of following grazing cattle and other large mammals, walking along and grabbing insects and small vertebrates that are flushed from their hiding places by the moving cattle. A typical pasture scene is a flock of these egrets intermixed with a cattle herd, several of the white birds perched atop the unconcerned mammals. In Hawaii, Cattle Egrets are most often seen following moving tractors and other heavy equipment. Many herons also spend much of their foraging time as *sit-and-wait* predators, standing motionless in or adjacent to the water, waiting in ambush for unsuspecting prey to wander within striking distance. Then, in a flash, they shoot their long, pointed bills into the water to grab or spear the prey. They take anything edible that will fit into their mouths and down their throats, and then some. One particular heron that I recall grabbed a huge frog in its bill and spent the better part of a half hour trying to swallow it. Most herons are day-active, but many of the subgroup known as *night-herons* forage at least partly nocturnally. Most herons are social birds, roosting and breeding in colonies.

Members of the rail family, such as the COMMON MOORHEN and HAWAIIAN COOT (Plate 28), swim and/or stalk through marshes, swamps, shallow ponds, grassy shores, and wet grasslands, foraging for insects, small fish and frogs, bird eggs and chicks, and berries. Typically they move with a head-bobbing walk. Many rails are highly secretive, being heard but rarely seen moving about in marshes; the moorhen is often secretive, but coots are usually conspicuous.

Ecological Interactions

Herons and egrets often lay more eggs than the number of chicks they can feed. For instance, many lay three eggs when there is sufficient food around to feed only two chicks. This is contrary to our usual view of nature, which we regard as having adjusted animal behavior through evolution so that behaviors are finely tuned to avoid waste. Here's what biologists suspect goes on: females lay eggs one or two days apart, and start incubating before they finish laying all their eggs. The result is that chicks hatch at intervals of one or more days and so the chicks in a single nest are different ages, and thus different sizes. In years of food shortage, the smallest chick dies because it cannot compete for food from the parents against its larger siblings, and also because, it has been discovered, the larger siblings attack it (behavior called *siblicide*). The habit of laying more eggs than can be reared as chicks may be an insurance game evolved by the birds to maximize their number of young; in many years, true, they waste the energy they invested to produce third eggs that have little future, but if food is plentiful, all three chicks survive and prosper. Apparently, the chance to produce three offspring is worth the risk of investing in three eggs even though the future of one is very uncertain.

Breeding

Many herons breed in monogamous pairs within breeding colonies of various sizes. A few species are solitary nesters and some are less monogamous than others. Herons are known for their elaborate courtship displays and ceremonies, which continue through pair formation and nest-building. Generally, nests are constructed by the female of a pair out of sticks procured and presented to her by the male. Nests are placed in trees or reeds, or on the ground. Both sexes incubate the three to seven eggs for 16 to 30 days, and both feed the kids for 35 to 50 days before the young can leave the nest and feed themselves. The young are *altricial* – born helpless; they are raised on regurgitated food from the parents.

COMMON MOORHENS build nests of reeds and aquatic plants hidden in dense vegetation; HAWAIIAN COOTS build floating nests of aquatic plants. In both of these rail relatives, both sexes incubate the 5 to 10 or more eggs for about 3 weeks and care for the *precocial* young – born ready to swim and escape danger – for 6 to 8 weeks until they are independent.

Notes

The CATTLE EGRET is a common, successful, medium-sized white heron that, until recently, was confined to the Old World, where it made its living following herds of large mammals. What is so interesting about this species is that, whereas many of the animals that have recently crossed oceans and spread rapidly into new continents have done so as a result of people's intentional or unintentional machinations, these egrets did it themselves. Apparently the first ones to reach the New World were from Africa. Perhaps blown off-course by a storm, they first landed in northern South America in about 1877. Finding the New World to its liking, during the next decades the species spread far and wide, finding abundant food where tropical forests were cleared for cattle grazing. Cattle Egrets have now colonized much of northern South America, Central America, all the major Caribbean islands, and eastern and central North America, as far as the southern USA. We must assume that they have Chicago and New York City in their sights. Cattle Egrets were introduced to Hawaii in the mid-20th Century in an attempt to reduce pest insects near cattle.

According to an early chronicler of Hawaii's birds, the Hawaiian legend of how the COMMON MOORHEN came to have a red forehead is this: "Fire was unknown to the (Hawaiian) people, hence they could neither cook their food nor warm themselves during the cold weather. The bird took pity on them and, flying to the home of the gods, stole a blazing brand and carried it back to Earth. On this return flight its formerly white forehead was scorched by the flames; hence its name *alae*, signifying a burnt forehead. The descendants of this valiant bird all bear the red mark of honor." (G. C. Munro 1944).

Status

CATTLE EGRETS and BLACK-CROWNED NIGHT-HERONS are both common birds with broad distributions. Indeed, herons and egrets are such ecologically successful birds that few species are threatened worldwide. On the other hand, a good number of rail species throughout the world are threatened or endangered. This is particularly the case for island-bound species, where introduced predators such as rats, cats, and mongooses find the adults, as well as their eggs and young, easy prey. The GUAM RAIL, endemic to Guam, is endangered (USA ESA listed) and the WAKE RAIL (endemic to the mid-Pacific's Wake Island) and HAWAII RAIL (a small brown marsh bird with yellow bill and orange legs that was endemic to Hawaii) have both become extinct during the past hundred years. Both the HAWAIIAN COOT and Hawaiian subspecies of the COMMON MOORHEN are endangered (USA ESA). Introduced predators and introduced wetland plants that render some wetlands unsuitable for foraging or breeding, take their toll on these birds. A main threat is the limited availability of their freshwater habitats. These habitats in Hawaii have long been drained and altered for housing and agricultural/business reasons; it's estimated that less than 30% of Hawaii's original freshwater wetlands still exist.

Profiles

Cattle Egret, *Bubulcus ibis*, Plate 27a
Black-crowned Night-Heron, *Nycticorax nycticorax*, Plate 27b
Hawaiian Stilt, *Himantopus knudseni*, Plate 27c
Common Moorhen, *Gallinula chloropus*, Plate 28a
Hawaiian Coot, *Fulica alai*, Plate 28b

5. Ducks, Geese, and a Grebe

One of Hawaii's most famous and most threatened animals, the NENE, or HAWAIIAN GOOSE (Plate 27), is a member of Family Anatidae, the group, about 150 species strong, that includes the *ducks, geese,* and *swans*. These water-associated birds are distributed throughout the world in habitats ranging from open seas to high mountain lakes. Although an abundant, diverse group throughout temperate regions of the globe, ducks, or *waterfowl,* (*wildfowl* to the British), have only limited representation in most tropical areas, and few representatives on small mid-ocean islands. A single, now-endangered duck is native to the main Hawaiian Islands, the KOLOA, or HAWAIIAN DUCK (Plate 28). It's closely related to the widely distributed, quintessential duck, the common MALLARD (Plate 28). (Another duck, the FULVOUS WHISTLING-DUCK, was introduced during the early 1980s to Oahu or arrived by itself and established itself naturally; but the population dwindled and may be extinct by the time this book is printed.) The remainder of the ducks you catch sight of in the islands (Plates 29, 30) are migratory, only passing through or wintering there; most are North American breeders, but some, such as the GARGANEY and EURASIAN WIGEON, are Eurasian. A few geese occur in Hawaii: the highly endangered, endemic Nene, and a few species that regularly visit the islands in winter, their non-breeding period, including a small subspecies of the broadly distributed CANADA GOOSE (Plate 27).

Ducks vary quite a bit in size and coloring, but all share the same major traits: duck bills, webbed toes, short tails, and long, slim necks. Plumage color and patterning vary, but there is a preponderance within the group of grays and browns, and black and white, although many species have at least small patches of bright color. In some species male and female look alike, but in others there are many differences between the sexes. Geese, usually larger than ducks, are also, in general, more terrestrial – they often graze on land. They have moderately long legs placed fairly centrally on the body (versus ducks with short legs placed near the back of the body), so geese, unlike ducks, with their waddling gait on land, can walk and run smartly. Male and female geese look alike. Nene, with a highly terrestrial life and non-migratory habits, have been modified through evolution from their Canada Goose-like ancestors (the Canada Goose is the Nene's closest living relative), and now have only partially webbed feet and relatively short wings.

Grebes, often mistaken for small ducks, are actually members of a different order (Podicipedidae), one only very distantly related. The 20 or so grebe species are distributed nearly globally (none occur in the high Arctic or Antarctic, or on some oceanic islands). They are fully aquatic, and built to dive and swim well underwater. They have short wings, very little tail, legs placed well back on their bodies, and lobed toes (but not fully webbed) that aid in propulsion underwater. Bills are usually sharp-pointed, not duck-like (although Hawaii's only grebe has a fairly broad bill). Male and female grebes look alike or nearly so.

Natural History

Ecology and Behavior

Ducks are birds of wetlands, spending most of their time in or near the water. Many of the typical ducks are divided into *divers* and *dabblers*. Diving ducks plunge underwater for their food; dabblers, such as Mallards, pintail, wigeon, and teal take food from the surface of the water or, maximally, put their heads down into the water to reach food at shallow depths. Ducks eat mostly aquatic plants or small fish, but some forage on land for seeds and other plant materials. The HAWAIIAN DUCK, a resident of ponds, marshes, and drainage ditches of Kauai, Oahu, and the Big Island, for instance, in addition to eating aquatic plants, snails, and insects, feeds on terrestrial foods including earthworms, some land snails, and insects.

Geese often spend a good portion of their foraging time on land, and the NENE is almost completely terrestrial (but will swim if water happens to be nearby). It mainly occupies rugged, rocky, sparsely vegetated regions of lava flows at middle and higher elevations on volcanic slopes (to 2400 m, 7900 ft), but also some dry scrublands, grasslands, and even golf courses. (see Chapter 5, p. 64). Nene are vegetarians, plucking fruits such as berries as well as plant leaves, buds, and flowers, and grazing on grasses and herbs.

Breeding

Ducks place their nests on the ground in thick vegetation or in holes. Typically nests are lined with downy feathers that the female plucks from her own breast. In many of the ducks, females perform most of the breeding duties, including incubation of the 2 to 16 eggs and shepherding and protecting the ducklings. Some of these birds, however, particularly among the geese and swans, have lifelong marriages during which male and female share equally in breeding duties. The young are *precocial*, able to run, swim and feed themselves soon after they hatch. HAWAIIAN DUCKS are usually monogamous, breeding mainly from December through May. Nests, consisting of a bowl of grass lined with feathers, are built on the ground, concealed in vegetation. The 2 to 10 eggs are incubated for 4 weeks, and fledging occurs when young are 50 to 60 days old. Breeding starts when these ducks are one year old. NENE breed monogamously in isolated pairs, usually from late October through June. The nest is a depression on the ground lined with feathers, often between slabs of lava or concealed under a bush. Two to five eggs are incubated for 29 days; fledging occurs at 70 to 80 days. Nene first breed when 2 or 3 years old.

Notes

Ducks, geese and swans have been objects of people's attention since ancient times, sometimes as cultural symbols (for instance, as a Chinese symbol of happiness), but chiefly as a food source. These birds typically have tasty flesh, are fairly large and so economical to hunt, and usually easier and less dangerous to catch than many other animals, particularly large mammals. Owing to their frequent use as food, several wild ducks and geese have been domesticated for thousands of years. Wild ducks also adjust well to the proximity of people, to the point of taking food from them – a practice that surviving artworks show has been occurring for at least 2000 years. Hunting ducks and geese for sport is also a long-practiced tradition. As a consequence of these long interactions between ducks and people, and the research on these animals stimulated by their use in agriculture and sport, a large amount of scientific information has been collected on the group; many of the ducks and geese are among the most well known of birds. The close association between ducks and people has even led to a long contractual agreement between certain individual ducks and the Walt Disney Company.

Apparently, neither the HAWAIIAN DUCK nor the NENE loomed large in ancient Hawaiian culture and folklore. Both species were undoubtedly hunted and eaten and, indeed, prior to the introduction to the islands of cats, large rats, and mongooses, people would have been these birds' main predators. According to one native-born Hawaiian writing during the early 1800s, Nene feathers were used to adorn accouterments of Hawaiian royalty, and "its body is excellent eating." Roast Nene, in fact, during the 1800s, was on the menu at Volcano House, the old hotel located within what is now Hawaii Volcanoes National Park.

The first PIED-BILLED GREBES (Plate 28) arrived on the Big Island during the early 1980s and set up shop. The single colony occupied a small pond area on the island's Kona (drier, western) side. It looked as if these grebes would be an example of a successful natural colonization by a bird. But the population was hard-hit during the 1990s by disease (botulism) and may not survive.

Status

Introduced mongooses, cats, and rats, along with sport hunting and habitat loss, severely reduced HAWAIIAN DUCK and NENE populations. During the late 1940s, the total Nene population, restricted to the Big Island, stood at about 30 individuals. Captive-breeding programs in Hawaii and England started, then eventually released into protected wild areas on Maui and on the Big Island, more than 2000 Nene. Predator eradication programs at Haleakala National Park and Hawaii Volcanoes National Park seek to remove the predation threat from adult Nenes and from their nesting efforts. Even though Nene are now a highly protected species (USA ESA listed, endangered), the beneficiary of much conservation attention (Chapter 5, p. 64), and Hawaii's state bird, the main current wild populations (several hundred individuals each on Maui and the Big Island) are not self-sustaining, and would steadily decline if not for the continued addition of captive-raised birds. Recently a third population was established in mongoose-free, lowland Kauai, and it is thriving.

Hawaiian Duck populations during the 1960s plunged to fewer than 3000 individuals, probably all located on Kauai (the bird was then extinct on the other islands). Breeding programs led to the reintroduction of populations on Oahu and the Big Island. The species is still considered endangered (USA ESA listed). In addition to increasing loss of its wetland habitats and its easy vulnerability to ground predators, these ducks face another threat: interbreeding with MALLARDS (which, in Hawaii, are usually feral domestic birds). Either because the Mallard males are very aggressive and forcibly mate with female Hawaiian Ducks, or because the Hawaiian Ducks sometimes cannot locate other members of their species with which to breed and so sometimes mate with Mallards, the result is hybrid young. If this happens enough, hybrids will come to characterize all the populations, and the Hawaiian Duck, as a species separate from the Mallard, will cease to exist. Pure-bred Hawaiian Ducks occur now only on Kauai. Another species closely related to the Mallard is the LAYSAN DUCK, endemic to Laysan Island, 1200 km, 700 miles, from Kauai, in the Northwestern Hawaiian Islands. It is considered endangered; its population size is about 500, which may be all the small island can support.

Profiles

Nene (Hawaiian Goose), *Branta sandvicensis*, Plate 27d
"Cackling" Canada Goose, *Branta canadensis minima*, Plate 27e
Pied-billed Grebe, *Podilymbus podiceps*, Plate 28c
Hawaiian Duck, *Anas wyvilliana*, Plate 28d
Mallard, *Anas platyrhynchos*, Plate 28e

Northern Pintail, *Anas acuta*, Plate 29a
Northern Shoveler, *Anas clypeata*, Plate 29b
Green-winged Teal, *Anas crecca*, Plate 29c
Blue-winged Teal, *Anas discors*, Plate 29d
Garganey, *Anas querquedula*, Plate 29e

American Wigeon, *Anas americana*, Plate 30a
Eurasian Wigeon, *Anas penelope*, Plate 30b
Lesser Scaup, *Aythya affinis*, Plate 30c
Greater Scaup, *Aythya marila*, Plate 30d
Tufted Duck, *Aythya fuligula*, Plate 30e
Ring-necked Duck, *Aythya collaris*, Plate 30f

6. Raptors

Raptor is another name for *bird-of-prey*, birds that make their living hunting, killing, and eating other animals, usually other vertebrates. When one hears the term raptor, one usually thinks of soaring *hawks* that swoop to catch rodents, and of speedy, streamlined *falcons* that snatch small birds out of the air. Although these *are* common forms of raptors, the families of these birds are large, the members' behavior, diverse. The two main raptor families are the Accipitridae, containing the *hawks*, *kites* and *eagles*, and the Falconidae, including the *true falcons*, *forest-falcons*, and *caracaras*. The reasons for classifying the two raptor groups separately mainly have to do with differences in skeletal anatomy and hence, suspected differences in evolutionary history. Raptors are common and conspicuous animals in most parts of the world, but only a few species occur in Hawaii, and most of them are seen only rarely. Some can occasionally be seen soaring during the day, using the currents of heated air that rise from the sun-warmed ground to support and propel them as they search for meals. But Hawaii's raptors, like raptors everywhere, are found in a broad variety of habitats, including dry woodlands and closed rainforests.

The *accipitrids* are a worldwide group of about 200 species; they occur everywhere but Antarctica. Two species occur in Hawaii, one, the HAWAIIAN HAWK (Plate 31), being native and endemic, and the other, the OSPREY (Plate 31), being an occasional visitor that breeds only outside the Hawaii region. *Falconids*, likewise, are worldwide in their distribution. There are about 60 species, but only one, the PEREGRINE FALCON (Plate 31) occurs in the Hawaii area, occasionally visiting the main islands but never nesting there. Some falcons have very broad distributions; indeed, the Peregrine is found almost everywhere (that is, its distribution is *cosmopolitan*) and may have the most extensive natural distribution of any bird.

Raptors vary considerably in size and in patterns of their generally subdued color schemes, but all are similar in overall form – we know them when we see them. They are fierce-looking birds with strong feet, hooked, sharp claws, or *talons*, and strong, hooked and pointed bills. Accipitrids vary extensively in size, from small hawks only 28 cm (11 in) long to one-meter-long (40 in) eagles. Females are usually larger than males, in some species noticeably so. Most raptors are variations of gray, brown, black, and white, usually with brown or black spots, streaks, or bars on various parts of their bodies. The plumages of these birds are actually quite beautiful when viewed close-up, which, unfortunately, is difficult to do. Males and females are usually alike in color pattern. Juvenile raptors often spend several years in *subadult* plumages that differ in pattern from those of adults. Falcons, like hawks, span a good range of sizes, from small species such as the *kestrels* familiar to North Americans and Europeans, to large birds such as Peregrines. They are usually distinguished from hawks by their long, pointed wings, which allow the rapid, acrobatic flight for which these birds are justifiably famous.

Owls. Owls, mostly nocturnal birds-of-prey, can also be considered raptors. Owls comprise Order Strigiformes, a worldwide group of about 150 species that lacks representation only in Antarctica and some remote oceanic islands. Two species occur in Hawaii. One, the SHORT-EARED OWL (also called HAWAIIAN OWL, or PUEO, Plate 31), is a naturally-occurring Hawaiian subspecies, or race, of a species that is broadly distributed over the northern hemisphere. (New research suggests that the Hawaiian subspecies is essentially indistinguishable from mainland Short-eared Owls.) The other, the BARN OWL (Plate 31), introduced from North America, has a cosmopolitan distribution on six continents.

Most people can always identify owls because of several distinctive features. All have large heads with forward-facing eyes, small, hooked bills, plumpish bodies and sharp, hooked claws. Most have short legs and short tails. Many owls are clad mostly in mixtures of gray, brown, and black, the result being that they usually are highly camouflaged against a variety of backgrounds. They have very soft feathers. Most are medium-sized birds, but the group includes species that range in length from 15 to 75 cm (6 to 30 in). Males and females generally look alike, although females are a bit larger.

Natural History

Ecology and Behavior

Raptors are meat-eaters. Most hunt and eat live prey. They usually hunt alone, although, when mated, the mate is often close by. HAWAIIAN HAWKS mainly take small vertebrate animals – rodents and small birds – but they take some invertebrates, such as large insects and spiders, as well. Prey is snatched with talons first, and then killed and ripped apart with the bill. OSPREY, sometimes called "fish-eagles," eat fish; they hover momentarily over freshwater or saltwater, spot a likely meal, then dive in feet-first and grab the fish in their talons, then fly off to a tree branch to feast. Falcons are best known for their remarkable eyesight and fast, aerial pursuit and capture of flying birds – they are "birdhawks." Most people are familiar with stories of PEREGRINE FALCONS diving through the atmosphere (*stooping*, defined as diving vertically from height to gain speed and force) at speeds approaching 320 kph (200 mph) to stun, grab, or knock from the sky an unsuspecting bird. (But some falcon species eat more rodents than birds, and some even take insects.)

Many raptors are territorial, a solitary individual or a breeding pair (as in the Hawaiian Hawk) defending an area for feeding and, during the breeding season, for reproduction. Displays that advertise a territory and may be used in courtship consist of spectacular aerial twists, loops, and other acrobatic maneuvers. Although many raptors are common birds, typically they exist at relatively low densities, as is the case for all *top predators* (a predator at the pinnacle of the food chain, preyed upon by no animal). That is, there usually is only enough food to support one or two of a species in a given area. For example, a typical density for a small raptor species, perhaps one that feeds on mice and other small rodents, is one individual per square kilometer (0.4 sq miles). A large eagle that feeds on rabbits and large rodents may be spaced so that a usual density is one individual per 500 sq km (190 sq miles).

Owls. In general, owls occupy a variety of habitats: forests, clearings, fields, grasslands, mountains, marshes. Most owls hunt at night, taking prey such as small mammals, birds (including smaller owls), and reptiles; smaller owls specialize on insects, earthworms, and other small invertebrates. But some owls, including the SHORT-EARED OWL and BARN OWL, hunt at twilight (*crepuscular* activity) and sometimes during the day. Owls hunt by sight and sound. Their vision is very good in low light, the amount given off by moonlight, for instance; and their hearing is remarkable. They can hear sounds that are much lower in sound intensity (softer) than most other birds, and their ears are positioned on their heads asymmetrically, the better for localizing sounds in space. This means that owls in the darkness can, for example, actually hear small rodents moving about on the forest floor, quickly locate the source of the sound, then swoop and grab. Additionally, owing to their soft, loose feathers, owls' flight is essentially silent, permitting prey little chance of hearing their approach. Owls swallow small prey whole, then instead of digesting or passing the hard bits, they regurgitate bones, feathers, and fur in compact *owl pellets*. These are often found beneath trees or rocks where owls perch and they can be interesting to pull apart to see what an owl has been dining on. For instance, one Short-eared Owl pellet in Hawaii was found to contain the remains of three Pacific Rats and one House Mouse (obviously, the owl had quite a good dinner that day).

Ecological Interactions

The hunting behavior of falcons has over evolutionary time shaped the behavior of their prey animals. Falcons hit perched or flying birds with their talons, stunning the prey and sometimes killing it outright. An individual bird caught unawares has little chance of escaping the rapid, acrobatic falcons. But birds in groups have two defenses. First, each individual in a group benefits because the group, with so many eyes and ears, is more likely to spot a falcon at a distance than is a lone individual, thus providing all in the group with opportunities to watch the predator as it approaches and so evade it. This sort of anti-predation advantage may be why some animals stay in groups. Second, some flocks of small birds, such as starlings, which usually fly in loose formations, immediately tighten their formation upon detecting a flying falcon. The effect is to decrease the distance between each bird, so much so that a falcon flying into the group at a fast speed and trying to take an individual risks injuring itself – the "block" of starlings is almost a solid wall of bird. Biologists believe that the flock tightens when a falcon is detected because the behavior reduces the likelihood of an attack.

Breeding

Hawk nests are constructed of sticks that both sexes place in a tree or on a rock ledge. HAWAIIAN HAWKS put their stick nests lined with leaves in trees 4 to 18 m (13 to 60 ft) above the ground. For hawks in general, only the female incubates the one to six eggs (only one or two in the larger species) for 28 to 49 days and gives food to the nestlings. The male frets about and hunts, bringing food to the nest for the female and for her to give to the nestlings. Both sexes feed the young when they get a bit older; they can fly at 28 to 120 days of age, depending on species size. After fledging, the young remain with the parents for several more weeks or months until they can hunt on their own. Female Hawaiian Hawks lay usually one but sometimes two or three eggs, which are incubated for 38 days; fledging occurs after about 60 days in the nest, and young are then cared for for another 7 or 8 months until they are fully independent.

Owls. Most owls are monogamous breeders. They do not build nests themselves, but either take over nests abandoned by other birds or nest in cavities such as tree or rock holes. Incubation of the one to ten or more eggs (often two to four) is usually conducted by the female alone for 4 to 5 weeks, but she is fed by her mate. Upon hatching, the female broods the young while the male hunts and brings meals. Young fledge after 4 to 6 weeks in the nest. SHORT-EARED OWLS nest on the ground, often in grass; three to six eggs are usual. Owls in general have a reputation for fierce, aggressive defense of their young; many a human who ventured too near an owl nest has been attacked and had damage done!

Notes

Large, predatory raptors have doubtless always attracted people's attention, respect, and awe. For instance, wherever eagles occur, they are chronicled in the history of civilizations. So it is not surprising that Hawaii's native raptors, limited essentially to one hawk and one owl, were part of the ancient Hawaiians' mythology. Io, the HAWAIIAN HAWK, in legend, was a symbol of Hawaiian royalty. Likewise, Pueo, the SHORT-EARED OWL, was worshipped as a god and revered as a guardian spirit. It is believed by some that human sacrifices were even offered to the owl gods. On Maui, legend had it that an owl god was able to bring back to life the souls of those people who died while lost in the wilds; on Oahu, the owl was considered a guardian of local families.

The forward-facing eyes of owls are a trait shared with only a few other animals: humans, most other primates, and to a degree, the cats. Eyes arranged in this way allow for almost complete binocular vision (one eye sees the same thing as the other), a prerequisite for good depth perception, which, in turn, is important for quickly judging distances when catching prey. On the other hand, owl eyes cannot move much, so owls swivel their heads to look left or right.

"Peregrine" means "tendency to wander" – and these large falcons certainly do: individuals that breed during summer in arctic Alaska and Canada have been found spending their winters in Central Argentina and Chile – quite the international travellers.

Status

HAWAIIAN HAWKS, limited for the most part to Hawaii's Big Island (they occasionally visit some of the other main islands), were considered endangered until recently; field studies indicated, however, that the hawk's population was stable, so its status was upgraded to threatened (USA ESA listed). The entire population is estimated to be about 2700 birds. Main threats are deforestation, hunting, and human interference with nests. Trade in almost all hawks, kites, eagles, and falcons is heavily regulated (most are CITES Appendix II listed). Conservation measures aimed at raptors worldwide are difficult to formulate and enforce because the birds are often persecuted for a number of reasons (hunting, pet and feather trade, ranchers protecting livestock) and they roam very large areas. Further complicating the picture, some breed and winter on different continents, and thus need to be protected in all parts of their ranges, including along migration routes.

The SHORT-EARED OWL is considered by the State of Hawaii to be endangered on Oahu, but the owl occurs, often in good numbers, on all of Hawaii's main islands, and has a broad distribution outside of the state; overall, the species is not threatened. Most currently threatened owls are Old World species. The endangered Northern Spotted Owl, of which so much is heard in the USA, is a subspecies (race) of Spotted Owl; because the other subspecies, the Southern Spotted Owl, which occurs in the southwestern USA and Mexico, is still fairly common, the species as a whole is not presently threatened.

Profiles

Osprey, *Pandion haliaetus*, Plate 31a
Hawaiian Hawk, *Buteo solitarius*, Plate 31b
Peregrine Falcon, *Falco peregrinus*, Plate 31c
Barn Owl, *Tyto alba*, Plate 31d
Short-eared Owl, *Asio flammeus*, Plate 31e

7. Chicken-like Birds (Pheasants, Quail, and Turkey)

Hawaii has a plethora of chicken-like birds, in a broad range of sizes, running and scratching their way around the islands. All were introduced by people – these are strongly terrestrial birds and poor flyers, not the kinds of birds that could cross thousands of kilometers of open ocean to colonize islands naturally. All of Hawaii's chicken-like birds can be considered members of Family Phasianidae, a globally distributed group of about 200 species included in Order Galliformes (along with, among others, *grouse* and *curassows*). The family includes the *pheasants, quails, bobwhites, francolins, partridges, junglefowl, peafowl,* and *turkeys*. For want of a better term, even professional ornithologists refer technically to this group as the "chicken-like birds," (but they can also be called *galliforms*, and are sometimes referred to as "game birds"). Indeed, roasting chickens are domesticated members of the group; the RED JUNGLEFOWL (Plate 33) is actually the wild form of the domesticated chicken. (Some recent classifications consider the turkeys and the New World group of quails to be separate families within Order Galliformes.) About 12 species occur on the main islands and some of them are widespread, abundant, and readily seen.

These are stocky birds with short, broad, rounded wings, shortish legs with heavy claws adapted for ground scratching, short, thick, chicken-like bills, and short or long tails – with some of the pheasants having tails to 1.5 m (5 ft) long. Some species, particularly among the pheasants, are exquisitely marked with bright colors and intricate patterns. The long tails of male peafowl (the "peacock") are, of course, among nature's most ornate and colorful constructions. Others, however, mostly the smaller and so more at-risk from predators, are duller and more camouflaged in their typical on-the-ground habitats. The sexes can look alike or different; males in some groups are a bit larger than females.

Natural History

Ecology and Behavior

The quails, francolins, pheasants, and turkeys are all terrestrial birds that feed and nest on the ground. Most flights are very brief; over short distances, such as when making a sudden, quick escape from a potential predator, these birds are powerful, swift aviators. But with their short, rounded wings, they are not built for sustained flight. In fact, few in the family are migratory to any significant degree. These birds are mostly vegetarians, eating seeds and other plant matter such as shoots, some green leaves, berries, roots, as well as some insects and other small invertebrate animals. They scratch the ground, clearing away vegetation and soil, looking for seeds and roots. Most species are gregarious, travelling in small family or multi-family groups called *coveys* – of perhaps 4 to 15 individuals; but some species are more solitary in their habits.

Breeding

Mating systems in this group are highly variable. Some, such as the New World quails (in Hawaii, represented by the CALIFORNIA QUAIL, Plate 32), are usually monogamous breeders, some species even maintaining their pairs throughout the year, but many others, such as many of the pheasants, are polygynous. In these systems, a male's contribution to breeding is limited to mating with several females. The females then incubate the eggs and rear the young by themselves. Nests are placed on the ground, either in a bare scrape or perhaps one lined with leaves or grass. The nest is often at the foot of a tree or next to a large, sheltering grass clump. The female incubates the usually 6 to 12 eggs by herself, and then leads and cares for the young (and, in monogamous species, may be assisted by the male in leading young). Young are *precocial* – born covered with downy feathers and ready soon to leave the nest, follow the parents, and feed themselves.

Notes

The 30 or so species of pheasants are largely confined in their natural distributions to Asia. So why do pheasants now run about almost everywhere else – for instance, the RING-NECKED PHEASANT (Plate 33) found all over Europe, the USA and even Hawaii? (and called, simply, Pheasant, by the British)? People like to hunt and eat pheasants, so these large birds have been introduced to many parts of the world, where, often, they thrive in their new environs. The Ring-necked Pheasant was long ago brought from Asia to Europe, then to North America and, during the late 1800s, to Hawaii. They are now resident breeders on all the main islands. In fact, many of the chicken-like birds – the partridges, quail, and pheasants – are considered to be among the most important game birds, pursued globally for food and sometimes for sport. All the birds shown in Plates 32 and 33, except for the peafowl and junglefowl, are classified as game birds in Hawaii, and may be hunted on many state lands, including forest reserves, natural area reserves, game management areas, and military training areas. Some daily bag limits: Ring-necked Pheasant: three (cocks only) per hunter; Kalij Pheasant: three birds of either sex per hunter; Chukar, Gray Francolin, Black Francolin: eight birds of either sex per hunter, which may be of any one species or a combination of these species.

Pheasants and peafowl, with their ornate plumage and long tails, have an associated rich mythology, appearing in ancient Greek, Buddhist, and Hindu legends, among others. The Buddha was sometimes shown riding on a displaying peacock. Early Christians apparently regarded the peacock as a symbol of immortality.

Status

Many of the chicken-like birds are common, abundant species and this is certainly the case for the ones that have been introduced to Hawaii. However, some species of galliform birds, such as several of the Asian pheasants found in tropical forests, are now threatened or endangered, mostly owing to over-hunting.

Profiles

Black Francolin, *Francolinus francolinus*, Plate 32a
Gray Francolin, *Francolinus pondicerianus*, Plate 32b
Erckel's Francolin, *Francolinus erckelii*, Plate 32c
Japanese Quail, *Coturnix japonica*, Plate 32d
California Quail, *Callipepla californica*, Plate 32e

Chukar, *Alectoris chukar*, Plate 33a
Kalij Pheasant, *Lophura leucomelana*, Plate 33b
Red Junglefowl, *Gallus gallus*, Plate 33c
Ring-necked Pheasant, *Phasianus colchicus*, Plate 33d
Common Peafowl, *Pavo cristatus*, Plate 33e
Wild Turkey, *Meleagris gallopavo*, Plate 33f

8. Shorebirds

Spotting *shorebirds* is usually a priority only for visitors to Hawaii who are rabid birdwatchers. The reason for the usual lack of interest is that shorebirds are often very common, plain-looking brown birds that most people are familiar with from their beaches back home. Still, it is always a treat watching shorebirds in their tropical wintering areas as they forage in meadows, along streams, on mudflats, and especially on the coasts, as they run along beaches, parallel to the surf, picking up food. Some of the small ones, such as SANDERLINGS (Plate 35), as one biologist wrote, resemble amusing wind-up toys as they spend hours running up and down the beach, chasing, and then being chased by, the outgoing and incoming surf. Shorebirds are often conspicuous and let themselves be watched, as long as the watchers maintain some distance. When in large flying groups, shorebirds such as *sandpipers* provide some of the most compelling sights in birddom, as their flocks rise from sandbar or mudflat to fly fast and low over the surf, wheeling quickly and tightly in the air as if they were a single organism, or as if each individual's nervous system were joined to the others'.

Shorebirds are traditionally placed along with the gulls and terns (p. 117) in the avian order Charadriiformes. They are global in distribution and considered to be hugely successful birds – the primary reason being that the sandy beaches and mudflats on which they forage usually teem with their food. About 35 shorebird species occur in Hawaii, all but one of them (the HAWAIIAN STILT, Plate 27) being winter visitors only that migrate to breed in arctic regions. Only five of these are more or less commonly seen: PACIFIC GOLDEN-PLOVER, WANDERING TATTLER, RUDDY TURNSTONE (Plate 34), SANDERLING, and BRISTLE-THIGHED CURLEW (Plate 36). There are several shorebird families, three of which require mention. The *sandpipers*, Family Scolopacidae, are a worldwide group of approximately 85 species, and include the typical sandpipers, Sanderling, turnstones, tattlers, dowitchers, yellowlegs, curlews, and Whimbrel (Plate 36). *Plovers,* Family Charadriidae, with about 60 species, likewise have a worldwide distribution. Three species occur in the Hawaiian region, all of which breed elsewhere. The broadly distributed Family Recurvirostridae consists of about seven species of *stilts* and *avocets*, one of which occurs in Hawaii.

All shorebirds, regardless of size, have a characteristic "look." They are usually drably colored birds (especially during the non-breeding months), darker above, lighter below, with long, thin legs for wading through wet meadows, mud, sand, or surf. Depending on feeding habits, bill length varies from short to very long. Most of the Hawaiian sandpipers range from 15 to 43 cm (6 to 17 in) long. They are generally slender birds with straight or curved bills of various lengths. Plovers, 15 to 30 cm (6 to 12 in) long, are small to medium-sized, thick-necked shorebirds with short tails and straight, relatively thick bills. They are mostly shades of gray and brown but some have bold color patterns such as a broad white or dark band on the head or chest. The HAWAIIAN STILT is a striking, mid-sized black and white bird with very long coral-pink legs and long, fine bill. The sexes look alike, or nearly so, in most of the shorebirds.

Natural History

Ecology and Behavior

Shorebirds typically are open-country birds, associated with coastlines and inland wetlands, grasslands, and pastures. Sandpipers and plovers are excellent flyers but they spend a lot of time on the ground, foraging and resting; when chased, they often seem to prefer running to flying away. Sandpipers pick their food up off the ground or use their bills to probe for it in mud or sand – they take insects and other small invertebrates, particularly crustaceans. They will also snatch bugs from the air as they walk and from the water's surface as they wade or swim. Larger, more land-dwelling shorebirds may also eat small reptiles and amphibians, and even small rodents; some of the plovers also eat seeds. Usually in small groups, HAWAIIAN STILTS wade about in shallow fresh and salt water, using their bills to probe the mud for small fish, insects, snails, crustaceans, and worms.

Many shorebirds, especially among the sandpipers, establish winter *feeding territories* along stretches of beach; they use the area for feeding for a few hours or for the day, defending it aggressively from other members of their species. Many of the sandpipers and plovers are gregarious birds, often seen in large groups, especially when they are travelling. Many species make long migrations over large expanses of open ocean, a good example being Hawaii's abundant PACIFIC GOLDEN-PLOVER, which flies apparently nonstop in May from Hawaii to its arctic breeding grounds in Alaska (and returns in August and September).

Breeding

Shorebirds breed in a variety of ways. Many species breed in monogamous pairs that defend small breeding territories. Others practice *polyandry*, the least common type of mating system among vertebrate animals, in which some females have more than one mate in a single breeding season. This type of breeding is employed at times by SANDERLINGS. The normal sex roles of breeding birds are reversed: the female establishes a territory that she defends against other females. More than one male settles within the territory, either at the same time or sequentially during a breeding season. After mating, the female lays a clutch of eggs for each male. The males incubate their clutches and care for the young. Females may help care for some of the broods of young provided that there are no more unmated males to try to attract to the territory.

Most shorebird nests are simply small depressions in the ground in which eggs are placed; some of these *scrapes* are lined with shells, pebbles, grass, or leaves. Sandpipers lay two to four eggs per clutch, which are incubated, depending on species, by the male alone, the female alone, or by both parents, for 18 to 21 days. Plovers lay two to four eggs, which are incubated by both sexes for 24 to 28 days. HAWAIIAN STILTS breed in small colonies on mudflats near water. The scrape, into which four eggs are placed, is lined with vegetation. Both sexes incubate for 22 to 26 days. Shorebird young are *precocial*, that is, soon after they hatch they are mobile, able to run from predators, and can feed themselves. Parents usually stay with the young to guard them at least until they can fly, perhaps 3 to 5 weeks after hatching.

Notes

The manner in which flocks of thousands of birds, particularly shorebirds, fly in such closely regimented order, executing abrupt maneuvers with precise coordination, such as when all individuals turn together in a split second in the same direction, has puzzled biologists and engendered some research. The questions include: what is the stimulus for the flock to turn – is it one individual within the flock, a "leader," from which all the others take their "orders" and follow into turns? Or is it some stimulus from outside the flock that all members respond to in the same way? And how are the turns coordinated? Everything from "thought transference" to electromagnetic communication among the flock members has been advanced as an explanation. After studying films of DUNLIN, a North American sandpiper, flying and turning in large flocks, one biologist has suggested that the method birds within these flocks use to coordinate their turns is similar to how the people in a chorusline know the precise moment to raise their legs in sequence or how "the wave" in a sports stadium is coordinated. That is, one bird, perhaps one that has detected some danger, like a predatory falcon, starts a turn, and the other birds, seeing the start of the flock's turning, can then anticipate when it is their turn to change course – the result being a quick wave of turning coursing through the flock.

The British often use the term "waders" instead of "shorebirds" to refer to this group.

Status

Of Hawaii's migratory shorebirds, only the BRISTLE-THIGHED CURLEW, which breeds in Alaska and winters on tropical Pacific islands, is in trouble (USA ESA listed, endangered); its worldwide population is perhaps below 10,000. One resident species, the endemic HAWAIIAN STILT, is endangered (USA ESA listed). Some authorities consider it a subspecies of the broadly distributed BLACK-NECKED STILT, which it closely resembles; others consider it a separate species. The main threat to it is the limited availability of the freshwater marshy wetlands in which it forages and adjacent to which it nests. These habitats in Hawaii have long been drained and altered for housing and agricultural/business reasons, and it is estimated that less than 30% of Hawaii's original freshwater wetlands still exist. Introduced predators and introduced wetland plants that render some wetlands unsuitable for stilt breeding, also take their toll on these beautiful birds. The current population of the stilt in Hawaii is probably 2000 or fewer individuals.

A major goal for conservation of shorebirds is the need to preserve critical migratory stopover points – pieces of habitat, sometimes fairly small, that hundreds of thousands of shorebirds settle into mid-way during their long migrations to stock up on food. For instance, one famous small patch of coastal mudflats near Grays Harbor, Washington State, USA, is a popular, traditional stopover point for millions of shorebirds. Its destruction or use for any other activity could cause huge losses to the birds' populations. Fortunately, it has been deemed essential, and protected as part of a national wildlife refuge. The destruction of shoreline and wetland habitats in Hawaii has apparently greatly reduced migratory shorebird numbers in the islands in recent years. The establishment of several national wildlife refuges that protect wetlands, such as Maui's Kealia Pond National Wildlife Refuge, may help reverse this trend.

Profiles

Hawaiian Stilt, *Himantopus knudseni*, Plate 27c
Pacific Golden-Plover, *Pluvialis fulva*, Plate 34a
Black-bellied Plover, *Pluvialis squatarola*, Plate 34b
Semipalmated Plover, *Charadrius semipalmatus*, Plate 34c
Ruddy Turnstone, *Arenaria interpres*, Plate 34d
Wandering Tattler, *Heteroscelus incanus*, Plate 34e

Sanderling, *Calidris alba*, Plate 35a
Pectoral Sandpiper, *Calidris melanotos*, Plate 35b
Sharp-tailed Sandpiper, *Calidris acuminata*, Plate 35c
Long-billed Dowitcher, *Limnodromus scolopaceus*, Plate 35d
Least Sandpiper, *Calidris minutilla*, Plate 35e
Lesser Yellowlegs, *Tringa flavipes*, Plate 35f

Bristle-thighed Curlew, *Numenius tahitiensis*, Plate 36a
Whimbrel, *Numenius phaeopus*, Plate 36b

9. Pigeons

The *pigeon* family is a highly successful group, represented, often in large numbers, almost everywhere on dry land, except for Antarctica and some oceanic islands. The family's continued ecological success must be viewed as at least somewhat surprising, because pigeons are largely defenseless creatures and quite edible, regarded as a tasty entrée by human and an array of nonhuman predators. They inhabit almost all kinds of habitats, from semi-deserts to tropical moist forests, to high-elevation mountainsides. Smaller species generally are called *doves*, larger ones, *pigeons*, but there is a good amount of overlap in name assignments. The family, Columbidae, includes approximately 250 species, four of which occur in Hawaii; all were introduced by people.

All pigeons are generally recognized as such by almost everyone, a legacy of people's familiarity with domestic and feral pigeons. Pigeons worldwide vary in size from the dimensions of a sparrow to those of a small turkey. They are plump-looking birds with compact bodies, short necks, and small heads. Legs are usually fairly short, except in the ground-dwelling species. Bills are small, straight, and slender. Typically there is a conspicuous patch of naked skin, or *cere*, at the base of the bill, over the nostrils. Hawaii's pigeons, as well as the ones with which most North Americans and Europeans are familiar, are generally colored in understated grays and browns, although some have bold patterns of black lines or spots; many have splotches of iridescence, especially on necks and wings. But some pigeons in other regions are among the most gaily colored of birds: *fruit doves*, for instance, genus *Ptilinopus*, which occur on many Pacific Islands, are small and mid-sized green pigeons with bold patches of red, yellow, and orange. Male and female of most pigeon species are generally alike in size and color.

Natural History

Ecology and Behavior

Most of the pigeons are at least partly arboreal, but some spend their time in and around cliffs, and still others are primarily ground-dwellers. They eat seeds, ripe and unripe fruit, berries, and the occasional insect, snail, or other small invertebrates. Even those species that spend a lot of time in trees often forage on the ground, moving along the leaf-strewn forest floor, for example, with the head-bobbing walk characteristic of their kind. Owing to their small, weak bills, they eat only what they can swallow whole; "chewing" is accomplished in the *gizzard*, a muscular portion of the stomach in which food is mashed against small pebbles that are eaten by pigeons expressly for this purpose. Pigeons typically are strong, rapid flyers, which, along with their cryptic color patterns, is their only defense against predation. Most pigeons are gregarious to some degree, staying in groups during the non-breeding portion of the year; some gather into large flocks. SPOTTED DOVES (Plate 36) usually forage on the ground for seeds and insects; ZEBRA DOVES (Plate 36), likewise, forage on the ground for seeds (or on outdoor restaurant tables, for crumbs!).

Ecological Interactions

The great success of the pigeon family – a worldwide distribution, robust populations, the widespread range and enormous numbers of *rock doves* (wild, domestic pigeons) – is puzzling to ecologists. At first glance, pigeons have little to recommend them as the fierce competitors any hugely successful group needs to be. They have weak bills and therefore are rather defenseless during fights and ineffectual when trying to stave off nest predators. They are hunted by people for food. In several parts of the world they compete for seeds and fruit with parrots, birds with formidable bills, yet pigeons thrive in these regions and have spread to many more that are parrot-less. To what do pigeons owe their success? First, to reproductive advantage. For birds of their sizes, they have relatively short incubation and nestling periods; consequently, nests are exposed to predators for relatively brief periods and, when nests fail, parents have adequate time to nest again before the season ends. Some species breed more than once per year. Also, the ability of both sexes to produce pigeon milk to feed young may be an advantage over having to forage for particular foods for the young. Second, their ability to capitalize on human alterations of the environment points to a high degree of hardiness and adaptability, valuable traits in a world in which people make changes to habitats faster than most organisms can respond with evolutionary changes of their own.

Breeding

Pigeons are monogamous breeders. Some breed solitarily, others in colonies of various sizes. Nests are shallow, open affairs of woven twigs, plant stems, and roots, placed on the ground, on rock ledges, or in shrubs or trees. Reproductive duties are shared by male and female. This includes nest-building, incubating the one or two eggs, and feeding the young, which they do by regurgitating food into the nestlings' mouths. All pigeons, male and female, feed their young *pigeon milk*, a nutritious fluid produced in the *crop*, an enlargement of the esophagus used for food storage. During the first few days of life, nestlings receive 100% pigeon milk but, as they grow older, they are fed an increasing proportion of regurgitated solid food. SPOTTED DOVES and ZEBRA DOVES breed year-round in Hawaii. Spotteds usually have two eggs, which are incubated for 14 to 16 days; fledging occurs 15 days after hatching. Zebras incubate their two eggs for 13 days, and nestlings fledge at 11 to 12 days old.

Notes

Intriguingly, although pigeons over evolutionary time have naturally colonized most tropical islands with forests, they never made it to the Hawaiian archipelago. With people's assistance, however, four species are now resident breeders in Hawaii; most were intended to be game animals. SPOTTED DOVES were introduced during the mid-1800s, ZEBRA DOVES during the 1920s. The MOURNING DOVE, a common North American resident, was introduced to the Big Island in the early 1960s, and occurs there in localized areas over the island's western sector. Common pigeons, technically ROCK DOVES, *Columba livia*, were first brought to Hawaii in the late 18th Century and, as true for nearly everywhere else on Earth, are now an occupying force in the state's cities and towns. A fifth species, the CHESTNUT-BELLIED SANDGROUSE (Plate 32), a pigeon relative, was introduced to the Big Island in 1961 as a game bird. All of Hawaii's pigeons, including the sandgrouse but excluding the rock dove, are classified as game birds; bag limit is 10 individuals for Spotted Doves, Mourning Doves, and sandgrouse, 20 for Zebra Dove.

Although many pigeons today are very successful animals, some species met extinction within the recent past. There are two particularly famous cases. The DODO was a large, flightless pigeon, the size of a turkey, with a large head and strong, robust bill and feet. Dodos lived, until the 17th Century, on the island of Mauritius, in the Indian Ocean, east of Madagascar. Reported to be clumsy and stupid (hence the expression, "dumb as a dodo"), but probably just unfamiliar with and unafraid of predatory animals, such as people, they were killed by the thousands by sailors who stopped at the island to stock their ships with food. This caused population numbers to plunge; the birds were then finished off by the pigs, monkeys, and cats introduced by people to the previously predator-free island. The only stuffed Dodo in existence was destroyed by fire in Oxford, England, in 1755.

North America's PASSENGER PIGEON, a medium-sized, long-tailed member of the family, suffered extinction because of overhunting and because of its habits of roosting, breeding, and migrating in huge flocks. People were able to kill many thousands of them at a time on the Great Plains in the central part of the USA, shipping the bodies to markets and restaurants in large cities through the mid-1800s. It is estimated that when Europeans first settled in the New World, there were 3 billion Passenger Pigeons, a population size perhaps never equalled by any other bird, and that they may have accounted for up to 25% or more of the birds in what is now the USA. It took only a little more than 100 years to kill them all; the last one died in the Cincinnati Zoo in 1914.

Status

Pigeons and doves in most parts of the world are common, abundant animals. Some species have even benefited from people's alterations of natural habitats, expanding their ranges, for example, where forests are cleared for agriculture – North America's MOURNING DOVE, for instance. But a number of New World and Old World pigeons are threatened or endangered, mostly from a combination of habitat loss (generally forest destruction), reduced reproductive success owing to introduced nest predators, and excessive hunting. Many threatened birds now are species restricted to islands (p. 69), and several pigeons fall into this category; in fact, of the 10 pigeon species known to have become extinct during the past 200 years, nine were island-bound.

Profiles

Spotted Dove, *Streptopelia chinensis*, Plate 36c
Zebra Dove, *Geopelia striata*, Plate 36d
Chestnut-bellied Sandgrouse, *Pterocles exustus*, Plate 32f

10. Parrots

Parrots may be the birds most commonly symbolic of the tropics. The 300+ parrot species that comprise the family Psittacidae (the P is silent; refer to parrots as *psittacids* to impress your friends) are globally distributed across the region, with some species extending into subtropical and even temperate zone areas. Although some parrots apparently occur naturally on some Pacific islands (it's also likely that in the distant past some had human help in spreading among islands), none made it to the highly isolated Hawaiian chain. During the past few decades, many parrot species have been spotted in Hawaii, but all are escaped pets or individuals intentionally (but illegally) released. Probably only one, the ROSE-RINGED PARAKEET (Plate 36), a common parrot native to parts of Africa and Asia, has succeeded in establishing itself as a permanent, if alien, breeding resident. A few other species may be in the process of becoming permanently established.

Consistent in form and appearance, all parrots are easily recognized as such. They share a group of traits that set them distinctively apart from all other birds. Their typically short neck and compact body yield a form variously described as stocky, chunky, or bulky. All possess a short, hooked, bill with a hinge on the upper part that permits great mobility and leverage during feeding. Finally, their legs are short and their feet, with two toes projecting forward and two back, are adapted for powerful grasping and a high degree of dexterity – more so than any other bird. The basic parrot color scheme is green, but some species depart from basic in spectacular fashion, with gaudy blues, reds, and yellows. Green parrots feeding quietly amid a tree's high foliage can be difficult to see, even for experienced birdwatchers. Ornithologists divide parrots by size: *parrotlets* are small birds (as small as 10 cm, or 4 in) with short tails; *parakeets* are also small, with long or short tails; *parrots* are medium-sized, usually with short tails; and *macaws* are large (up to 1 meter, or 40 in) and long-tailed. Male and female parrots often look alike or nearly so.

Natural History

Ecology and Behavior

Parrots are incredibly noisy, highly social seed- and fruit-eaters. They are almost always encountered in flocks of four or more, and smaller parrots are often in large groups. The ROSE-RINGED PARAKEET, in its native India and Central Africa, is seen in flocks of 60 to 70 individuals. In Hawaii, where it occurs over small, localized areas of Kauai, groups are much smaller. (Rose-ringed Parakeets are also occasionally spotted on Oahu and the Big Island.) They prefer drier sites – edges of dry woodlands, dry scrub areas – and feed on fruit and seeds, often in agricultural lands; in fact, they may become economically harmful farm pests, as they are within their native Old World haunts.

Breeding

ROSE-RINGED PARAKEETS are monogamous breeders. They nest in holes in trees, usually 3 to 10 m (10 to 33 ft) from the ground. The female only incubates the three to five eggs for 22 to 28 days; the male brings food to her at the nest. Young fledge from the nest 6 to 7 weeks after hatching.

Profile

Rose-ringed Parakeet, *Psittacula krameri*, Plate 36e

11. Crow, Thrushes, and Elepaio

Crows are members of the Corvidae, a passerine family of a hundred or so species that occurs just about everywhere in the world – or, as ecologists would say, *corvid* distribution is cosmopolitan. Corvids are known for their adaptability and for their seeming intelligence; in several ways, the group is considered by ornithologists to be the most highly developed of birds. The group includes the crows, *ravens, magpies,* and *jays.* A single species, the HAWAIIAN CROW (Plate 37), or ALALA (rhymes with tra-la-la), highly endangered, occurs in Hawaii, on the Big Island. Members of the family range in length from 20 to 71 cm (8 to 28 in), many near the higher end – large for passerine birds. Corvids have robust, fairly long bills and strong legs and feet. Many corvids (crows, ravens, rooks, jackdaws) are all or mostly black, but the jays are different, being attired in bright blues, purples, greens, yellows, and white. The COMMON RAVEN, all black, ranges over most of the northern hemisphere and is the largest passerine. The Hawaiian Crow, a dull black bird fairly closely related to the Common Raven, is one of the larger corvids, to 51 cm (20 in) long.

More than 300 species of *thrushes* inhabit most terrestrial regions of the world and include some of the most familiar park and garden songbirds. The thrush family, Turdidae, has few defining, common features that set all its members apart from other groups, as perhaps could be expected; so large an assemblage of species is sure to include a significant amount of variation in appearance, ecology, and behavior. Thrushes as a group are tremendously successful birds, especially when they have adapted to living near humans and benefitting from their environmental modifications. Most obviously, on five continents, a thrush is among the most common and recognizable garden birds, including North America's

AMERICAN ROBIN and Europe's REDWING and BLACKBIRD. Today in Hawaii it's possible to see three members of the thrush family, two native and one alien; several others recently went extinct (see *Status*, below). Hawaii's native thrushes (the OMAO and PUAIOHI, Plate 37), closely related to the New World's *solitaires*, are mid-sized songbirds with smallish, slender bills, drably turned out in brown, with gray chests and bellies. The sexes are very similar in appearance. During their first year of life, young Hawaiian native thrushes are distinctive, with striking scalloped or spotted patterns on their chests. Hawaii's alien thrush, the WHITE-RUMPED SHAMA (Plate 38), is a striking mid-sized bird imported from southeast Asia.

One of the few native forest songbirds that a Hawaiian visitor with a little time and effort can locate and see is a small flycatcher called the ELEPAIO (Plate 37). It's an endemic representative not of the large group of American flycatchers (Family Tyrannidae), but of an Old World group, Family Monarchidae, the *monarch flycatchers*. (In fact, the Elepaio is sometimes called the Hawaiian Flycatcher.) This group, of about 130 species, occurs over large sectors of Africa, Asia, and the Australia/Pacific region. They are mostly small forest and woodland birds, with many of them, like the Elepaio, being chestnut-colored above and white below. Although the latest classification schemes indicate all Hawaii's Elepaios comprise a single species, there are three to five different forms (or sub-species, or races; in this book, three separate species are recognized) that each look slightly different: a rare Elepaio at lower and middle elevations in drier forested valleys and slopes on Oahu; a fairly abundant Elepaio at a range of elevations on Kauai (but mostly above 1200 m, 4000 ft); and one, two or three different forms, quite common, at a variety of elevations on the Big Island (some more adapted to wet forest, some to dry forest).

Natural History

Ecology and Behavior

Crow. Bright and versatile, corvids are quick to take advantage of new food sources and to find food in agricultural and other human-altered environments. Crows use their feet to hold food down while tearing it with their bills. Hiding food for later consumption, *caching*, is practiced widely by the group. Most corvids are fairly omnivorous, taking bird eggs and nestlings, carrion, insects, and fruits and nuts. HAWAIIAN CROWS are different in that they eat mostly fruit, but also insects, spiders, and land snails that they take from trees and shrubs, and some carrion. Corvids are usually quite social, often living in family groups. Hawaiian Crows apparently occupy small (1 sq km, 0.4 sq miles), permanent territories that are defended by a breeding pair and their non-breeding offspring. (In 1999, there were only two pairs of Hawaiian Crows left in the wild.) Corvids are usually quite noisy, and the Hawaiian Crow is no exception, giving a variety of harsh calls, eerie growls, and screams.

Thrushes. Among the thrushes are species that employ a variety of feeding methods and that take several different food types. Many eat fruits, some are primarily insectivorous, and most are at least moderately omnivorous. Although arboreal birds, many thrushes frequently forage on the ground for insects, other arthropods, and, a particular favorite, delicious earthworms. These birds are residents of many kinds of habitats – forest edge, clearings, and other open sites such as shrub areas and grasslands, gardens, parks, suburban lawns, and agricultural lands. Many thrushes are quite social, spending their time during the non-breeding season in flocks of the same species, feeding and roosting together. OMAO are fairly common, if hard to see, residents of the Big Island's higher elevation, wetter native forests. They eat fruits such as berries, seeds, and, less frequently, land snails and insects such as caterpillars. PUAIOHI are very rare birds of high elevation dense native forests of Kauai's Alakai Swamp region. Relatively little is known of their natural history. They eat fruit and insects. Both native thrushes have a habit of quivering, or vibrating, their wings when perched, and the Omao, at least, is considered an excellent singer. The WHITE-RUMPED SHAMA, also a fine singer, with loud, clear, melodies, is common on Oahu and Kauai, where it prefers lushly vegetated forested sites, but also occurs in semi-open habitats including residential areas and gardens. It eats mostly insects.

Elepaio. Elepaio are birds predominantly of higher elevation native forests, but they are apparently extremely adaptable little birds, occurring in a wide range of habitats, including wet and dry forests on the Big Island (the different habitats they occupy probably accounting for the variation in coloring of the subspecies, Plate 37) and, on Kauai and Oahu, in both native- and introduced-tree forests. Elepaio eat mainly insects and other small arthropods such as spiders, which they catch in the air, pick off the bark of trees and pull from leaves. They are monogamous breeders, pairs staying on their nesting territories throughout the year. Predators on adult and fledgling Elepaio include Hawaiian Hawks, Short-eared Owls, and Barn Owls.

Breeding

Crow. HAWAIIAN CROWS breed monogamously. The breeding pair's offspring from past nests stay on the breeding territory perhaps helping to defend it. In other corvid species, the non-breeding group members often serve as *helpers*, assisting in nest construction and feeding the young, but it is not known if this occurs in the Hawaiian Crow. The open nests, built of twigs and branches and lined with grass and other vegetation, are in wet forest, usually in upper branches of ohia trees. One to five eggs are incubated for 19 to 20 days, the young then being fed in the nest by parents for about 40 days. Young often fledge before they are fully flight-certified, flopping around on the ground below the nest for a few days, being fed by parents, and being easy prey for ground predators.

Thrushes. Thrushes breed monogamously, male and female together defending exclusive territories during the breeding season; pairs may associate year round. Nests, usually built by the female and placed in tree branches, shrubs, or crevices, are cup-shaped, made of grass, moss and like materials, and often lined with mud. Two to six (usually two or three) eggs are incubated by the female only for 12 to 14 days. Young are fed by both parents for 12 to 16 days prior to their fledging. OMAO nests, usually placed in nooks and crannies of rock walls or ledges, or in the tops of treeferns, are constructed of grasses, ferns, leaves, bark, and other plant materials. PUAIOHI have been found nesting in cliff cavities near streams. WHITE-RUMPED SHAMA nest in tree holes and other natural and artificial cavities.

Elepaio. Elepaio nest in trees. Male and female construct the cup-like nest of vegetation (grasses, rootlets, bark, leaf pieces), using spider-webbing to help hold it all together. One to three eggs, usually two, are laid. Both sexes incubate eggs (18 days) and feed and brood nestlings; young fledge after about 16 days in the nest.

Notes

Crow. Although considered by many to be among the most intelligent of birds, and by ornithologists as among the most highly evolved, corvid folklore is rife with tales of crows, ravens and magpies as symbols of ill-omen. This undoubtedly traces to the group's frequent all-black plumage and habit of eating carrion, both sinister traits. Ravens, in particular, have long been associated in many Northern cultures with evil or death, although these large, powerful birds also figure more benignly in Nordic and Middle Eastern mythology. Several groups of indigenous peoples of northwestern North America consider the COMMON RAVEN sacred and sometimes, indeed, as a god. In native Hawaiian culture, the black feathers of HAWAIIAN CROWS were used for adorning royal thrones and for dressing idols. One of the first books about Hawaiian birds, written during the first half of the 20th Century, noted that the antics of captive Hawaiian Crows were quite entertaining and that, in fact, one reason for protecting the crows from farmers that killed them as pests was their "potential entertainment value."

Elepaio. "Of all the (Hawaiian) birds, the most celebrated in ancient times was the Elepaio, and for this reason. When the old natives used to go up into the forest to get wood for their canoes, when they had felled their tree the Elepaio would come down to it. If it began to peck it was a bad sign, as the wood was no good, being unsound; if, on the contrary, without pecking, it called out 'ona ka ia,' 'Sweet the fish,' the timber was sound." (Perkins 1893, quoted in Berger, 1981). A possible explanation for this bit of folklore: Elepaio eat insects, including ones that bore into tree trunks. An Elepaio that began pecking a fallen tree would be indicating the presence of boring insects in the wood, meaning that the tree was riddled with small insect holes and tunnels; such a tree would provide wood unsuited to boat-making.

"Elepaio are often the first birds to sing in the morning, and their songs were thought to warn spirits of the night that their work must end because dawn was approaching." (Pukui 1983, quoted in VanderWerf 1998.)

Status

Crow. Most corvids are common or very common birds. Some adjust well to people's activities, indeed often expanding their ranges when they can feed on agricultural crops. In fact, only a few corvids worldwide are threatened; the two most endangered are the two Pacific island species, the HAWAIIAN CROW and the MARIANA CROW (both USA ESA listed). The Hawaiian is close to extinction with only a few individuals left in the wild (confined to the Big Island's McCandless Ranch, 26,000 hectares, 64,000 acres, of private property on the forested western slopes of Mauna Loa), but the species is currently the object of an intense captive breeding program and other conservation efforts (p. 62). The Mariana Crow, with very small populations, occurs on only two islands (Guam, with fewer than 100 individuals, and Rota, with perhaps 500) in the Western Pacific's Mariana Islands. The Brown Tree Snake (p. 11), via its predation on crow eggs, nestlings, and fledglings, is believed by many to have pushed the crow to near extinction on Guam.

Thrushes. Hawaii's native thrushes have suffered significantly since people colonized the islands. Habitat reduction, severe reductions in some of their food plants caused by introduced mammals, the introduction of mammals that prey on bird nests, and the arrival in Hawaii of alien diseases such as avian malaria and pox have taken a huge toll. Two species, Kauai's KAMAO and Molokai's

OLOMAO, survived in small numbers until just recently (both were probably last seen during the late 1980s), but both are now feared extinct or reduced to just a few individuals; the Kamao, as late as 1910, had been one of Kauai's commonest forest birds. PUAIOHI (USA ESA listed, endangered) survive now in the wild only in remote high elevation areas of Kauai's Alakai Swamp Wilderness Reserve. The total population from 1976 to 1983 was estimated to be only about 20 individuals, but the population is much larger now (helping to increase the wild population is a conservation program that breeds Puaiohi in captivity and then releases young birds into the wild in the Alakai Swamp area; for example, 14 birds were released in early 1999). About 30 other members of the thrush family, worldwide, are threatened or endangered, including one of the shamas, the Philippines' BLACK SHAMA.

Elepaio. Elepaio, with large populations on Kauai and the Big Island, are not, in general, considered threatened. (But the Oahu Elepaio has been proposed for listing as an endangered species, USA ESA.) Biologists believe that Elepaio have been able to maintain healthy populations on some islands because, compared with many other of Hawaii's native forest birds, which are now threatened or endangered, they are less specific in their habitat requirements and more flexible in their foraging techniques.

Profiles

Hawaiian Crow, *Corvus hawaiiensis*, Plate 37a
Omao, *Myadestes obscurus*, Plate 37b
Puaiohi, *Myadestes palmeri*, Plate 37c
Kauai Elepaio, *Chasiempis sclateri*, Plate 37d
Oahu Elepaio, *Chasiempis ibidis*, Plate 37e
Hawaii Elepaio, *Chasiempis sandwichensis*, Plate 37f
White-rumped Shama, *Copsychus malabaricus*, Plate 38a

12. Mocker, Myna, Skylark, and Meadowlark

Four fairly common to very common, easily seen birds of Hawaii's more open areas comprise a group of introduced, unrelated brown or gray birds. The NORTHERN MOCKINGBIRD (Plate 38), introduced from North America to Oahu during the early 1930s, is a representative of Family Mimidae, a New World group that consists of about 30 species of mid-sized, often slender songbirds with characteristically long tails. The family includes the *mockingbirds, thrashers,* and *catbirds*. Most are brown or gray (like the Northern Mocker) with lighter chests, which are often streaked or spotted; within a species, the sexes generally look alike.

Considered by most locals to be a pest, the hugely abundant COMMON MYNA (Plate 38), introduced from India during the 1860s to control crop-damaging worms, is actually a beautiful bird that North Americans and Europeans, unfamiliar with it, will appreciate at least initially. (After you've seen your millionth myna, however, the novelty begins to wear thin.) The chocolate-brown mynas, seen in trees and on roads seemingly everywhere, are members of the *starling* family, Sturnidae, an Old World group of about 110 species. Most of the family consists of sturdily built brown or blackish birds, many with patches of iridescence in their plumage. They are mainly arboreal birds, but some, such as the Common Myna, are also at home on the ground. Male and female look alike.

The EURASIAN SKYLARK (Plate 38), introduced to Hawaii from Europe (via

New Zealand) in 1865, is a member of the *lark* family, Alaudidae, an almost entirely Old World group of about 75 species (for some reason, one species, the HORNED LARK, occurs over much of North America). Larks are smallish ground-dwelling, open-country birds, outfitted in dull brown and buffy streaked plumages that allow them to meld well into their grassland habitats. Male and female larks usually look alike.

The WESTERN MEADOWLARK (Plate 38), introduced from North America to Oahu and Kauai in 1931 (it managed to permanently establish itself, however, only on Kauai) is not a lark at all, but a member of a large, successful family of 95 or so species, Icteridae, the New World blackbirds and orioles. This is a highly diverse family, its species varying extensively in size, coloring, ecology, and behavior. It includes birds called *blackbirds, caciques, cowbirds, grackles, meadowlarks, orioles,* and *oropendolas*. The group occurs throughout North, Central, and South America, but most species occur south of the USA. Meadowlarks are mid-sized yellow, black, and brown grassland birds with long, conical, sharply pointed bills. Meadowlark sexes look alike.

Natural History

Ecology and Behavior

Mockingbird. Mockingbirds, such as the NORTHERN MOCKINGBIRD, are mostly birds of the ground, shrubs, and low trees. They forage on the ground in dry woodlands, parklands, open areas, and gardens for insects and other small invertebrate animals, and they also take some fruit and seeds. These birds, as a group, are known for their virtuoso singing performances, their highly intriguing ability to closely mimic the songs of other species, and their aggressive territoriality during breeding seasons – many a person who wandered innocently across a mockingbird territory during nesting has been hit on the head by the swooping mockers.

Myna. The COMMON MYNA is one of the land birds seen most often by visitors to Hawaii, occuring as it does over a broad range of habitats, including city parks and other settled areas. They spend a lot of their time on the ground, walking rather than hopping, looking for insects and other things to eat; they are considered fairly omnivorous. During the day they are often seen in mated pairs, but large numbers assemble by evening at communal tree roosts.

Skylark. The EURASIAN SKYLARK is a ground-dwelling bird of open fields, grasslands/savannah, roadsides, and the edges of woodlands. They mostly stay rooted on the ground, and they walk, rather than hop. They eat insects and some plant materials.

Meadowlark. Meadowlarks are ground birds of open fields, grasslands, and pastures. To broadcast their high-volume melodious songs, however, they often fly up to perch on shrubs, fences, or telephone wires. They eat mostly insects and some seeds, foraging for both on the ground. Distinguishing meadowlarks and other members of the *icterid* family from other birds is a particular feeding method not widely used by other birds, known as *gaping* – a bird places its closed bill into crevices or under leaves, rocks or other objects, then forces the bill open, exposing the previously hidden space to its prying eyes and hunger.

Ecological Interactions

The COMMON MYNA has benefitted enormously, directly and indirectly, from people's actions. Within its native range, India and southeast Asia, it expanded its distribution where people modified the environment for agriculture, adapting well to new habitats such as cleared areas, settled areas, and croplands. Furthering the species' expansionist plans, people began introducing the myna to other regions of the world, primarily to try to control crop pests such as locusts (to little positive effect). The Common Myna is now abundant not only in Hawaii, but in South Africa and other Pacific islands, including Fiji, the Cook Islands, and the Society Islands. The species recently colonized Midway Island (first seen there in 1974); most likely the myna initially reached Midway with human help. The Common Myna could be considered to have a *commensal* relationship with people – the myna benefits by the association, and people, more or less, are unaffected. (The myna may have played a small role in the decline of some of Hawaii's native forest birds by competing for food and habitat with some of them – at one time the myna was more of a forest resident than it is now – and in perhaps being a reservoir for and transmitting avian diseases and parasites.)

Breeding

Mockingbird. NORTHERN MOCKINGBIRDS are monogamous. Cup nests in trees or shrubs are built of sticks and leaves by both sexes or by the female alone; she also incubates the three to five eggs for about 12 days. Young are fed in the nest by both sexes for 11 to 13 days until they fledge.

Myna. The COMMON MYNA is a cavity nester, especially partial to holes in trees, but also places its nests in holes in buildings and sometimes in dense foliage. Two to five eggs are incubated by both sexes for about 13 days, and both sexes feed the chicks for up to 30 days, until they fledge.

Skylark. EURASIAN SKYLARKS are monogamous. The female alone incubates the two to four eggs in a grass nest on the ground. Both sexes feed the chicks for 9 to 10 days until fledging.

Meadowlark. WESTERN MEADOWLARKS, monogamous (sometimes polygynous) breeders, nest on the ground. The female builds a grass nest in a natural depression, usually hidden in tall grass. Three to six eggs are incubated by the female for 13 to 15 days; young, fed mostly by the female, fledge after about 12 days in the nest.

Notes

It is not the COMMON MYNA that imitates human speech quite well; that dubious achievement is accomplished by the HILL MYNA, *Gracula religiosa*, also native to India and southeast Asia.

Status

Mockingbird. The NORTHERN MOCKINGBIRD is an abundant North American species, but a few mockingbirds are threatened. One of Mexico's, the SOCORRO MOCKINGBIRD, endemic to one of Mexico's Pacific islands, is endangered; probably only a few hundred individuals survived as of the early 1990s.

Myna. Mynas, as well as other members of the starling family, are often very common, abundant birds within their native ranges. A few starlings, such as Indonesia's BALI STARLING and Micronesia's POHNPEI MOUNTAIN STARLING, are critically endangered. The EUROPEAN STARLING, spread from the Old World to the New by people during the late 1800s, is now probably one of the most numerous birds on Earth.

Skylark. EURASIAN SKYLARKS, as recently as 1972, were considered well-established and fairly common on all six main islands. Since then they disappeared from Kauai and their population has plunged on Oahu. The species is common and abundant in its native lands. About eight species within the lark family are considered threatened or endangered, most of them African inhabitants.

Meadowlark. Most meadowlark species are not threatened; they are abundant New World birds. But one, southern South America's PAMPAS MEADOWLARK, is endangered.

Profiles

Northern Mockingbird, *Mimus polyglottos*, Plate 38b
Common Myna, *Acridotheres tristis*, Plate 38c
Western Meadowlark, *Sturnella neglecta*, Plate 38d
Eurasian Skylark, *Alauda arvensis*, Plate 38e

13. White-eye, Bush-warbler, and Babblers

Five species of some of Hawaii's more nondescript songbirds are members of three separate avian families. If you turn your eyes away from the beaches and ocean views for even a short period, one you are almost sure to see is the JAPANESE WHITE-EYE (Plate 39). An alien species, but essentially the most common land bird in the islands, the white-eye is seen almost everywhere, from lowland forests and gardens to the high slopes, at treeline, of Hawaii's volcanoes. White-eyes comprise a family, Zosteropidae, of about 85 species of small greenish birds with small, slightly down-curved, pointed bills and (usually highly conspicuous) white eye-rings composed of fine white feathers. Male and female look alike. White-eyes are distributed over Africa, Asia, Australia, and New Guinea, and often have representatives on Pacific islands – some there naturally, some, such as the Japanese White-eye in Hawaii, introduced by people.

Another small, drably colored alien bird very common on several of the main islands is the JAPANESE BUSH-WARBLER (Plate 39). Bush-warblers, native to the China/Japan region, are members of a huge assemblage (about 350 species) of tiny or small, usually inconspicuously marked birds, mainly arboreal insect eaters with small, slender, pointed bills, known as the Old World Warblers (Family Sylviidae). (In the New World the group is represented by a small number of species of tiny birds known as *gnatcatchers* and *gnatwrens*.) Bush-warbler sexes look alike.

The RED-BILLED LEIOTHRIX and two species of *laughing-thrushes* (Plate 39) in Hawaii are also introduced members of a large assemblage of Old World birds, the *babblers*, Family Timaliidae. The group, about 250 species in all (with a single native North American representative, the WRENTIT of the USA's western coast), consists mostly of small and mid-sized nondescript songbirds clad in dull browns and grays (but some species, particularly ones that tend to inhabit wet forests, such as the Red-billed Leiothrix, are quite colorful). The sexes in babblers often look alike.

Natural History

Ecology and Behavior

White-eye. White-eyes are birds of forest canopies, woodlands, forest edges, and gardens. They eat insects and spiders, fruit such as berries, and sometimes, nectar from flowers. They are fast-moving birds, alighting in a tree, searching leaves for bugs, probing into crevices for spiders, then moving on to another tree. During breeding periods white-eyes defend small territories in monogamous pairs; during non-breeding periods they are usually seen in small flocks.

Bush-warbler. Bush-warblers in Hawaii are usually seen in the dense undergrowth of forests, both of native and introduced trees, and in scrubby/high grass areas and thickets. They eat mainly insects, which they pull off foliage, but also some fruit, and nectar from flowers.

Babblers. Babblers occupy a wide range of habitat types, from tropical rainforests to marshes to desert scrub areas. Hawaii's babblers – the laughing-thrushes and the RED-BILLED LEIOTHRIX – are mostly forest birds, but occur in other habitats as well. These birds, as a group, eat mostly insects and other small invertebrate animals, but also fruit and some seeds. Many of the babblers, including the Red-billed Leiothrix and the GREATER NECKLACED LAUGHING-THRUSH, spend most of the year in small parties of 3 to 30 individuals; in some species, group territories are defended, and in some, the group breeds cooperatively within the territory.

Ecological Interactions

You might think that the introduction of small, non-native songbirds to Hawaii would hardly have major harmful effects on native wildlife. But some of these birds, such as the JAPANESE WHITE-EYE, RED-BILLED LEIOTHRIX, and MELODIOUS LAUGHING-THRUSH, may indeed have harmed native songbirds – not by eating them or their young, but by the simple act of competing with them for food. They occupy some of the same habitats as do some of Hawaii's native forest birds (especially the honeycreepers, p. 151), and eat some of the same foods (for instance, the Japanese White-eye eats plant nectar, among other things, a mainstay food of the now highly threatened Hawaiian honeycreepers). While these alien birds are not regarded as a chief cause of the decline of Hawaii's native forest birds, they may have been a contributing factor (Chapter 5, p. 56).

Breeding

White-eye. White-eyes build cup-shaped nests of grasses and other vegetation, suspended in a tree fork. Two to four eggs are incubated for 10 to 12 days; nestlings fledge after 10 to 13 days in the nest. Young are fed insects at first by the parents, then, as they get older, mostly plant materials.

Bush-warbler. Bush-warblers construct cup-like nests woven from vegetation and placed usually in a shrub. Two to five eggs are incubated for 12 to 14 days; chicks fledge after 11 to 15 days in the nest.

Babblers. Babblers usually build their nests in shrubs or trees. Two to six eggs are incubated for 14 to 15 days; young fledge after 13 to 16 days in the nest.

Notes

White-eye. Hawaii's first white-eyes were first imported from Japan in 1929, apparently an official act by the Hawaii Territorial Board of Agriculture and Forestry, and released on Oahu; later releases occurred on the Big Island and probably on Maui as well. The bird spread to all the main islands and is now considered the most abundant land bird in the state. It occurs in essentially all terrestrial habitats from sea level to treeline. In an exhaustive field census of Hawaii's forest birds conducted from 1976 through 1983, researchers, in their species-by-species report of their results, state that the JAPANESE WHITE-EYE was "ubiquitous in our study areas;" they estimated that there were 1.3 million (give-or-take 25,000) individual white-eyes in the areas they surveyed. We actually know quite a bit about the ecology and behavior of white-eyes because one species, the GRAY-BREASTED WHITE-EYE, has been intensively studied for more than 20 years on Heron Island, a small island in Australia's Great Barrier Reef; all birds on the small island are color-banded for individual identification, and all nests are monitored to find out about breeding success and causes of nesting loss. White-eyes are cage birds in some areas of Asia, prized for their singing.

Babblers. The MELODIOUS LAUGHING-THRUSH, a cage bird kept by many Chinese citizens of colonial-era Hawaii, supposedly was first released into the wild in Hawaii during the pandemonium at the time of the great Honolulu fire of 1900.

Status

White-eye. Most white-eye species are abundant birds, but a few species, particularly some that are restricted to small islands, are in trouble; one, the WHITE-CHESTED WHITE-EYE, endemic to Norfolk Island, east of central Australia, with no confirmed sightings since about 1980, is CITES Appendix I listed. The IUCN Red List contains about 20 white-eye species listed as vulnerable or endangered, with about half of them being critically endangered.

Babblers. Several laughing-thrush species of China and southeast Asia, and the ASH-HEADED LAUGHING-THRUSH of Sri Lanka, are considered vulnerable to threat, mostly owing to habitat destruction and alteration.

Profiles

Greater Necklaced Laughing-thrush, *Garrulax pectoralis*, Plate 39a
Melodious Laughing-thrush, *Garrulax canorus*, Plate 39b
Red-billed Leiothrix, *Leiothrix lutea*, Plate 39c
Japanese Bush-Warbler, *Cettia diphone*, Plate 39d
Japanese White-eye, *Zosterops japonicus*, Plate 39e

14. Cardinals and Bulbuls

Hawaii's cardinals and bulbuls comprise a group of striking aliens, most with conspicuous sharp crests, that for the most part stick to open, scrubby sites, parks, some cultivated regions and settled, residential areas. Family Pycnonotidae, the *bulbuls*, consists of about 120 species of mostly tropical, forest-dwelling birds, distributed by nature throughout Africa and southern Asia. But people, often through accidental introduction via the release of pet cage birds (as apparently happened in Hawaii), have spread bulbuls far and wide, so that there are now established breeding populations in such far-flung locations as New Zealand, Australia, Florida (USA mainland), and Hawaii and other Pacific island groups. Bulbuls are usually crested, small to mid-sized songbirds, with slender, often slightly down-curved bills, and longish tails. Most are dully turned out in subdued grays, browns, or greens; the sexes look alike.

Cardinals are handsome, mostly crested, often colorful mid-sized birds with short but sturdy "finch-like" bills, adapted for seed-crunching. They are members of a large, diverse, worldwide family, Emberizidae, which includes the North American *sparrows, towhees, grosbeaks*, the North American *buntings*, and some North American *finches*. But the family's classification is controversial, modified every 10 years or so by ornithologists who study these birds (a recent classification scheme places the NORTHERN CARDINAL, Plate 40, into a separate family, Cardinalidae, along with grosbeaks and some of the buntings). *Emberizids* are mostly small birds, such as most sparrows and finches (but Hawaii's SAFFRON FINCH, Plate 41, is mid-sized) and, although some, such as the cardinals, are brightly or strikingly colored, most are fairly drably attired in grays, browns, and dark or brownish greens, often with streaking; sexes may look alike or quite different.

Natural History

Ecology and Behavior

Bulbuls. Although within their native ranges bulbuls are most often birds of forests or forest edges, in Hawaii they have adapted to other habitats, and now are mostly birds of open areas, usually with scattered trees: parks, residential areas, agricultural sites. Bulbuls eat fruits, including many kinds of berries, and also some insects, perhaps the occasional tiny lizard. In their native lands, within forests, bulbuls are known as shy, skulking birds, but in habitats near people, such as parks and garden areas, they become quite bold. Many species, such as the RED-VENTED BULBUL (Plate 40), particularly following the breeding season, form fairly large flocks. When in flocks (which are not common in Hawaii), they are considered to be potentially harmful fruit-crop pests.

Cardinals. Emberizids are birds of diverse habitats. Most eat seeds and grain that they find on the ground. Although they eat seeds during most of the year, many switch to insects and other small arthropods during the breeding season, and so the group is considered fairly *omnivorous*. Cardinals are perhaps more arboreal than many other emberizids, which spend less time in trees and more in bushes and thickets. Cardinals eat mainly seeds, but also some soft fruit and insects. Most are monogamous breeders, male and female pairing and together defending a territory from others of their species in which they will place their nest and on which they will find all or much of the food they will feed to their young. The SAFFRON FINCH, a bird of open and semi-open sites, is often seen on the Big Island in urban areas, on or near lawns, golf courses, etc. After breeding, it often forms large, conspicuous flocks.

Breeding

Bulbuls. Mating systems among the bulbuls apparently are quite variable, but little research has been done with them, so facts are few. Nests consist of loosely woven twigs, leaves, and other plant materials formed into a cup, placed in the fork of a tree or bush. Two to five eggs are incubated for 12 to 14 days; young fledge after 13 to 14 days in the nest.

Cardinals. Cardinal nests, cup-shaped, are usually placed in shrubs or small trees. Nests are made of various kinds of vegetation such as twigs, bark, weed stems, grass, and usually lined with fine materials like grass, hair, moss, and feathers. Two to five eggs are incubated, often by the female only but sometimes by the male also, for 12 to 13 days; both sexes feed the chicks, which fledge from the nest when about 10 to 11 days old.

Notes

The Latin root of the word "cardinal" means "important," and cardinals, of course, are important members in the Catholic Church hierarchy (the Pope being the Cardinal of Rome). Church cardinals often wear red robes; hence the name "cardinal" for a red bird such as *Cardinalis cardinalis*, the NORTHERN CARDINAL. Speaking of the Northern Cardinal: this species is a bit of a troublemaker in Hawaii, in such places as the Kokee State Park area of Kauai. You see a streak of red in the trees and think "Ah – a native Iiwi or Apapane" – small red birds that visitors to the islands who are birdwatchers love to see – but no, it turns out to be a very non-native Northern Cardinal, utterly familiar to so many North Americans from their home backyards.

Status

Bulbuls. Most bulbul species are fairly common to abundant birds; about 12 species, most from Indonesia, Africa, or Madagascar, are considered to be threatened or endangered.

Cardinals. About 30 species within the large family Emberizidae are threatened or endangered in various parts of the world; no cardinals are on that list.

Profiles

Northern Cardinal, *Cardinalis cardinalis*, Plate 40a
Red-crested Cardinal, *Paroaria coronata*, Plate 40b
Yellow-billed Cardinal, *Paroaria capitata*, Plate 40c
Red-vented Bulbul, *Pycnonotus cafer*, Plate 40d
Red-whiskered Bulbul, *Pycnonotus jocosus*, Plate 40e
Saffron Finch, *Sicalis flaveola*, Plate 41a

15. Fringillid Finches (House Finch, Canaries) and House Sparrow

This group of introduced Hawaiian birds consists mainly of species that "make their livings" in association with people. That is, they live in parks, villages, towns, and cities, nest in these places, sometimes in buildings and other structures, and eat a variety of foods, including what we can refer to euphemistically as "table scraps" (and perhaps less charitably as garbage). The HOUSE SPARROW (Plate 41), one of the globe's most widespread and well known birds, is a member of Family Passeridae, the Old World *sparrows*, a group of about 35 species of small open-country birds. Many of them, including House Sparrows, are drably costumed in brownish streaked plumages; male and female sometimes look alike, sometimes not. *Canaries* and *finches* (together with other types of birds known as *siskins* and *seedeaters*) are representatives of Family Fringillidae, which contains about 125 species. The group is broadly distributed across the New World, Eurasia and Africa. *Fringillids* occupy a great variety of habitats, from forests and grasslands to deserts and arctic tundra. But several have adapted to live with or around people, and so the group is best known for a few very familiar species, such as the HOUSE FINCH (Plate 41), which tend to live in settled areas, especially around farms and ranches and in parks and gardens. Fringillids range in size from very small to mid-sized songbirds (11 to 19 cm, 4 to 7.5 in); their coloring ranges from brownish and highly cryptic to quite colorful, and many are heavily streaked. The chief family trait is probably the strong, stout, conical bill used to crush seeds (see below).

Natural History

Ecology and Behavior

HOUSE SPARROWS, having been introduced from their native Eurasia and North Africa to most other parts of the world, now flourish over vast stretches of the terrestrial Earth, essentially wherever there are people, excepting rainforests, deserts, and arctic tundra areas. They live in and near cities and towns at elevations from sea level to 4500 m (14,700 ft); they have been seen feeding on upper floor ledges of New York City's Empire State Building. They eat mostly seeds, but will eat or attempt to eat a great variety of things, including insects, fresh plant buds and leaves, and discarded people food (french fries being a perennial favorite). Most live and breed in small colonies. After breeding, House Sparrows form into flocks sometimes in the hundreds, sometimes the thousands (at which point they can become significant agricultural pests on such crops as cereals and grains).

Fringillids such as the HOUSE FINCH, YELLOW-FRONTED CANARY (Plate 41) and ISLAND (or COMMON) CANARY (Plate 41) are generally gregarious flocking birds during non-breeding periods, roaming widely, seeking their favorite food, seeds, which are usually taken from the ground. Other foods, such as new buds, blossoms, and some fruit, are taken from small trees; some fringillids also dabble in insects. House Finches eat mostly grass and weed seeds, but also some tree buds and fruit (in fact, they are considered fruit crop pests in some orchard areas of the western USA). Canaries eat mainly grass and weed seeds. Fringillid finches are specially adapted for seed-eating: they actually crush a small seed by wedging it into a groove on the roof of their mouth, then raise the lower jaw into it; the broken husk is then peeled off with the tongue and discarded, the bare seed swallowed. When breeding, fringillids usually do not defend large territories from which they will obtain most of their food, but they do defend smaller areas around their nests.

Breeding

HOUSE SPARROWS nest in trees or building cavities, constructing messy-looking, ball-shaped nests of straw, grass, and pieces of trash. Two to four eggs are incubated by the female for 10 to 13 days; young fledge after 14 to 17 days in the nest. Fringillids build their nests, usually of grass, moss, and other vegetation, in trees or bushes. The female alone usually builds the nest and incubates the three to five eggs for 12 to 14 days; nestlings fledge when 11 to 17 days old.

Notes

Because almost everyone is familiar with HOUSE SPARROWS (whether they know it or not) and because they're one of the world's commonest birds, it might seem a waste here to devote to them illustration space and a paragraph or two. But if you think about it, the lowly House Sparrow is actually nothing short of amazing. It's a 30-g (1-oz) dynamo of vigorous, competitive energy. These sparrows have probably been in a close commensal relationship with people since the development of agriculture in the Middle East, 10,000+ years ago, feeding on cereal and grain crops and on seeds left in fields. Ornithological historians suspect that these small brown birds would then have prospered and multiplied during the age of non-mechanized cities, when beasts of burden (I've always wanted to use the phrase "beast of burden," and this is my chance) and transport such as horses perpetually filled streets with their seed-rich droppings. Europeans brought House Sparrows with them when they colonized great sections of the planet during the 17th through 19th Centuries. They brought many more of their native bird species also, but relatively few made "successful" colonizations (meaning, the first few individuals released into new lands survived, thrived, reproduced, and initiated self-perpetuating, range-expanding populations), and fewer still colonized with the swiftness and eventual pervasiveness of the House Sparrow. Its first successful North American introduction evidently occurred during the mid-1850s on New York's Long Island, along the Atlantic Coast, in a place called Brooklyn (there were a few releases before then, but those birds failed to establish themselves). It took only 50 years for the birds to reach the Pacific Ocean, essentially colonizing and occupying all suitable habitat on the entire USA mainland. They now also occur over much of sub-arctic Canada, Mexico, and Central and South America, and they are still spreading.

The YELLOW-FRONTED CANARY in its native Africa is often trapped and sold for the pet trade; it is a popular cage bird (sometimes known as the Green Singing Finch) in Europe and North America. The ISLAND CANARY, which does not occur in the main Hawaiian Islands, was introduced to Midway Atoll during the early 1900s.

Status

Only four or five species of the canary/finch family, and probably none of the Old World sparrows, are currently endangered. Conservation of finches and sparrows really concerns conservation of their habitats; the destruction and alteration of natural habitats, mostly for crop farming and ranching, is generally agreed to be the prime threat to their populations.

Profiles

House Sparrow, *Passer domesticus*, Plate 41b
House Finch, *Carpodacus mexicanus*, Plate 41c
Yellow-fronted Canary, *Serinus mozambicus*, Plate 41d
Island Canary, *Serinus canaria*, Plate 41e

16. Hawaiian Honeycreepers

The *Hawaiian Honeycreepers* form a group, now much reduced in diversity, of stunning little birds with an incredible story. Most of them, because they have small and declining populations and inhabit only remote mountain forests, are difficult to find. Yet because of their story and their current status, they are also the birds that visitors to the islands who are birdwatchers most want to search out and see. The story is simply this: a large group of small birds, perhaps 50 species or more, developed in the Hawaiian Islands from a single species that colonized naturally perhaps 3.5 million years ago. These species existed only in Hawaii; they thrived there, protected, in a real sense, by their mid-ocean isolated location. But from people's first colonization of the islands 1600 years ago to the present day, these birds were persecuted: hunted for their feathers, their forest homes cleared for agriculture, their eggs and young eaten by introduced nest predators such as rats and mongooses, and their immune systems overwhelmed by introduced diseases (p. 58, and Close-up, p. 158). Only a few remain, most of them endangered, confined now mainly to patches of protected high-elevation wet forests on Kauai, Maui, and the Big Island. They are among the islands' most famous animal residents, they represent a truly amazing group of birds, and now some of them are among the Earth's most critically endangered animals. If you have some interest in animals and nature, it would be worth your while to seek out one or more of the few final refuges of these little birds (p. 110) and take a look at them; few may still exist a generation hence.

The Hawaiian Honeycreepers, also called the Hawaiian Finches (because they all evolved from a single finch or finch-like ancestor that long ago colonized Hawaii), now comprise perhaps 17 species (the ones shown in Plates 42, 43, 44 and two that occur on small islands in the Northwestern Hawaiian Islands chain); a few more species may still exist in the wild (in the remotest part of the Alakai Swamp region, for instance), but have been reduced to such small populations (20 individuals? 10 individuals?) that without massive conservation efforts, they are doomed. The honeycreepers' classification is controversial, but they are usually considered to be a separate passerine family, Drepanididae, closely related to the finch family Fringillidae (p. 149), or a subfamily (Drepanidinae) within the Fringillidae.

Drepanidids are small birds, most being from 10 to 13 cm (4 to 5 in) long, but a few are as large as about 18 cm (7 in). Most are greenish or green and yellow, but some are bright red or orangish, or brownish, and at least one surviving member, the AKOHEKOHE, or CRESTED HONEYCREEPER (Plate 44), is black. Within a species, male and female are usually similarly colored and marked, although females are often a bit duller than males. As a group, the birds' most distinctive physical trait is their bills, which vary from the short and straight, such as that of the MAUI ALAUAHIO (Plate 43) to the very long, highly down-curved, as in the IIWI (Plate 44).

Natural History

Ecology and Behavior

Honeycreepers primarily occupy Hawaii's native forests, although a few of them also move into areas with mixed native/introduced forests and occasionally into shrublands. But most are limited to higher elevation wet forests, usually those dominated by ohia (p. 232) or koa (p. 226) trees. On the Big Island, honeycreepers are also found in drier high elevation forests, mostly with mamane (p. 242) trees (mamane-naio forests). The honeycreepers now occur only in high- and some middle-elevation sites because their low elevation wooded habitats were mostly cleared for agriculture during the past 1600 years and because, with the introduction to the islands during the 19th Century of mosquitos that can carry bird diseases such as avian malaria, the birds can survive now only where these mosquitos cannot – in high, and therefore cool, places.

Probably the most-studied and so best-known aspect of honeycreeper ecology is the obvious functional relationship between the birds' bills and their respective feeding habits. The bills have been adapted through evolution to closely reflect feeding ecology, or feeding "niches."

(1) Some of the birds, seed-eaters, have short, stout, strong "finch-like" bills to crush seeds, such as the PALILA (Plate 44), which feeds mostly on seedpods from the mamane tree.
(2) Some, such as the MAUI ALAUAHIO, HAWAII CREEPER, AKIKIKI (Plate 43), and ANIANIAU (Plate 42), have short, fairly narrow, sharply pointed, slightly down-curved "warbler-like" bills – the better to quickly aim at and grab insects from tree trunks and branches.
(3) Other insect-eaters, such as the amakihis (which also, at times, feed extensively on nectar; Plate 42), have mid-sized, slim, down-curved bills to probe for insects in mosses and tree crevices and under bark.
(4) Some, such as the APAPANE (Plate 44) eat mostly nectar (and some insects), and have mid-sized, slightly down-curved bills specialized to take nectar from flowers.
(5) The IIWI (Plate 44) has a very long, strongly down-curved bill, which permits

it to take nectar from long, tubular flowers (many native plants have flowers of this type).

(6) Some have highly specialized bills in which the top and bottom parts differ significantly. For instance, the bottom part of the AKIAPOLAAU'S (Plate 43) bill is short, thick, and straight, used to dig small holes into soft tree wood and bark like a woodpecker does; the top part, fine and down-curved, is nearly twice as long as the bottom part and is used to probe into the holes and extract insect larvae such as grubs.

Below are details on the ecology and behavior of the three most commonly seen honeycreepers.

Apapane. These small red and black birds, the most abundant of the honeycreepers, occur in native wet forests above 1000 m (3300 ft; often only above 1250 m, 4000 ft). They are common on Kauai, Maui, and the Big Island (and can be thick in the trees at Hawaii Volcanoes National Park), uncommon on Oahu, and rare or extinct on Molokai and Lanai. They occupy mostly ohia forests, feeding at the red ohia flowers on the nectar in them and insects around them; on the Big Island they are also found in mamane and koa forests and some shrublands. They are predominantly *nectarivores*, but also eat insects and spiders. Apapane are often seen flying over forests during their daily search for flowering trees at which they can feed. The ohia trees, for instance, are not all in bloom in the same month, so the birds must constantly search for patches of trees with flowers, and may have to cover many kilometers each day to find enough food. Some apparently make daily *altitudinal migrations*, moving to somewhat lower elevations during the day to feed – for example, at the Big Island's Hawaii Volcanoes National Park – then returning to higher sites in late afternoon, where they will roost for the night. During nonbreeding parts of the year, Apapane are seen usually in small flocks. Although Apapane themselves are surviving well, their abundance may have a negative effect on some other honeycreeper species: Apapane, although sometimes immune to the diseases themselves, can apparently be carriers of avian malaria and avian pox, diseases partly responsible for the decline of many of these birds.

Amakihis. The three species of Amakihis are common honeycreepers, the most abundant honeycreepers after Apapane. They occur now on all the main islands but Lanai, generally above 600 m (2000 ft) elevation, but occasionally lower. They occur in wet and some drier forests ranging up to 3000 m (10,000 ft), to treeline, and sometimes higher, into drier scrublands. Amakihi eat insects and spiders, which they take from tree banches and from high leafy areas in the treetops, flower nectar from a host of native and introduced trees and other plants, plus some fruit and fruit juice. Amakihi are considered by ecologists to be particularly adaptable, able to live at middle, high, and even some lower elevations, able to feed at an array of plant species (that is, not as specialized in their feeding habits as many of the other honeycreepers), and with immune systems that are apparently sometimes able to fight off diseases such as avian malaria, diseases that kill most other honeycreepers.

Iiwi. The IIWI (eee-eee-vee), the unofficial poster bird for the Hawaiian honeycreepers, with its striking red body, black wings and tail, and long, drooping, reddish-orange bill, is one of the most exotic-looking animals that people who live out their lives in Europe or North America will ever see. It occurs on five of the main islands (but is very rare and endangered on Oahu and Molokai), in wet forests generally above 1500 m (4900 ft). The species is highly susceptible to avian malaria, so individuals venturing to lower elevations, where there are disease-carrying mosquitos, die quickly. Iiwi eat mainly nectar from flowers of ohia (about 90% of their foraging time is spent at this nectar source) and mamane trees and from lobelia plants, and also some insects and spiders that they take from the tree leaves. They forage for these foods mainly in the middle and upper forest levels. Iiwi are known to be strong fliers, and are often seen flying rapidly over the forest, making long daily flights of many kilometers, in their perpetual search for flowering ohia trees from which to feed.

Ecological Interactions

Mutualisms are ecological interactions in which both participants benefit (p. 71). As such, we usually have very positive impressions of these interactions – animals, or animals and plants, essentially "helping each other." But when the interacting parties become so specialized that they depend absolutely on one another, mutualisms show a dark side and can lead to ecological disasters. Nectar-eating honeycreepers such as the IIWI have mutualistic interactions with the plants from which they gather nectar. The birds benefit because they get from the plants the food they need to live. The plants make the nectar for the birds because they benefit by using the birds to transfer their pollen, thus achieving successful reproduction. The bird accidentally picks up pollen on its body when it feeds at the flowers of the plant, then drops some of the pollen at flowers of other individuals of the same plant species, and cross-pollination is accomplished. In mainland areas, plants that use pollinators (insects, birds, or bats) almost always have more than one species carrying pollen, and nectar-eating insects, birds, or bats almost always feed at more than one species of plant. But on islands, where the ecological communities are much less diverse, often plants are pollinated by single species, and a species of bird, for example, may feed exclusively at a single plant species. Thus, bird and plant depend on each other entirely – and on islands undergoing rapid ecological changes as people alter habitats, this is a recipe for disaster.

One group of plants, the *lobelioids*, is in particular trouble in Hawaii. (The lobelioids comprise a subfamily of the bellflower family; most are in genus *Cyanea* or genus *Clermontia*; Plate 3.) More than a hundred species of these plants exist in Hawaii, probably all descended from a single species that colonized long ago and diversified. Some evolved along with the long-billed honeycreepers, the birds' bills coming to resemble in shape the elaborate, long, curved, flower tubes through which they gained access to nectar. Many of these honeycreeper species are now extinct or exceedingly rare. The effect on the lobelioids that depended on the little birds for reproduction? Nothing less than extinction. Some of the lobelioid species that depended on extinct honeycreepers are still alive only because individual plants, even if they cannot reproduce, can live a long time. Some species, however, have been reduced to just a few surviving individual plants, and indeed, one species exists now as a single individual. Plant conservation biologists work with some of these species, cross pollinating them by hand, hoping to maintain viable populations. Mutualisms of this type work both ways: some of the birds that depended on specific tree species for their food are now partly endangered because their food trees are increasingly scarce.

Breeding

Most honeycreepers breed within the period from January through July. Most or all are monogamous breeders, male and female building a nest after such courtship behaviors as rapid and repeated chasing. Nests are shallow cups placed in trees or cavities, built of twigs, grasses, mosses, leaves, and other plant materials. One to four eggs (usually two or three) are incubated for 13 to 14 days, usually only by the female; nestlings fledge from nests after being fed by parents for 15 to 22 days.

Apapane. Apapane breed monogamously, nesting mainly from February through June. Male and female defend a small area around the nest, which is placed in an ohia or other tree or in a tree cavity. Both sexes build the cup nest of mosses, twigs, leaves, bark, and lichens; the nest is lined with fine grass. The female only incubates the one to four eggs (usually three) for about 13 days; Both sexes feed the chicks (caterpillars being a primary food) for about 16 days until they fledge.

Amakihi. Amakihi, also monogamous breeders, nest mainly from April through July. A cup nest of vegetation is built usually near the tip of a small tree branch (which may help prevent nest predation by rats and other mammal predators). The female only incubates the two to four eggs for about 14 days. The nestlings, fed by both parents, fledge when they are 15 to 20 days old.

Iiwi. Iiwi are monogamous breeders, usually nesting from February through June. Male and female defend a small area around the nest, which is placed an average of 7 m (23 ft) above the ground in an ohia tree. The nest, built by both sexes, is an open cup constructed of twigs, mosses, lichens, and bark. The female only incubates the one to three eggs for about 14 days; both parents feed the nestlings for about 21 days, until they fledge.

Notes

IIWI and APAPANE are known to have been trapped and killed in large numbers by early Hawaiians for their feathers and for eating. The birds' bright red feathers were often used in feather artwork and feather adornments. A royal cape or cloak – a symbol of power – included literally hundreds of thousands of these feathers. Iiwi feathers, being brighter red, were probably considered more desirable than the duller red Apapane feathers. James Cook, the first Westerner to arrive in the Hawaiian Islands, during the late 1700s, wrote that in markets he saw large numbers of dead Iiwi for sale in bundles of 20.

The Hawaiian honeycreepers are often called "dreps" (for the family, Drepanididae) by knowledgeable birdwatchers, but the term is only for the initiated; don't be surprised if personnel at your hotel return only blank stares when you ask them for directions to places to see "dreps."

Status

Of the 47 to 50 honeycreeper species that occurred in Hawaii prior to human settlement, only about a third still exist, and probably half the remainder are threatened or endangered. These birds occupy more slots on the USA's endangered species list (for species inhabiting USA territory) than members of any other bird family. The causes for their past declines and present problems are now generally known:

(1) *Habitat destruction/alteration.* Their low elevation forest habitats were cleared for agriculture by early Polynesian settlers and by later-arriving Europeans and Americans; and many of their habitats at all elevations were altered and degraded by feral pigs and feral grazers – cattle, sheep, goats.
(2) *Introduction of nest predators.* Small mammals that attack the nests of small

birds, such as rats, cats, and the mongoose, probably strongly reduced the nesting success of many species in areas into which the introduced predators penetrated; previous to the introductions, the honeycreepers would have had few, if any, nest predators, and so no experience in dealing with such threats.

(3) *Introduction of bird diseases.* Diseases, such as avian malaria, to which the honeycreepers had never previously been exposed, and so to which they had no natural immunity, arrived with the introduction of mosquitos (which could transmit the disease organisms to the birds) starting during the early 1800s. Most of the honeycreepers at lower, warmer elevations, where the mosquitos could breed, died of these diseases, and they are a continuing threat.

(4) *Competition with introduced birds.* Some of the introduced land birds, such as the JAPANESE WHITE-EYE, occupy some of the same forests as do the honeycreepers, and probably compete with them for foods such as tree flower nectar.

The current status of the Hawaiian honeycreepers can only be said to be precarious and highly threatened. Details on threats to honeycreepers and on some of the programs attempting to save them are discussed in Chapter 5 and in Close-up 4, p. 158.

Profiles

Common Amakihi, *Hemignathus virens*, Plate 42a
Kauai Amakihi, *Hemignathus kauaiensis*, Plate 42b
Oahu Amakihi, *Hemignathus flavus*, Plate 42c
Anianiau, *Hemignathus parvus*, Plate 42d
Akekee, *Loxops caeruleirostris*, Plate 42e

Akiapolaau, *Hemignathus munroi*, Plate 43a
Maui Parrotbill, *Pseudonestor xanthophrys*, Plate 43b
Maui Alauahio, *Paroreomyza montana*, Plate 43c
Akikiki, *Oreomystis bairdi*, Plate 43d
Hawaii Creeper, *Oreomystis mana*, Plate 43e

Palila, *Loxioides bailleui*, Plate 44a
Akepa, *Loxops coccineus*, Plate 44b
Apapane, *Himatione sanguinea*, Plate 44c
Iiwi, *Vestiaria coccinea*, Plate 44d
Akohekohe, *Palmeria dolei*, Plate 44e

17. Estrildid Finches: Waxbills and Mannikins

Estrildid finches are small, common birds of Hawaii's grassy, brushy open-country habitats; all of the 10 or so species that occur here were introduced. Their classification is controversial, but we can consider them members of an Old World family, Estrildidae, which contains about 130 species of small or very small seed-eating birds known as *waxbills*, *grassfinches*, and *mannikins*. They occur broadly over Australia (grassfinches, in particular), the New Guinea region, southern Asia, and sub-Saharan Africa. Many are drably marked in shades of brown, perhaps with barring patterns, but some have large bold patches of black and/or white and others are quite colorful (culminating in northern Australia's GOULDIAN FINCH, with green back, purple chest, yellow belly, and red, blue, and black head). Bills of some species appear disproportionately large for such tiny birds – the big, powerful bills are required to handle and crush the diet staple, seeds. One group usually has red waxy-looking bills, so are called – you guessed it – waxbills (the name originally referred to the resemblance between the birds' bills and the red wax used to seal documents and letters – this before the advent of e-mail, of course). Another group (genus *Lonchura*), slightly heavier-looking than the others and perhaps overall the most dully-colored of the family, are generally called mannikins (the African and New Guinea species) or *munias* (Asian species); the former name, inevitably, causes these birds to be confused with the *manakins*, a Central and South American group of small, colorful songbirds. Estrildid finches vary in length from 9 to 15 cm (3.5 to 6 in). The sexes look alike in some species, different in others.

Natural History

Ecology and Behavior

Estrildid finches are mainly birds of open and semi-open habitats – savannahs, arid grasslands, brush and scrublands, forest edges and clearings (but there are also some that dwell in open woodlands and even dense forests). Most are seedeaters (but two African species eat ants), with grass seeds, collected from the ground or pulled from grass stalks, being the primary diet item; many species switch to an at least partial insect diet during breeding seasons. Estrildid finches are often monogamous breeders, a male and a female maintaining a stable pair-bond for one or more breeding seasons; some may mate for life. Perhaps the group's outstanding feature is the extreme gregariousness of the estrildid lifestyle. Various species are either in flocks all year, breeding in colonies when the reproductive urge hits, or flock together after pairs nest solitarily. Large numbers of individuals spend the entire day together, feeding, bathing, and roosting in flocks.

Breeding

Most estrildid finches build untidy domed nests of grass and other vegetation that they place in shrubs, trees, or sometimes on the gound in tall dense grass. Both sexes construct the nest (the male often doing more shopping for and delivering materials, the female doing more actual building), the sexes take turns incubating the three to six eggs, and both parents regurgitate food to nourish the nestlings for 15 to 17 days until they fledge; fledglings, out of the nest, are fed by parents for another week or two before they are fully independent.

Notes

When estrildid finch flocks swell into the thousands, they can become significant agricultural pests. Species such as the NUTMEG MANNIKIN (Plate 46), CHESTNUT MANNIKIN (Plate 46) and JAVA SPARROW (Plate 46) harm crops in their native lands and also sometimes in Hawaii. In fact, the Nutmeg Mannikin, known locally in Hawaii as the "Ricebird," was largely responsible for eliminating rice as an agriculural crop in the islands.

Status

Estrildid finches are some of the world's most popular cage birds and some areas in which they are heavily hunted and trapped for the pet trade have experienced significant population declines for these birds. The JAVA SPARROW, for instance, is now scarce over parts of its native range in Bali and Java, the result of its being trapped intensively for the pet trade and as food; the species is considered vulnerable to threat by the IUCN Red List. Conservation of estrildid finches mainly concerns conservation of their habitats; the destruction and alteration of natural habitats, mostly for crop farming and ranching, is generally agreed to be the prime threat to their populations. Added to this is the persecution of the birds for the international pet trade, which is ameliorated slightly by official (but widely flouted) international controls on wildlife shipments and by increased abilities of aviculturalists, with more experience, to breed these birds in captivity to meet the demands for pet cage birds. Three estrildid species (one each from Australia, Fiji, and the Philippines) are considered endangered.

Profiles

Red-cheeked Cordonbleu, *Uraeginthus bengalus*, Plate 45a
Lavender Waxbill, *Estrilda caerulescens*, Plate 45b
Orange-cheeked Waxbill, *Estrilda melpoda*, Plate 45c
Common Waxbill, *Estrilda astrild*, Plate 45d
Black-rumped Waxbill, *Estrilda troglodytes*, Plate 45e
Red Avadavat, *Amandava amandava*, Plate 46a
Warbling Silverbill, *Lonchura malabarica*, Plate 46b
Nutmeg Mannikin, *Lonchura punctulata*, Plate 46c
Chestnut Mannikin, *Lonchura malacca*, Plate 46d
Java Sparrow, *Padda oryzivora*, Plate 46e

Environmental Close-up 4
Restoring Palila and other Hawaiian Forest Birds: Understanding Problems and Developing Solutions

by Paul C. Banko
Pacific Island Ecosystems Research Center
US Geological Survey – Biological Resources Division
Hawaii National Park

When the first Polynesians discovered and colonized the Hawaiian Islands, they encountered some of the most unusual birds in the world. Because of Hawaii's remote location in the Pacific, many colonizing bird species became isolated from the continental and oceanic populations from which they originated. Consequently, they evolved into species unique to Hawaii (*endemic* species). As a result of evolving in this remote setting, Hawaiian birds became adapted to a variety of habitat types, foraging techniques, and food resources. In some instances, a spectacular radiation of forms was derived from only one or a few colonizing species. The most extraordinary bird group to radiate was the Hawaiian honeycreepers, an endemic family or subfamily of finches (p. 151; the group's precise classification is controversial). Reflecting the many opportunities for exploiting nectar, fruit, seeds, and insects available in the Hawaiian Islands, the variety of bill forms which evolved in the Hawaiian honeycreepers is unmatched by any other single group of birds, including the finches of the Galápagos Islands which Charles Darwin collected and observed with keen interest. Other colonizing birds, including geese, ducks, rails, ibises, and owls, also evolved into strikingly different species or groups of species. In addition to the dramatic radiation of honeycreepers and other birds, some species of ducks and shorebirds established migratory connections to their places of origin and did not evolve into new species. The early Hawaiians, therefore, found birds in virtually all available habitats, including alpine lava fields.

Unfortunately, this remarkable assemblage of island-adapted birds could not withstand the many changes brought by Polynesians and the other cultures which followed many centuries later. As in most other archipelagos, Hawaiian bird communities have been greatly altered and many species have become rare or extinct as a result of human activities. Discoveries of fossil remains reveal that many more bird species were known to the early Hawaiians than were described by early Western naturalists. Populations of endemic species declined markedly in numbers and distribution after human colonization began around 400 AD. About 88 (66%) of the approximately 133 endemic Hawaiian bird species and subspecies known from collected specimens (71 species or subspecies) or fossil remains (62 species or subspecies) have become extinct within the past 1600 years. About 47% (62 of 133) of these endemic species or subspecies were eliminated during Polynesian colonization and were unknown to 19th Century naturalists, and an additional 20% (26 of 133) were eliminated after 1825 due to Western influences. About 62% (28 of 45) of the remaining endemic species or subspecies are listed as endangered or threatened, and others are candidates for listing by the US Fish and Wildlife Service. In addition, at least nine species or subspecies listed as endangered are presently too rare to be recovered or are extinct already.

Early Hawaiians restricted their permanent settlements primarily to the large, high islands in the southern portion of the island chain. They established thriving communities from Niihau Island (Map 1, p. 13) to the Big Island, and even managed to maintain a small settlement on tiny Nihoa Island in the northwestern chain. Although they utilized resources from deep offshore waters to the summits of the great volcanoes, they lived and farmed mainly in lowland areas below about 750 m (2500 ft) elevation in leeward areas and below about 450 m (1500 ft) on the wetter, more densely forested windward slopes. During the past 200 years of Western occupation, however, birds and their habitats at mid and high elevations have been greatly affected by human activity.

Cultures arriving early or late in Hawaii affected bird communities in three major ways. First, they preyed directly on some species, particularly larger species which nested on the ground and occurred in easily accessible places. The early Hawaiians would have been especially attracted to the gooselike, flightless "moa-nalo" (or ancient fowl) which were distributed in many lowland areas. Moa-nalo is the term recently coined to describe these large birds which are known only from bones found in lava tubes and other places where remains of birds were naturally preserved. We know from written accounts that Hawaiians ate and maintained semi-domesticated flocks of NENE (*Branta sandvicensis*). The Nene survived into modern times probably because it was the only goose that was capable of strong flight, and it was able to exploit a variety of habitats, including some that were less accessible to Hawaiians and the Westerners that followed.

People also affected bird populations by removing or altering the native vegetation that sustained them. Polynesians burned much of the lowland vegetation for agriculture, and Westerners later converted additional areas by livestock grazing, sugar cultivation, logging, and other agriculture. Few bird species tolerated the gross transformation of lowland vegetation communities to agricultural uses. At mid-elevations, where rainfall is higher, forests were relatively little disturbed by early Hawaiians. However, large areas of these forests were cleared for sugarcane and other agriculture during the past 150 years. Even habitats above

1200 m (4000 ft) elevation, where rainfall again declines to lower levels, were significantly modified by agriculture during the historic period.

The third major way in which people affected populations of Hawaiian birds was to introduce alien species of plants and animals to the archipelago. Crop species brought by Polynesians had no obvious impact on native ecosystem function or structure, but the Pacific Rat (*Rattus exulans*) and possibly their dogs and pigs may have affected some bird species, especially those nesting on the ground. Although rats undoubtedly spread into native habitats, there is no evidence suggesting that pigs and dogs established feral populations. However, Westerners brought to Hawaii many alien species that had profound consequences for native ecosystems and the birds in them. For example, they introduced cattle, goats, sheep, and the European pig, a larger strain of pig which may have been more prone to disturbing the soil than the Polynesian strain. Populations of these ungulates spread quickly into native habitats on most of the large, high islands, destroying native plants communities and even whole watersheds in many areas. Small mammalian predators, such as feral house cats, Black Rats, and Mongooses, also had major impacts on birds. A host of introduced invertebrates (mosquitos, ants, parasitic and predatory wasps, predatory snails, to name a few) spread avian diseases and decimated native arthropod and snail populations, including many species that were important food resources for native birds.

The magnitude of the decline and extinction of Hawaiian birds and the resources they depend upon makes restoration of species and ecosystems exceptionally challenging. The PALILA (*Loxioides bailleui*, Plate 44), a finch-billed Hawaiian honeycreeper that relies primarily on the immature seeds of Mamane (*Sophora chrysophylla*) trees (p. 242), provides an example of the difficulty of restoring an endemic forest bird that has suffered from human activity over the course of human contact. As is the case with many other Hawaiian birds, the Palila now occupies only a small fraction of its prehistoric and historic ranges and holds precariously to existence at the upper extreme of its habitat. Fossil evidence suggests that Polynesians found Palila living on at least two islands, Oahu and the Big Island, although the Oahu population apparently disappeared as a result of forest clearing to make way for agriculture. Early Western naturalists found Palila only on the drier slopes of Mauna Loa, Hualalai, and Mauna Kea volcanoes on the Big Island. More recently, Palila have become dangerously restricted to the upper slopes of Mauna Kea, where the population is increasingly concentrated on the western slope. Contraction of the Palila's range on Hawaii has been due to many factors, including: exposure to avian malaria and pox at the lower limits of its historical range where mosquitos occur; predation by feral house cats and introduced rats; habitat destruction and alteration by feral cattle, sheep, goats, and introduced game animals, which eat Māmane seedlings and strip foliage from the lower branches of trees; and threats to important insect foods (such as caterpillars) by parasitic wasps and flies which were introduced to control modern agricultural pests (themselves introduced).

The reliance by Palila on Māmane seeds for a major portion of its diet makes the species particularly susceptible to habitat degradation. For example, Palila are disappearing from forests on Mauna Kea, which have been almost eliminated or greatly damaged below around 1800 to 2400 m (6000 to 8000 ft) elevation by cattle grazing. The altitudinal width of Mamane forests is important in the foraging ecology of Palila, because the timing of flowering and seed set is affected by elevation (and rainfall). Mamane trees flower and set seed throughout the year on

Mauna Kea, but there are seasonal peaks in the availability of these important foods. Generally, flowers and seeds are produced first at higher elevations, and slowly this process moves downhill until these resources are most abundant in lower elevations. Palila move in response to changes in food availability, although they do not seem inclined to move many miles from where they are reared. If the Mamane forest extends over a wide range of elevations, there is sufficient food to sustain the Palila population. Where the band of forest is narrow, Palila struggle to find sufficient food throughout the year. Restoring forests at lower elevations will require many decades of seedling regeneration after cattle have been removed.

Herds of feral cattle, sheep, and goats and wild Mouflon Sheep (*Ovis musimon*), introduced for sport hunting, have been eliminated or reduced in much of the Palila's habitat on Mauna Kea, and Mamane has regenerated dramatically in many areas. The increased density of Mamane, however, has not yet benefitted Palila. Palila mainly forage and nest in large Mamane trees, and another decade or so of growth is needed before saplings grow sufficiently large to attract Palila. Nonetheless, forest recovery has begun and birds eventually will benefit. Palila more quickly benefit from the regrowth of low branches which were removed by browsing on Mamane trees. Regrowth of branches increases the production of flowers and seeds on trees, and Palila respond by foraging for seeds on branches which are touching the ground as well as on higher branches.

Caterpillars are another food resource of great importance to Palila. The most heavily exploited species are Mamane Coddling Moths (*Cydia* spp.), small moths whose larvae somehow gain access into Mamane pods to eat the immature seeds within. Adult Palila find *Cydia* caterpillars inside pods which they remove from branches and tear open. The caterpillars are eaten by adult birds and are an important source of protein for nestlings. Because Palila are the only bird able to tear open the tough pods, they have almost exclusive access to the caterpillars – except, unfortunately, for tiny wasps which somehow parasitize *Cydia* caterpillars (or their eggs). The larval young of the parasitic wasp live on the tissues of the caterpillar but eventually consume the host and emerge from its carcass as adult wasps. Parasitism rates of *Cydia* by four species of wasps introduced to Hawaii to control agricultural pests are extremely high. In some sites at lower elevations, the parasites consume virtually all *Cydia* caterpillars. In addition, other native species of caterpillars that are eaten by Palila are also heavily parasitized by introduced wasps.

Palila and other native birds evolved in the absence of mammalian predators but were preyed upon by native raptors, including hawks and owls. Today, however, Palila are vulnerable mainly to the depredations of introduced mammals, especially rats and feral house cats. These mammals kill Palila during nesting and roosting. Palila seem highly traditional about their choice of roost trees (where they sleep at night). In fact, some individuals return to the same perch within their roost tree for weeks at a time. Fecal accumulation below their perches may attract rats and cats. In addition, rats seem particularly attracted to the hard seeds of Naio (p. 232) trees, and Palila roosting or nesting in Naio may be preyed upon more often than birds which select Mamane trees.

In order to begin restoring Palila populations, biologists and resource managers of the Pacific Island Ecosystems Research Center (US Geological Survey – Biological Resources Division), Hawaii State Division of Forestry and Wildlife, US Fish and Wildlife Service, US Army, The Peregrine Fund, and others have been studying Palila, managing habitat, controlling predator populations, and developing new recovery techniques. Some lands may be withdrawn from cattle grazing to allow forest recovery at lower elevations. This may result in greater year-round availability of Mamane seeds and other foods. Feral sheep and Mouflon Sheep are being removed from most areas on Mauna Kea and Mamane regeneration is occurring. Mamane seedlings are being planted in areas where natural recovery of the trees is slow. As forests are managed for the benefit of Palila and other native species, alien weeds will be monitored and controlled. There presently is no practical management for parasitic wasps and other threats to caterpillar foods of Palila, but additional research will explore possible techniques. Methods for widespread cat and rat control are being developed and populations of these introduced predators are being reduced in a few important areas by trapping.

Methods for reintroducing Palila from their stronghold on the western slope of Mauna Kea to areas where they have disappeared are also being developed. Wild birds have been translocated to the northern slope of Mauna Kea, but most have returned to their original home ranges 16 km (10 miles) away. Experiments with younger birds may result in a greater number of birds remaining in the reintroduction area. Alternatively, some Palila may occasionally travel between the source (original) and reintroduction areas. If so, they may be able to escape unfavorable habitat conditions in one area or the other. Efforts to propagate Palila in captivity may result in stock which will remain permanently in reintroduction areas; however, they may not survive as well as wild-reared birds. Trials to determine what methods will be most effective in repopulating areas may begin soon. Given the slow progress of habitat recovery, Palila restoration requires active management (as opposed to simply fencing off habitat and leaving the birds to fend for themselves).

These and many other activities represent the beginning stages of recovery of Palila and its habitat on Mauna Kea. Recovery throughout its historic range requires effort and resources far surpassing what is available today. Furthermore, there is little prospect that Palila could be reintroduced to Oahu or other parts of their prehistoric range. Recovering Palila and other Hawaiian birds requires ecosystem management over large areas. That will be the greatest challenge facing biologists and resource managers in Hawaii during the 21st Century.

Chapter 9

Mammals

Mammals of Hawaii

Unless you make a special effort to do some off-shore whale-watching, mammal sightings will not contribute much to your Hawaiian wildlife experience. The reason is simple: there are few mammal species on the islands and of those, most are secretive and/or nocturnal and so difficult to see. Furthermore, almost all were introduced to the islands by people, which, to most viewers, lessens an animal's attractiveness – the experience of watching is reduced because you feel you are not seeing the animal in its native wild habitat. Also, several are fairly common types of animals introduced for sport or agriculture (goat, sheep, deer, pig) or they are common rodents (mice, rats), wildlife that most people have little interest in seeking out. There is only one native non-aquatic mammal, a bat. Although it is now thought to be at least somewhat common, most people will never see it owing to its nocturnal and solitary habits. There's a highly endangered native seal, the HAWAIIAN MONK SEAL, which is seen very occasionally hauled out on beaches. In my opinion, the most interesting Hawaiian mammal on land, one that is very common and most everyone will see, is the Small Indian Mongoose (called in Hawaii, simply, MONGOOSE). It's of interest because it's a member of a mammalian family (Viverridae) unrepresented in North America, northern

Europe, or Japan, so it's highly exotic to the great majority of visitors to the islands; many visitors remember childhood stories involving mongooses, but have never seen a real one. Mongooses are also of interest because of the ecological trouble they've caused – chiefly their habit of preying on bird nests and on birds incubating eggs (p. 173).

Probably the most important thing to know about terrestrial mammals in Hawaii is that, in large measure, they are unwanted. If you asked knowledgeable local people whether, given the opportunity to eliminate all non-captive land mammals in Hawaii with the push of a button, they would do so, most would yell a resounding "YES!". (Feral Pigs are the main exception: pig hunting for food and sport is quite popular in the islands and is considered part of the local culture.) The reason is that all of these mammals were introduced to the islands by people and almost without exception, they have caused and will continue to cause ecological harm. The family profiles below that deal with rodents, ungulates, and the Mongoose detail some of the negative effects these animals have had on native Hawaiian habitats and wildlife.

Characteristics of Mammals

If birds are feathered vertebrates, mammals are hairy ones. The group first arose, so fossils tell us, approximately 245 million years ago, splitting off from the primitive reptiles during the late Triassic Period of the Mesozoic Era, before the birds did the same. Four main traits distinguish mammals and confer upon them great advantage over other types of animals that allowed them in the past to prosper and spread and continue to this day to benefit them: hair on their bodies which insulates them from cold and otherwise protects from environmental stresses; milk production for the young, freeing mothers from having to search for specific foods for their offspring; the bearing of live young instead of eggs, allowing breeding females to be mobile and hence, safer than if they had to sit on eggs for several weeks; and advanced brains, with obvious enhancing effects on many aspects of animal lives.

Mammals are quite variable in size and form, many being highly adapted – changed through evolution – to specialized habitats and lifestyles, for example, bats specialized to fly, marine mammals specialized for their aquatic world. The smallest mammals are the *shrews*, tiny insect eaters that weigh as little as 2.5 g (a tenth of an ounce). The largest are the *whales*, weighing in at up to 160,000 kg (350,000 lb, half the weight of a loaded Boeing 747) – as far as anyone knows, the largest animals ever.

Mammals are divided into three major groups, primarily according to reproductive methods. *Monotremes* are an ancient group that actually lays eggs and still retains some other reptile-like characteristics. Only three species survive, the platypus and two species of spiny anteaters; they are fairly common inhabitants of Australia and New Guinea. The *Marsupials* give birth to live young that are relatively undeveloped. When born, the young crawl along mom's fur into her *pouch*, where they find milk supplies and finish their development. There are about 240 marsupial species, including kangaroos, koalas, wombats, and opossums; they are limited in natural distribution to Australia and South and Central America (the industrious but road-accident-prone VIRGINIA OPOSSUM also inhabits much of northern

Mexico and the USA). Surprisingly, there is a Hawaiian member of the kangaroo family (Macropodidae, with about 55 species distributed throughout the Australian region), the BRUSH-TAILED ROCK WALLABY (Plate 48).

I mention the wallaby here because, even though it occurs only over a small portion of Oahu (the Kalihi Valley and environs, north of Honolulu; often along the Likelike Highway), some travellers and tours seek it out as a biological curio. It's a 60-cm (2-ft) tall grayish brown kangaroo look-a-like with a long, bushy, cylindrical tail. The present restricted population, probably numbering fewer than 200 individuals, is descended from a single pair, presumably male and female, which escaped captivity in 1916. Herbivores, the wallabies found the vegetation around the cliffs and rocky slopes of the Kalihi Valley to their liking, and have been there ever since, living in rock crevices and caves (which is also their preferred habitat in their native haunts). Wallabies eat a variety of plant materials, including grasses, bark, and roots. As long as this wallaby population remains small and causes little ecological damage, Hawaiian biologists and wildlife managers seem content to leave the marsupials alone; the species is apparently fairly scarce in Australia, so another population existing elsewhere may eventually prove beneficial. Some biologists also believe that the Hawaiian subspecies of the wallaby may be unique – it's a slightly different color from others – owing either to the elimination in Australia of the subspecies from which it descended, or from rapid evolution in Hawaii; and this is another reason that preservation of the Oahu 'roos, or, minimally, a "hands-off" attitude toward them, is probably a good idea. For readers who have not seen kangaroo-like animals in the wild and who have little prospect of a trip to Australia anytime soon, a search for the not-very-easy-to-see wallabies may be fun.

The majority of mammal species are *Eutherians*, or *true* mammals. These animals are distinguished from the other groups by having a *placenta*, which connects a mother to her developing offspring, allowing for long internal development. This trait, which allows embryos to develop to a fairly mature form in safety, and for the female to be mobile until birth, has allowed the true mammals to be rather successful, becoming, in effect, the dominant vertebrates on land for millions of years. The true mammals include those with which most people are intimately familiar: rodents, rabbits, cats, dogs, bats, primates, elephants, horses, whales – everything from house mice to ecotravellers.

The 4600 species of living mammals are divided into about 20 orders and 115 families. Approximately 40 mammal species occur in and around the main Hawaiian islands and their surrounding seas: one bat, one seal, about 20 whales and dolphins, and 18 terrestrial, non-flying hairy vertebrates.

Seeing Mammals in Hawaii

Except for the Mongoose, which you will see, albeit usually briefly, almost everywhere, your Hawaiian encounters with non-domesticated mammals will be few. Axis Deer are frequently spotted crossing roads and in more open areas of Molokai and Lanai. Feral Goats are commonly seen along trails at Kauai's Waimea Canyon State Park, and Feral Pigs are seen on or near forest trails on most of the main islands. Feral Donkeys are seen on the western side of the Big Island, in the South Kohala district, among others. Hawaii is the easiest place in the world to identify bat species – there is only one, the HAWAIIAN HOARY BAT, which occurs on all main islands.

Whale-watching day cruises, on which whales and dolphins are usually spotted, leave daily from most of the main islands, particularly from Maui and the Kona coast of the Big Island. Humpback Whales are what most visitors want to see. A thousand or more of the Northern Pacific's population of 2000+ Humpbacks spend the winter near the Hawaiian Islands, mostly from November through May, with peak densities from December to March. Best sighting areas are often the relatively narrow, shallow channels between and among the close-grouped islands of Maui, Lanai, and Molokai. In fact, because the endangered Humpbacks dwell in this area, it was in 1992 declared a protected federal sanctuary (the Hawaiian Islands Humpback Whale National Marine Sanctuary), which protects plant and animal life in the ocean's waters from the high-tide lines around Maui, Lanai, and Molokai to a sea depth of 183 m (600 ft); Humpbacks apparently prefer these shallow waters, often about 100 m (330 ft) deep, for their breeding. Also seen in these waters are Hawaiian Monk Seals, Short-finned Pilot Whales, False Killer Whales, Bottle-nosed Dolphins, Spinner Dolphins, and Green and Hawksbill Sea Turtles. For the landlubbers among you, whale and dolphin watching can also be done from shore. Although the whales are often not very close to shore, they can be seen breaching, swimming, and cavorting. Popular viewing spots are along elevated areas of Maui's western coast, along the Big Island's western (Kona) coast, and from Kauai's Kilauea Point; whales can even be seen from Oahu (for instance, from Kaena Point, Makapuu Point, and the Pupakea Beach area).

Spinner Dolphins are often seen off South Point and along the Kona coast of the Big Island. In fact, a long-term population-monitoring study of Spinners took place in this location, which was selected by researchers because of the good visibility of the dolphins provided by the site as well as the calm waters off the Kona coast that allowed for good visibility and safe, comfortable boating for seeking out the dolphins. Why the dolphins like this area so much probably has to do with the abundant food resources it provides, its close proximity to safe, deep water, and the protected stretches of shoreline and coves with calm waters in which they can rest when not feeding.

Family Profiles

1. Bats, the Flying Mammals

Bats are flying mammals that are active only at night. Like birds, they engage in sustained, powered flight – the only mammals to do so. Bats navigate the night atmosphere chiefly by "sonar," or *echolocation*: not by sight or smell but by broadcasting ultrasonic sounds – extremely high-pitched chirps and clicks – and then gaining information about their environment by "reading" the echos. Because of these characteristics, bats are quite alien to people's primate sensibilities; but precisely because their lives are so very different from our own, they are increasingly of interest to us. In the past, of course, bats' exotic behavior, particularly their nocturnal habits, engendered in most societies not ecological curiosity but fear and superstition.

Bats have true wings, consisting of thin, strong, highly elastic membranes that extend from the sides of the body and legs to cover and be supported by the elongated fingers of the arms. (The name of the order of bats, Chiroptera, refers to the wings: *chiro*, meaning hand, and *ptera*, wing.) Other distinctive anatomical features include bodies covered with silky, longish hair; toes with sharp, curved claws that allow the bats to hang upside down and are used by some to catch food; scent glands that produce strong, musky odors; and, in many, very odd-shaped folds of skin on their noses (nose-leaves) and prominent ears that aid in echolocation. Like birds, bat's bodies have been modified through evolution to conform to the needs of energy-demanding flight: they have relatively large hearts, low body weights, and fast metabolisms.

Bats are widely distributed, inhabiting most of the world's tropical and temperate regions except for some oceanic islands. With a total of about a thousand species, bats are second in diversity among mammals only to rodents. Ecologically, they can be thought of as night-time equivalents of birds, which dominate the daytime skies. A single, native species, the HAWAIIAN HOARY BAT (Plate 47) occurs in Hawaii. It's a member of a large family, Vespertilionidae, which contains more than 350 species distributed over temperate and tropical areas of the globe. The genus to which it belongs, *Lasiurus*, known as the hairy-legged bats, contains 13 species and in North and South America is one of the most widespread bat groups. The Hoary Bat actually occurs widely in the New World; the Hawaiian Hoary Bat is considered an endemic subspecies. It's a fairly small bat, dark gray or reddish-gray, weighing 14 to 18 g (half an ounce), with a wingspan of 27 to 35 cm (10.5 to 14 in). "Hoary," meaning gray or white with age, refers to the fact that the body hair often has a "frosted" appearance.

Natural History

Ecology and Behavior

Most bats, including the Hoary Bat, specialize on insects. They use their sonar not just to navigate but to detect insects, which they catch on the wing, pick off leaves, or scoop off the ground. Bats use several methods to catch flying insects. Small insects may be captured directly in the mouth; some bats use their wings as nets and spoons to trap insects and pull them to their mouth; and others scoop bugs into the fold of skin membrane that connects their tail and legs, then somersault in mid-air to move the catch to their mouth. Small bugs are eaten immediately on the wing, while larger ones, such as large beetles, are taken to a perch and dismembered.

Bats spend the daylight hours in day roosts, usually tree cavities, shady sides of trees, caves, rock crevices, or, these days, in buildings or under bridges; Hoary Bats roost mostly in trees and on rock ledges. For most species, including the Hoary, the normal resting position in a roost is hanging by their feet, head downwards, which makes taking flight as easy as letting go and spreading their wings. Many bats leave roosts around dusk, then move to foraging sites at various distances from the roost. Night activity patterns vary, perhaps serving to reduce food competition among species. Some tend to fly and forage intensely in the early evening, become less active in the middle of the night, then resume intense foraging near dawn; others are relatively inactive early in the evening, but more active later on. Hoary bats usually become active in late afternoon and continue foraging into the night. They are strong, fast flyers, and are commonly seen, at a range of elevations, flitting about open fields, lava flows, woodland clearings, and over water, including ocean bays near shore.

Many bats are highly social animals, roosting and often foraging in groups, but Hoary Bats are predominantly solitary, male and female coming together only to mate. In North America, where the species migrates seasonally, large groups of several hundred are known to coalesce for mass journeys.

Ecological Interactions

Bats eat a variety of insects and some of the prey species have responded evolutionarily to this predation. For instance, several groups of moth species can sense the ultrasonic chirps of some echolocating insectivorous bats; when they do, they react immediately by flying erratically or diving down into vegetation, decreasing the success of the foraging bats. Some moths even make their own clicking sounds, which apparently confuse the bats, causing them to break off approaches. The interaction of bats and their prey animals is an active field of animal behavior research because the predators and the prey have both developed varieties of tactics to try to out-maneuver or outwit the other.

Breeding

Bat mating systems are diverse, various species employing monogamy (one male and one female breed together), polygyny (one male and several females), and promiscuity (males and females both mate with more than one individual); the breeding behavior of many species, such as the Hawaiian Hoary Bat, has yet to be studied in detail. It apparently gives birth from May to July. Most bats produce a single pup at a time, but bats of the genus *Lasiurus* are known for having litters of two, three, or even four young; HAWAIIAN HOARY BATS apparently often give birth to twins. Females probably leave their young at a perch while they forage.

Notes

Bats are appreciated by knowledgeable people for their insect-eating ways. Individual bats can snap up thousands of small, pesky bugs per night. Owing to this facility, a few bat species were brought from mainland regions and released in Hawaii, with the objective of controlling insect populations. A few BRAZILIAN FREE-TAILED BATS, for example, were brought to Hawaii at the end of the 19th Century. None of these introductions were successful; that is, none of these released bat species managed to establish self-perpetuating populations.

Bats have frightened people for a long time. The result, of course, is that there is a large body of folklore that portrays bats as evil, associated with or incarnations of death, devils, witches, or vampires. Undeniably, it was bats' alien lives – their activity in the darkness, flying ability, and strange form – and people's ignorance of bats, that were the sources of these myriad superstitions. Many cultures, worldwide, have legends of evil bats, from Australia to Japan and the Philippines, to Europe, the Middle East, and Central and South America.

Status

Determining the statuses of bat populations is difficult because of their nocturnal behavior and habit of roosting in places that are hard to census. The HAWAIIAN HOARY BAT is a case in point. Long believed to be rare and possibly threatened (in fact, the species is still USA ESA listed as endangered), recent research using sophisticated equipment that detects the bat's ultrasonic echolocating clicks indicates healthy populations; in fact, bats were detected at a range of elevations almost everywhere in Hawaii researchers looked for them. Their nocturnal behavior and usually secretive, solitary existence had led most people, even experts, to believe the bats were rare. Also boding well for the bats is that they apparently thrive in various types of habitats, including human-altered ones such as agricultural areas. If the Hawaiian Hoary Bat has declined significantly during the past 100 years, the main cause would probably be habitat loss and the resulting reduced availability of roosting sites.

Because many bats worldwide roost in hollow trees, deforestation is obviously a primary threat. Further, many bat populations in temperate regions in Europe and the USA are known to be declining and under continued threat by a number of agricultural, forestry, and architectural practices (about five bat species or subspecies on the USA mainland are endangered, USA ESA listed). Traditional roost sites have been lost on large scales by mining and quarrying, by the destruction of old buildings, and by changing architectural styles that eliminate nooks and crannies such as building overhangs and church belfries. Many forestry practices advocate the removal of hollow, dead trees, which frequently provide bats with roosting space. Additionally, farm pesticides are ingested by insects, which are then eaten by bats, leading to death or reduced reproductive success.

Profiles

Hawaiian Hoary Bat, *Lasiurus cinereus*, Plate 47a

2. Rodents, the Gnawing Undesirables

Ecotravellers discover among *rodents* an ecological paradox: although by far the most diverse and successful of the mammals, rodents are, often with a few obvious exceptions in any region, relatively inconspicuous and rarely encountered. The number of living rodent species globally approaches 2000, more than 40% of the approximately 4600 known mammalian species. Probably in every region of the world save Antarctica (where they do not occur), rodents – including the *mice*, *rats*, *squirrels*, *chipmunks*, *marmots*, *gophers*, *beaver*, and *porcupines* – are the most abundant land mammals. More individual rodents are estimated to be alive at any one time than individuals of all other types of mammals combined. Rodents' near-invisibility to people derives from the facts that most rodents are very small, most are secretive or nocturnal, and many live out their lives in subterranean burrows. That most rodents are rarely encountered, of course, many people do not consider much of a hardship.

Rodent ecological success is likely related to their efficient, specialized teeth and associated jaw muscles, and to their broad, nearly omnivorous diets. Rodents are characterized by having four large incisor teeth, one pair front-and-center in the upper jaw, one pair in the lower (other teeth, separated from the incisors, are located farther back in the mouth). With these strong, sharp, chisel-like front teeth, rodents "make their living": gnawing (*rodent* is from the Latin *rodere*, to gnaw), cutting, and slicing vegetation, fruit, and nuts, killing and eating small animals, digging burrows, and even, in the case of beaver, imitating lumberjacks.

Hawaii has four rodent species, all introduced by people. They are three species of rats, PACIFIC, BLACK, and NORWAY RATS (Plate 47), and the HOUSE MOUSE (Plate 47). All four species are members of the largest rodent family, Muridae, which encompasses about 65% of rodent species, including rats, mice, voles, lemmings, muskrat, gerbils, and mole rats, and has a nearly worldwide distribution. In this series of guidebooks I usually refrain from discussing and illustrating rats and mice on the grounds that they are relatively rarely seen, most people actually prefer not to interact with them, and there are other, more interesting rodents to describe. But here I profile rats and mice because they are the islands' only rodents, they are common, and they probably played a significant role in the decline of Hawaii's native birdlife.

Most of the world's rodents are small mouse-like or rat-like mammals that weigh less than a kilogram (2.2 lb); they range, however, from tiny pygmy mice that weigh only a few grams to South America's pig-like CAPYBARA, behemoths at up to 50 kg (110 lb). The three rats under our consideration here are small rodents, all usually weighing less than 280 g (10 oz), and one, the small Pacific Rat, usually less than 90 g (3 oz). HOUSE MICE, the smallest Hawaiian rodents, weigh between 10 and 20 g (four-tenths to eight-tenths of an ounce). Rats and mice are usually clad in various shades or gray and/or brown, but some are all black.

Natural History

Ecology and Behavior

PACIFIC and BLACK RATS (also called ROOF RATS because they like to climb and most often are not black) are found in similar types of vegetation at low (Pacific Rat) and low and middle (Black Rat) elevations, although both are also occasionally spotted in high elevation locations (such as at 2950 m, 9700 ft, on Maui and the Big Island). They frequent wooded areas of both wet and dry forest, wooded gulches, brushy areas, and fields such as sugarcane fields and abandoned pineapple plantations. The Pacific Rat particularly appreciates agricultural sites, and is considered a major farm pest. Black Rats are common rats of wooded areas, agricultural fields, but also houses, walls, and gardens. Unlike Black and NORWAY RATS, Pacific Rats rarely invade buildings or live in close association with people. Pacific and Black Rats have a varied diet, eating such things as insects, earthworms, snails, green plants, soft parts of sugarcane plants, fruit, contents of birds' nests (eggs and young), and adult birds. Norway Rats usually are found living only where there are people, mostly in lower elevation sites; they are not animals of forest, grassland, or open agricultural fields. They feed on human food, on garbage, on stored agricultural products, and on anything else edible they find.

The HOUSE MOUSE can be considered nearly ubiquitous on the main islands of Hawaii, from sea level to 3000 m (10,000 ft), and has sometimes been spotted at even higher elevations on volcanic mountainsides. It occurs in association with people wherever they live, but also in agricultural fields (sugarcane, pineapple) and a host of wild habitats such as beaches, grasslands, scrub areas, and forests. Populations are usually most dense, however, in low elevation drier areas. In and around buildings they eat almost anything edible; in the wild they munch insects and other small invertebrates, seeds, and fruit.

All Hawaiian rodents make for tasty meals for Hawaiian Hawk, owls, cats, and the Mongoose.

Ecological Interactions

Introduction of three species of rats had dire consequences for Hawaii's native birds. The main negative interaction has been direct predation by the rats on bird nests and on adult birds, most often on females incubating their eggs. Black Rats are often seen in trees, and are thought to have had enormous impacts on native forest birds by eating their eggs and nestlings. Wherever rats have been introduced to Pacific islands, dramatic reductions in the health of native forest bird populations have been recorded. For example, the elimination of at least two species from Midway Island, the LAYSAN RAIL and LAYSAN FINCH, are traceable to rat predation by introduced Black and Pacific Rats. Pacific Rats apparently prey with some frequency on adult ground and burrow-nesting seabirds such as albatrosses, petrels, and shearwaters; they may be at least in part responsible for low population numbers of HAWAIIAN PETRELS (Plate 22) at Maui's Haleakala National Park and for the extinction of the HAWAIIAN RAIL in the late 1800s. Black Rats, in particular, are also environmentally destructive in other ways: they strip bark from native trees, inhibiting their growth; they damage flowers and fruit; they eat green plant materials and seeds; they may even compete with birds for some fruits birds use as food. Furthermore, rats and mice exacerbate other kinds of damage to native ecosystems by transporting the seeds of alien plants, facilitating the spread of these aliens.

Rodents also have their beneficial aspects: burrowing is an aspect of their behavior that has significant ecological implications because of the sheer numbers of individuals that participate. When so many animals move soil around (rats and mice, especially), the effect is that over several years the entire topsoil of an area is turned, keeping soil loose and aerated, and therefore more suitable for plant growth.

Breeding

Rats and mice, of course, are prolific breeders. The three Hawaiian rat species breed essentially throughout the year (Black Rats may not breed in mid-winter). They all construct underground burrows, where females give birth to up to eight or more young at a time after pregnancies of about 3 weeks. Females begin breeding when they are 3 to 6 months old. Likewise, House Mice may breed all year, females giving birth to three to five or more young after a pregnancy of about 3 weeks; females can breed before they are teenagers – actually at 6 weeks of age.

Notes

The PACIFIC RAT, which may have originated in southeast Asia, was probably brought to the Hawaiian and other Pacific islands long ago by early Polynesian colonizers from the central Pacific, as much as 1500 years ago. The introduction may or may not have been intentional (there is some indication these rats may have been hunted for food or sport, which, if true, suggests the rats may have been transported from island to island intentionally). NORWAY RATS, which may have first evolved in the northern China region, were distributed essentially worldwide by the ships of Western commerce; they reached Hawaii during the 1800s on ships from either Europe or the Americas. The BLACK RAT, with a probable origin in the Malaysian region and a natural range in Asia, was also spread worldwide by ocean-crossing ships; they now occur in probably all temperate and tropical land areas; they reached Hawaii during the late 1800s. Both Norway and Black Rats live in close association with people, and are known globally as threats to human health because they often carry and spread diseases, such as bubonic plague and typhus. The HOUSE MOUSE, native to western Europe and the Mediterranean region, and now spread throughout the world's temperate and tropical regions, reached Hawaii during the late 1700s or early 1800s.

Through the animals' constant gnawing, rodents' chisel-like incisors wear down rapidly. Fortunately for the rodents, their incisors, owing to some ingenious anatomy and physiology, continue to grow throughout their lives, unlike those of most other mammals.

Status

PACIFIC and BLACK RATS cause millions of dollars of losses annually to the Hawaiian sugarcane industry and to other sectors of the agricultural economy. Efforts are made at rat management, both with traps and poisons (rodenticides), but rats are notoriously difficult to control. In areas with remaining good populations of native forest birds, especially Maui's Haleakala National Park and the Big Island's Hawaii Volcanoes National Park, efforts are underway to study the impacts of rats on bird populations and to try to reduce those negative impacts through fencing, trapping, and poisoning.

Profiles

Norway Rat, *Rattus norvegicus*, Plate 47b
Black Rat, *Rattus rattus*, Plate 47c
Pacific Rat, *Rattus exulans*, Plate 47d
House Mouse, *Mus musculus*, Plate 47e

3. Mongooses

As you drive around Hawaii, the small brown mammal that often darts across roads and parking lots and that resembles nothing so much as a cross between a weasel and a squirrel, is a *mongoose*, the SMALL INDIAN MONGOOSE (Plate 47), to be precise. How this animal got to Hawaii, what it was supposed to do when it arrived, and what it actually did, is a cautionary ecological tale; but more on that later. Mongooses (no, you may *not* use the cuter word, mongeese, as the plural) are members of the Family Viverridae – the approximately 70 species of civets, genets, and mongooses, an Old World mammal group with no close relatives in the Americas. They are the only *viverrids* in Hawaii and, in fact, aside from cats and dogs, the only local representative of Order Carnivora, which includes the weasel, raccoon, and bear families, as well as the cat and dog families. Viverrids such as the Small Indian Mongoose (known simply as MONGOOSE in Hawaii) are small, usually short-legged, long-tailed mammals with small, rounded ears. Various species range in weight from 0.5 to 4 kg (1 to 9 lb); the Hawaiian mongoose weighs from 0.5 to 1.3 kg (1.1 to 3 lb), with females being concentrated toward the lower end of the range. The Small Indian Mongoose is native to India and other parts of central Asia, but it has been transported by people to other regions, such as some Caribbean islands; in Hawaii, it occurs on all main islands save for Kauai and Lanai.

Natural History

Ecology and Behavior

The Hawaiian mongoose occurs in many habitat types at low, middle and some high elevations, to about treeline. But it prefers, and is most densely distributed in, lowland areas, generally below 600 m (2000 ft), on the islands' windward (and so wetter) coasts. They are common beach and plantation denizens. They shelter and sleep in crevices in the ground or among rocks, and they move easily over the ground and in trees. Adults are apparently fairly solitary animals, coming together only to breed. Home ranges, the area over which an individual lives and seeks food, are small, perhaps about 1.5 km (1 mile) in diameter for males, and about half that for females; home ranges of individuals overlap; that is, these mammals do not defend exclusive territories from which all other individuals of the species are actively excluded. The Mongoose is considered *omnivorous*, eating fruit such as berries, carrion, and eggs, and attacking and eating insects such as beetles and cockroaches, land snails, frogs, lizards, marine animals such as fish and crabs caught in tidepools, sea turtle hatchlings, birds, and rodents. They are known for their predatory aggressiveness, even sometimes attacking mammals much larger than themselves.

Ecological Interactions

The Hawaiian mongoose is a voracious predator on birds, at all stages of birds' life histories; they take both eggs and nestlings from nests, they are fearsome stalkers of fledglings – birds just out of the nest but not yet efficient fliers, and they routinely catch and eat adult female birds as they sit on their nests to incubate eggs or brood young. Mongooses take many birds that are non-native and plentiful, such as domestic poultry, Red Junglefowl, pheasants, and quail, but they also take many ground-nesting, colonial seabirds and some ground and tree-nesting native land birds, and have contributed significantly to the drastic declines in populations of some of these. Most well-known are the probable effects mongooses had on reducing NENE (p. 64) and HAWAIIAN CROW (p. 62) populations. Nene breeding in higher-elevation subalpine areas may suffer relatively little from mongooses because these predators are scarce at such altitudes, but at lower elevations, where Nene could be transplanted to establish new populations, mongooses are the prime threat. Indeed, one study showed that unsuccessful Nene nests were caused 62% of the time by egg predation and 10% of the time by death of the incubating female, and that mongooses were responsible for essentially all these losses. Hawaiian Crows, which exist in the wild now only over a single small portion of the Big Island, are particularly vulnerable to mongooses just after they leave the nest; fledglings are poor fliers and spend several days on the ground or in low vegetation, being fed by their parents. To control mongooses near the Hawaiian Crows and at other sites, these mammals are trapped and also killed with poison baits.

Mongooses also prey on and continue to threaten endangered species such as Hawaiian Coot, Hawaiian Duck, Hawaiian Stilt, Dark-rumped Petrel, and Newell's Shearwater. It is now thought that mongooses had relatively little effect on reducing populations of small native forest songbirds.

Breeding

Mongooses give birth in dens in crevices in the ground and in rocky areas. Usually two or three pups (up to five) are born after a pregnancy of 42 to 50 days. Females may produce two litters per year, generally between February and August.

Notes

The Hawaiian mongoose had its origins in the Middle East, India, and central Asia. They were brought to the New World, particularly to islands in the Caribbean and elsewhere, because plantation owners believed they could be effective predators on, and so control, venomous snakes that killed farm workers and rodents that ate and damaged crops. Mongooses, about 70 of them from Jamaica, were first brought to the Big Island by sugarcane growers in 1883 and released; others were eventually released on Maui, Molokai, and Oahu. Although the mongooses prospered and multiplied, they never lived up to their rodent-controlling reputations. They do eat rats and mice, particularly in agricultural areas, but mongooses are chiefly day-active mammals. The rats they were meant to control, however, are mainly nocturnal or active both day and night, thus reducing the effect of mongoose predation on the rodent populations. The mongooses found other foods in daytime, such as native Hawaiian birds, the use of which caused, and is still causing, great ecological harm. These mongooses have also been implicated for causing the extinction of, or severely reducing the populations of, many species of Caribbean birds, lizards, and snakes.

Status

Many species of mongooses are very common animals within their native Old World ranges, as they are on Caribbean and Hawaiian islands where they have been introduced. About six species of Indian mongooses, including the SMALL INDIAN MONGOOSE, are regulated in India for conservation purposes, CITES Appendix III listed.

Profiles

Mongoose, *Herpestes auropunctatus*, Plate 47f

4. Ungulates, the Hoofed Mammals

Ungulate is a broad, vaguely scientific term for hoofed mammals. Many of these species are very familiar to us as four-legged farmyard inhabitants, providers of dairy and meat products, and beasts of burden. Aside from hoofed feet, they have in common a *cursorial* mode of locomotion (that is, they run along the ground) and herbivorous eating habits. There are two orders of ungulates, both of which have introduced representatives in Hawaii. Order Artiodactyla is a globally distributed group of mammals that have an even number of toes on each foot. Artiodactyls include antelope, bison, buffalo, camels, cattle, deer, gazelles, giraffes, goats, hippos, pigs, and sheep. In general, the group is specialized to feed on leaves, grass, and fallen fruit. Order Perissodactyla, which refers to the fact that all of its members have an odd number of toes on each foot, contains the horses, rhinoceroses, and tapirs (the last being a small family of largish, stocky animals with short legs and a long snout somewhat reminiscent of a horse's), which also eat grasses and other plants.

Although Hawaii has a definite dearth of hippos and rhinos, it does support a variety of feral ungulates: pigs, goats, sheep, deer, and donkeys. They were brought to the islands by people as food animals or later, for sport; or, in the case of the donkey, as a pack animal. The main things to know about them now, and the reason to include them here, are that these feral mammals (1) cause severe environmental damage and are responsible for many ecological changes in the islands (see below); and (2) could probably be eliminated from Hawaii, thus significantly stemming further damage, but are permitted to stay because local people enjoy hunting some of them.

FERAL PIGS and FERAL GOATS (Plate 48) occur on all main islands except Lanai; AXIS DEER (Plate 48) occur on Maui, Lanai, and Molokai; and the FERAL DONKEY (Plate 48) occurs in isolated parts of the Big Island. The pig is a member of Family Suidae, an Old World group of nine species, all of which are robust, large-headed animals that pretty much resemble pigs. The goat, along with African and Asian antelopes, bison, sheep, and cattle, is part of Family Bovidae, the hollow-horned ungulates, which contains about 125 species worldwide. And the Axis Deer is one of about 35 deer species, Family Cervidae, with representatives throughout the New World, Eurasia and northwestern Africa. (Another species, the BLACK-TAILED DEER, occurs in small numbers in remote sections of northwestern Kauai.) Feral Donkeys, restricted to small portions of the Big Island, are members of the horse family, Equidae, which has nine species in Africa and Asia, including zebras.

Also, two species of SHEEP occur in Hawaii, feral domestic sheep, at higher elevations in the Mauna Kea region of the Big Island, and some MOUFLON perhaps there and on Lanai; hybrids between these two also occur. The domestic sheep are descendants of animals ranched on the islands during the late 1800s and early 1900s. At one time there were thousands of these feral sheep roaming the central, saddle, portion of the Big Island. They did considerable damage to native vegetation, grazing native grasses down to stubble and browsing on shrubs and young trees; they damage trees by stripping bark, particularly from native Māmane (Plate 14). The Big Island sheep population was reduced by hunting during the 1950s, and was further reduced in the 1980s and 1990s after conservation groups undertook successful legal action against the state, showing that the sheep significantly harmed habitats necessary for the continued existence of native birds in the Mauna Kea area, particularly the PALILA (Plate 44; Close-up, p. 158). A few sheep may still roam the area, but conservationists are working to eliminate them entirely; few ecotravellers see them. The Mouflon is a wild mountain sheep native to the Mediterranean region that was introduced to several of the Hawaiian Islands during the 1950s for sport-hunting (males have large, backwards-curling horns that render them impressive trophies). Also ecologically damaging, these sheep have been largely eliminated from most islands; some remain on Lanai, and a few may roam the high reaches of the Big Island's volcanic peaks.

Natural History

Ecology and Behavior

FERAL PIGS in Hawaii occupy various habitats at a range of elevations, but are most often found at middle elevations in forests, including wet forests, and open, mountain pasture areas. Day- and night-active, they travel as solitary animals or in small family groups (usually females with piglets), moving along paths and trails, foraging for vegetation such as grasses and ferns, fruit, and animals such as snails, slugs, and especially, earthworms. They dig into the ground with their snouts, *rooting* for vegetation and worms, often doing substantial damage to habitats (see below). Home ranges, the area within which individuals live and roam in search of food, generally are less than 10 sq km (4 sq miles). FERAL GOATS, likewise, occur in a variety of habitats, but prefer drier, middle and higher-elevation sites, particularly rocky slopes and open lava flow and range areas; but they also are seen in forests, including rainforests. They eat grasses and other vegetation, particularly native plants, and apparently can survive without drinking much water; the water they need is extracted metabolically from their food. These goats are most often seen in small family groups, which apparently roam over fairly limited home ranges. AXIS DEER also occur at low to high elevations, probably being most common in wet forests and drier woodlands. They are seen singly or in small groups, and give peculiar barking sounds when alarmed. Like the pigs and goats, they do considerable damage to native vegetation. Pigs, goats, and deer are preyed on by wild dogs and killed by human hunters. FERAL DONKEYS are occasionally seen in groups on the western, or Kona, side of the Big Island.

Ecological Interactions

These feral ungulates – primarily pigs and goats – are responsible for a good deal of ecological damage in Hawaii, directly killing or damaging native plants with their feeding, spreading non-native plants by transporting seeds on their coats and in their droppings, and indirectly causing declines and perhaps extinctions of native animals – particularly birds. Feral pigs, after people, are currently considered the greatest despoilers of Hawaiian forests. They eat vegetation, sometimes selectively munching native rare plants (not because they are evil, or out of spite, but because, presumably, they prefer the taste) and cause further damage when they trample vegetation and root about in the forest floor for worms, roots, and to topple plants. For instance, they root up and eat native treeferns. In fact, it is estimated that within areas pigs occupy, they each year succeed in rooting about in, and so turning over, about half the area of topsoil.

This constant aeration of the soil, combined with the pigs' destruction of the plant cover over the forest floor, creates much open habitat with fertile soils – conditions ripe for colonization by alien plants.

The significant but indirect effect of pigs on native forest birds demonstrates the inter-connectedness of ecosystems: the pigs create ideal mosquito breeding habitat when rainwater collects in the small depressions in the forest floor caused by their rooting behavior and also in treeferns they knock down. The mosquitos carry avian malaria and other avian diseases that have eliminated several species of endemic small forest birds from all but high-elevation sites, where the coolness inhibits the mosquito life cycle (see p. 58). Pigs are now controlled in certain sites, for instance in Haleakala and Volcanoes National Parks, by fencing large areas and then hunting out all pigs contained within the enclosures. By 1993, for example, about 65 km (40 miles) of anti-pig fence at Volcanoes National Park prevented pig access to about a third of the park's area.

Goats also do great damage to native vegetation, as do deer; in fact, on Lanai, the Axis Deer are now considered the most serious threat to the small amount of remaining native forest.

Breeding

Feral Pig females, breeding at any time of year, produce usually four to eight young at a time after pregnancies of 3.5 to 4 months; they can breed twice a year. The female gives birth in a nest in a crevice or one she digs in the ground; the nest is often lined with vegetation. Young are weaned at 3 to 4 months of age. Females begin breeding at 12 to 18 months old. Female goats give birth usually to one or two young after 5-month pregnancies. The precocial young (they walk well and can run a bit 24 hours post-birth) are weaned at 4 to 5 months of age. Females begin breeding when about a year old or slightly younger. Axis Deer may be born at any time of year, females giving birth to a single fawn after pregnancies of about 8 months; young are weaned at about 3 months. Females are sexually mature at about 16 months.

Notes

Polynesians introduced pigs to Hawaii. They were apparently an important source of food in the ancient Hawaiian civilization, had religious and ceremonial significance, and probably caused relatively little ecological damage because they were not free to penetrate into forests. FERAL PIGS today have the smaller Polynesian pigs as partial ancestors, but the Polynesian pigs interbred with the larger domestic pigs brought to the islands by Europeans, and it is the larger European strains that today's Feral Pigs resemble. GOATS, with origins in the Mediterranean region, were brought to Hawaii by Europeans during the late 1700s. AXIS DEER, native to India, were brought to Hawaii as a diplomatic gift in 1868.

Status

All of the species considered here are common animals. Some of their close relatives, however, are threatened; there are several species of deer, mostly in Asia, as well as at least a single goat and a single pig species, that are currently endangered.

Profiles

Feral Pig, *Sus scrofa*, Plate 48b
Feral Goat, *Capra hircus*, Plate 48c
Axis Deer, *Axis axis*, Plate 48d
Feral Donkey, *Equus asinus*, Plate 48e

5. Seals

Seals are familiar marine mammals with front and back legs modified as flippers, and noses adapted to balance large beach balls. The seal order, Pinnipedia, is divided into three families, the *true* or *earless seals* (Family Phocidae), the *walrus* (Family Odobenidae), and the *eared seals* (Family Otariidae). *Pinnipeds* probably evolved from terrestrial mammals such as the weasels and bears (Order Carnivora). Most seal species occur in the Arctic and Antarctic regions, but one of the earless seals is a native Hawaiian: the HAWAIIAN MONK SEAL (Plate 49). Earless seals, such as the monk seal, have ears that are simply holes on each side of the head, and front legs that are held to their sides as their hind flippers propel them through the water (whereas eared seals, such as *fur seals* and *sea lions*, have small external ears and powerful front flippers that help with propulsion during swimming). The Hawaiian Monk Seal occurs only in the Hawaiian islands (*endemic*) and although it breeds and spends most of its on-shore time on tiny islands of the Northwestern chain, it is occasionally seen hauled out on secluded sandy beaches on some of the main islands. The monk seal is gray or brownish, and about 2 m (6.5 ft) long; adults weigh 180 to 270 kg (400 to 600 lb). Females are usually a bit bigger than males.

Natural History

Ecology and Behavior

Monk seals eat animals they find on or near the sea floor – bottom-dwelling fish, eels, lobsters, squid, octopus. Seals are usually great divers, often better at it, going down deeper and for longer periods, than many of the whales and dolphins. Although most of the monk seals' dives are in shallow lagoons around the islands they inhabit, they sometimes dive to depths of 500+ m (1640+ ft). They can stay under for nearly 20 minutes, but most dives last 5 to 10 minutes. Often these seals forage at night and spend the hotter parts of the day being relatively inactive, perhaps lying on a beach under cover of vegetation. Many seal species are quite social, often staying in groups. But monk seals are usually seen alone, although small groups gather at times at preferred haul-out beaches. Long-term studies of marked individuals show that about 90% of the monks spend their whole lives on and around the islands of their birth, with about 10% taking trips, exploring other sites and eventually changing islands. The main predator on adult monk seals appears to be the Tiger Shark, but many also die when they are entangled in plastic debris in the ocean, such as pieces of fishing nets. Some monk seals live 25+ years in the wild.

Breeding

During spring and summer, male monk seals cruise past seal beaches, seeking females ready for mating. Sometimes males fight over females. Mating occurs in the water. Young are born from February through July, but usually in April or May, the female giving birth to a single pup far up on a sandy beach, away from the water. Pups, initially glossy black but turning to gray, are nursed for 5 to 6 weeks, growing during this period from about 15 kg (33 lb) at birth to about 50 to 75 kg (110 to 165 lb). The mothers teach the young to swim. Then the mother, which does not eat during the nursing/training period, departs to feed, and the young seal is on its own. Monk seals attain full size at about 4 years old and begin to breed at 5 to 10 years. Females tend to breed every other year.

Notes

HAWAIIAN MONK SEALS were probably abundant in the Hawaii region when Europeans arrived in the late 1700s. Commercial hunting nearly wiped them out and, in fact, their were believed to be extinct by 1825. Some survived, however, and the population rebounded somewhat, until disturbances caused by World War II and military occupation of the seal's breeding islands during the 1940s and 1950s led to breeding failure and steep population declines. Since the late 1970s, the USA National Marine Fisheries Service (Marine Mammal Research Program) and USA Fish and Wildlife Service have, to varying degrees, monitored populations and breeding of the monk seals, with the objective of gathering information on behavior, ecology, natural history, and population sizes so that species recovery plans can be instituted.

Most of the seal breeding occurs at five sites in the Northwestern Hawaiian Islands: French Frigate Shoals, Laysan and Lisianski Islands, Pearl and Hermes Reef, and Kure Atoll (smaller groups breed at and inhabit three smaller islets). The long-term study of the seals continues. Each spring researchers begin 5-month-long field research seasons, living on these tiny, remote, sand islands (such as on French Frigate Shoals, an atoll with several small islets and sandbars on which seals can haul-out and give birth), counting breeding seals without disturbing them (much), and marking newborns with tags so they can be followed for years, their ecology and behavior studied in detail. Also, in an effort to learn more about foraging behavior, researchers glue satellite tracking devices to the backs of some of the seals. Using this method, they have gathered extensive information on what they eat at sea and where. In fact, with additional funds provided by the National Geographic Society, they recently attached a video camera to the back of a seal (a seal-cam?) so they could actually watch the kind of fish the seals hunted and where they did so (the seals often chased quite small fish on the deep ocean bottom, which was surprising; large fish had been thought to be a main diet item). This is a challenging way to study seal behavior because the animal must be caught once to attach the camera, and then again whenever the videotape needs to be removed or replaced (and don't think the seals are any too cooperative!).

The research teams also promote seal population recovery by:

(1) rehabilitating underweight and sick pups (they are taken to a center in Honolulu, fed, treated, and released);
(2) actively removing dangerous trash and debris from breeding beaches; divers even remove tangled fishing nets and lines from reefs and rocks – debris that may cut seals or entangle them, killing them;

(3) keeping people – boaters, beach-combers, etc. – away from breeding beaches; just the presence of the federal scientists helps keep some people away;
(4) on a small scale, translocating some seals, particularly to alleviate "skewed" sex ratios in some of the populations; for example, if there are 30 males and two females in one area, capturing some of the males and taking them elsewhere makes sense (especially because so many males with so few females can lead to the females being hurt during repeated matings).

Status

All three species of monk seals, genus *Monachus*, are listed as endangered, CITES Appendix I and USA ESA, with one of them, the CARIBBEAN MONK SEAL, not seen since 1952 and now presumed extinct. All three were hunted to extinction or near-extinction during the 18th and 19th Centuries by commercial seal-hunting operators primarily for their furry skins (pelts), but also for oil and food. The MEDITERRANEAN MONK SEAL, which ranges through the Mediterranean Sea, the Black Sea, and off the northwestern coast of Africa, now probably numbers no more than a thousand individuals. Likewise, the HAWAIIAN MONK SEAL over the last few years has numbered probably between one and two thousand; both the Hawaiian and Mediterranean Monk Seals can be considered critically endangered. A few monk seals bred recently on secluded beaches on Oahu, Kauai, Maui, and Molokai, but because of all the human disturbance on the main islands, researchers do not think these will develop into flourishing populations.

Profiles

Hawaiian Monk Seal, *Monachus schauinslandi*, Plate 49a

6. Whales and Dolphins

The approximately 75 species of *dolphins*, *porpoises*, and *whales* belong to the Order Cetacea, and almost all of them are found only in sea water in the world's oceans (but a few of the smaller dolphins inhabit larger rivers and estuaries in Asia, Africa, and South America). *Cetaceans* never leave the water and generally come to the surface only to breathe. Their hind legs have been lost through evolution (they evolved from terrestrial mammals with legs) and their front legs modified into paddle-like flippers. Their tails have become broad and flattened into paddles called *flukes*. A single or double nostril, called a *blowhole*, is on top of the head. Up to a third of an individual's weight consists of a thick layer of fat (*blubber*) lying under the hairless skin. Although cetacean eyes are relatively small, hearing is well developed.

Whether a given cetacean species is called a whale or a dolphin has to do with length: whales generally are at least 4.5 to 6 m (15 to 20 ft) long, while dolphins and porpoises are smaller. The differences between dolphins and porpoises? Dolphins have a beak-type nose and mouth, a backwards-curving dorsal fin, and sharp, pointed teeth; porpoises are more blunt-nosed with a triangular dorsal fin and blunt teeth.

Cetaceans are often divided into two broad categories. One is a group of large whale species that have mouths that look like immense car radiator grills, filled with long, vertical, brownish strands of *baleen*, or *whalebone*. What is probably the largest animal that ever lived, the BLUE WHALE, is a baleen whale. Blue Whales, fairly regular in Hawaiian waters but rarely seen, grow to 30+ m (100 ft) in length and 160 tons in weight. The other group, known as the *toothed whales*, includes a few whales and all the porpoises and dolphins; they have mouths with teeth and not baleen.

The three most commonly seen dolphins (family Delphinidae) in Hawaiian waters are the SPINNER, SPOTTED, and BOTTLE-NOSED DOLPHINS (Plate 49), all members of groups that are distributed throughout the world's tropical and warm-water seas. These dolphins are cigar-shaped, long and thin, and tapered at the ends. They have smooth, hairless skin and prominent beaks and dorsal fins, and range in length from about 2 m (6.5 ft; Spinner) to almost 4 m (13 ft; Bottled-nosed). (The largest member of the dolphin family is the KILLER WHALE, *Orcinus orca*; occasionally seen around Hawaii, males grow to 9.5 m (31 ft) long, and weigh as much as 8 tons.)

Whales profiled here range in length from giant SPERM WHALES (to 20 m, 65 ft) and HUMPBACK WHALES (to 15 m, 50 ft) to relatively small PYGMY SPERM WHALES (to 3.7 m, 12 ft). Around Hawaii, the baleen whales are represented most frequently by the Humpback Whale, and this is also the top species on most whale-watchers' viewing wish-lists. The front flippers (or *pectoral fins*) of the Humpback are huge (as long as a third of the body length), white, and wing-like. (In fact, the Humpback's genus, *Megaptera*, means "large wing," referring to these flippers.) The dorsal fin is small and shark-like. It is placed two-thirds of the way back on the body and mounted on a fleshy pedestal, a trait that distinguishes it from all other baleen whales. When a Humpback initiates a deep dive (*sounds*), its large scalloped flukes (the entire tail can be 4.5 m, 15 ft, across) come well up off the water's surface to expose the underside of the flukes, which are mottled white and black (a character so variable and personalized that scientists use it to recognize and track individuals reliably).

Natural History

Ecology and Behavior

Dolphins. Because BOTTLE-NOSED DOLPHINS were the first to be kept in captivity for long periods (they are the species often seen in aquarium shows and achieved fame in the entertainment industry under their stage name, "Flipper"), and because they are often found close to shore and so are easily observed, more is known of their biology than of other species. SPOTTED DOLPHINS apparently lead lives that are very similar to those of Bottle-nosed Dolphins. (SPINNER DOLPHINS are discussed in a Close-up on p. 184). Although sometimes found as solitary animals, these dolphins usually stay in groups, sometimes of up to a thousand or more. Large groups apparently consist of many smaller groups of about two to six individuals, which usually are quite stable in membership for several years. There are dominance hierarchies within groups, the largest male usually being top dolphin. Large schools are believed to aid the dolphins in searching for and catching food, and to decrease the likelihood of the dolphins themselves becoming food for such enemies as large sharks. They eat primarily fish and squid, which they catch by making shallow dives into the water. They are fast swimmers, routinely jumping clear of the water when feeding or travelling. Dolphins use sounds as well as visual displays and touching to signal each other underwater; they also use high frequency sound (mostly clicking and popping sounds) for *echolocation*, like bats, for underwater navigation and to locate prey. Dolphins are considered highly intelligent and sometimes develop close affinities with people. Bottle-nosed Dolphins mature at about 6 years of age and live to be 25 years old. They feed at depths of up to 600 m (2000 ft) and eat a wide variety of seafood, from bottom-dwelling fish, eels, small sharks, and crabs, to tuna.

Whales. HUMPBACK WHALES migrate to feed in polar waters and return to equatorial waters to breed. Calves have no blubber so must remain in warmer waters until they have fed sufficiently to put on a layer of fatty insulation for the cold polar waters. Humpbacks are usually found in family groups of three or four individuals. Greatly paradoxical is that the baleen whales, behemoths so large they can only be measured in tons and tens of meters, feed mainly on planktonic crustaceans, small shrimp-like animals barely 5 to 10 cm (2 to 4 in) long. They swim through food-rich layers of water, especially in polar regions, with their mouths wide open. Then they close their mouths and use their immense tongues (some weighing 4 tons) to push water out through their 300+ baleen plates, straining the small shrimp, which stay in the mouth and are swallowed in a monumental gulp. Humpbacks also feed on fish and squid. However, in their Hawaiian winter breeding grounds, Humpbacks may not feed at all, surviving instead on energy stored during their summer feeding in polar regions. Consequently, a spectacular fishing behavior developed by Humpbacks is not seen in Hawaii: a Humpback locates a fish school, then swims under and around the fish, all the while releasing a stream of air bubbles. The stream of bubbles serves as a *bubble net* that frightens and concentrates the fish and forces them to rise as the bubbles rise. The whale then dives and turns to swim vertically up through the bubble net, its wide-open mouth closing only as it reaches the surface, entrapping then swallowing large numbers of fish.

Humpbacks also produce some of the most complex and fascinating songs of any animal. Each geographic group has its own song, or *dialect*, that all the individuals there copy and use, but the songs change from year to year. When given in the right layer of water – the appropriate depth and temperature – these songs can travel hundreds of kilometers. Because they apparently can communicate over very long distances, Humpbacks' social interactions – including mate attraction, group behavior, and territorial behavior – may be quite complex and difficult for us to understand. Humpbacks also frequently jump completely out of the water (*breaching*), usually in an arching back flip. This behavior may be associated with mate attraction and courtship or it may be to knock off parasitic barnacles that grow on the whales' skin.

Sightings of SPERM WHALES, historically rare in the area, are increasingly reported by whale-watchers, so these giants may be staging a comeback in Hawaiian waters. They feed on fish, lobsters, and sharks, but mainly on squid. An adult can eat up to a ton of squid a day. These whales regularly dive to depths of 1000 m (3300 ft) and individuals have been recorded descending to 2250 m (7400 ft). They use a sonar system at these depths to locate prey, and can stay down for 45 minutes and descend at a speed up to 8 kph (5 mph). Many individuals have scars produced by suckers on the tentacles of giant squid. By measuring the diameter of the sucker marks and calculating the body length from that information, researchers have determined that some of these squid may be up to 45 m (150 ft) long – making you wonder if it's the whale or the squid who is the aggressive predator down there.

Breeding

Dolphins. Dolphins usually produce a single young after pregnancies of about 12 months. When born, dolphins are about a meter (3 ft) long. The mating systems of dolphins in the wild are not well known, for the obvious reason that it is difficult to observe underwater courtship and mating behaviors; also complicating observation is that males and females look much alike. BOTTLE-NOSED DOLPHINS mate near the surface, and their courtship involves elaborate stroking, nuzzling, and posturing. Pregnancy lasts 12 months; birth is often attended by several female "midwives," which help nudge the newborn to the surface for its first breath. The calf accompanies its mother for about 2 years. SPINNER DOLPHIN pregnancies last about 11 months. When born they average 77 cm (30 in) in length; young Spinners nurse from their mothers for 15 to 18 months.

Whales. HUMPBACK WHALES reach sexual maturity at 9 to 10 years. Courtship is in shallow waters near the equator and involves a lot of splashing, churning and breaching. Pregnancy is about a year and calves nurse for an additional year. Courtship in SHORTFIN PILOT WHALES begins with loud head-butting underwater and then proceeds quickly to mating. Pregnancy is 15 months, and the calf may be in care of its mother for up to 2 years. Male and female SPERM WHALES lead very different lives. During June to September males leave tropical waters and live at the edge of the pack ice in polar regions, often in large bachelor groups. Females and young remain in warmer waters, including off Hawaii and New Zealand. In November to February all congregate in tropical waters, where the biggest males (usually those more than 25 years old) fight violently to gather harems of 20+ females. Males can be very aggressive toward ships and people during this period, if provoked. Of a male's harem of 20 or so females, only a few are ready to mate in any one year. Pregnancy lasts 14 to 16 months and mothers nurse young for a year or two after birth. A single female may bear young only once every 4 years. Several females serve as midwives during birth, helping to push the calf to the surface right after birth. Nursing females and their calves form groups called *nursery schools*, the adults of which have been seen swimming in circles to protect calves in the center of the circle from marauding KILLER WHALES.

Notes

Dolphins' intelligence and friendliness toward people have inspired artists and authors for thousands of years. Images of dolphins appear frequently on artworks and coins from at least 3500 years ago, and from both ancient Greece and Rome. Aristotle, 2300 years ago, noted that dolphins were mammals, not fish, and remarked on their intelligence and gentle personalities. Many other ancient writings tell stories of close relationships between people and dolphins. These animals are considered the only group, aside from humans, that regularly assists members of other species that are in distress. There have been many reports of dolphins supporting on the water's surface injured members of their own and other dolphin species, as well as helping people in the same way.

BOTTLE-NOSED DOLPHINS, among other claims to fame, were among the first species to be studied using photographs that allowed biologists to track and study individual animals. In this case, close photos of dorsal fins permitted researchers to identify individuals and follow their activities for extended periods. This method of identifying individuals by photographs is now widely used to study the long-term behavior and movements of such other animals as whales, elephants, and lions.

Many dolphins will approach moving sea vessels to "ride" the bow pressure wave (the wave produced at the front of the boat as it slices through the water).

They sometimes persist in "hitchhiking" in this way for 20 minutes or more, jostling and competing with each other for the best spots – where, owing to the water's motion, they need exert little energy to swim; they are essentially taking a "free ride."

Status

All marine dolphins are CITES Appendix II listed as species not currently threatened but certainly vulnerable if protective measures are not taken. SPOTTED DOLPHINS and COMMON DOLPHINS are among the dolphins most frequently caught accidentally in the nets of tuna fisherman, and hundreds of thousands have been killed in this way. Dolphins in some regions of the world are also sometimes killed by fishermen who consider them to be competitors for valuable fish, or to be used as bait – for instance, for crab fishing.

Many whale species were hunted almost to extinction during a 200-year period that ended in the early 1960s (when international controls and sanctions were placed on commercial whaling). For instance, whales, particularly SPERM WHALES, were heavily hunted in Pacific waters; during the 1830s, the American whaling fleet alone numbered more than 700 ships. Whales were killed by the thousands for the thin, transparent oil that is stored in a reservoir in the forward part of their heads; the oil was used in lamps. Sperm Whales also produce *ambergris*, a terrible-smelling black residue found in the intestines, which was used in making expensive perfumes. As late as 1963, more than 30,000 Sperm Whales were killed in a single year. However, this species, now under international protection as endangered (CITES Appendix I and USA ESA listed), has been recovering (population estimates range up to a half-million). Likewise, the HUMPBACK WHALE (CITES Appendix I and USA ESA listed), with a worldwide population now of perhaps 10,000, seems to be doing well. BLUE WHALES (CITES Appendix I and USA ESA listed), however, have not yet recovered from commercial whaling and have disappeared over much of their former range; worldwide population may be less than a thousand. Only a few countries continue to hunt whales (Japan, Norway, Iceland), but they also pressure others to rescind international rules against whaling or otherwise circumvent the rules. Also, several groups of indigenous peoples around the world, including within the USA, are permitted each year to kill a small number of whales to continue their cultural traditions.

Profiles

Pacific Bottle-nosed Dolphin, *Tursiops truncatus gillii*, Plate 49b
Spinner Dolphin, *Stenella longirostris*, Plate 49c
Spotted Dolphin, *Stenella attenuata*, Plate 49d
Rough-toothed Dolphin, *Steno bredanensis*, Plate 49e
Humpback Whale, *Megaptera novaeangliae*, Plate 50a
False Killer Whale, *Pseudorca crassidens*, Plate 50b
Short-finned Pilot Whale, *Globicephala macrorhynchus*, Plate 50c
Sperm Whale, *Physeter macrocephalus*, Plate 50d
Pygmy Sperm Whale, *Kogia breviceps*, Plate 50e

Environmental Close-up 5
Spinners in Paradise: Sleepy Dolphins and Pesky People

by Bernd Würsig
Marine Mammal Research Program
Texas A & M University
Galveston, Texas

Spinner Dolphins (Plate 49) are sleek, slender gray dolphins that generally occur in deep waters of the world's tropical oceans. They get their name from their unusual aerial behavior: "... school members burst from the water and rotate rapidly about their longitudinal axis (that is, they spin!) for as many as about four revolutions before falling back into the water. On re-entry after these leaps, some part of the body is slapped against the water – the dorsal fin, the flukes, or the back. This produces a sharp, smacking sound that ... (may mark) the location of the spinning animals to schoolmates swimming nearby." Spinners often travel with other species, such as Spotted or Common Dolphins, and researchers surmise (but have no proof) that this habit of *multispecies association*, a technical mouthful, takes place so that one species can pay attention to another's communication signals. For example, in the open ocean of the Eastern Tropical Pacific (ETP; east of Hawaii), Spinner Dolphins tend to feed on fish and squid that travel toward the ocean's surface at night. Spinners work hard feeding (and socializing) at night and rest during the day. Spotted Dolphins, on the other hand, feed more often on near-surface prey during day-time and rest at night. So, the thought goes, in mixed groups of Spinners and Spotteds, one species is on the alert for predatory sharks while the other rests, and vice versa. If true, this would be a good example of a *mutualistic* relationship (p. 71), beneficial to both parties. Furthermore, again in the ETP, both species often associate with Yellowfin Tuna. The tuna may benefit by "hiding" below the dolphins, but the dolphins may benefit by using the tuna as an early warning system – the tuna probably first detect a shark coming from below and fan out away from the predator. This type of rapid dispersal, when first the tuna and then the dolphins reacted dramatically to danger from below, has actually been witnessed from airplanes.

Spinner Dolphins also occur around tropical islands and atolls, especially those formed volcanically that rise steeply from deep water (such as the Hawaiian Islands). In these locations, Spinners feed on mid-water-depth (*mesopelagic*) lanternfishes and squid at night, but they do not form multispecies associations. Instead, during the day, they rest close to shore or actually within atoll lagoons, probably to avoid deep water shark predation (and perhaps also to avoid the often turbulent open seas, which can give young calves especially a difficult time). With much leaping and spinning, dolphins slip into these near-shore areas in the morning, and quickly descend into what for them must pass for sleep: in a tight group or school, of a few dozen to a few hundred individuals, they synchronously and slowly surface and dive. When you approach them during this rest phase (in a boat or as a diver), you get the impression that they have almost "descended evolutionarily" into a more primitive, "fish-like" state. There is little social activity at these times, and the school changes shape in response to objects in the environment

– a buoy for example – as one integrated system; no one individual is very alert, but the combined alertness of neighbors to the side allow all to deviate course, the school to change shape around the object, and then to coalesce again with the obstruction behind them. They are using what might be called an "integrated sensory system," paying attention to each other; but no single individual appears to be dominant in the school or to be leading it. They are all, in effect, on "automatic pilot." This day-time rest-phase, alternating with bouts of higher activity when animals periodically "wake up," is probably of critical importance to the Spinners' health after a hard night feeding out at sea. This night–day/feed–rest pattern is the way it has been for tens of thousands of years, and the way it could continue for ages.

But there is trouble in paradise. Some Spinner Dolphin bays are becoming heavily used by humans, and some bays – now important harbors for people – formerly occupied for day-time dolphin rest, have been abandoned by the marine mammals. Not much can be done about these particular bays, short of shutting down the harbors – a decidedly impractical solution. But the relatively few busy harbors that no longer have dolphins entering deep into their bays are not the real problem. The more important and insidious danger, but one that can be reduced with knowledge and education, is the problem of too many tourists loving the dolphins too much.

The situation, one of competing usage, is easy to understand. Both dolphins and tourists wish to use Hawaii's picturesque bays to rest and relax after strenuous activities elsewhere. The beauty and grace of dolphins is continually displayed on television and film the world over, and so it is no surprise that people who see the exceptionally graceful Spinners want to be with them. So, the tourists (and some local residents) swim, snorkel, kayak, or boat out to the resting dolphin groups; and just about every time they do, they wake up at least a portion of the group, which then swing back and forth between rest and high levels of activity. The tourists see the forceful tailslaps, sure signs of disturbance, and say, "How cute, they are performing for us." Some people race through the groups by kayak or motorboat, or swim and dive forcefully after animals, in order to see them turn, leap, and dash ahead instead of proceeding in their normal, slow meander. You see, dolphins, like humans, are relatively uninteresting when asleep.

Now, imagine being asleep after a long day at the office or at school, and a kind-eyed person sits by your bed, refuses to leave, and repeatedly says, "I really like you." You wake up and then drift off again to sleep, only to hear the "I really like you" again and again. This persistence would not only be disturbing, it would probably make you either mad or perhaps – if the person will not go away – resigned to the interruptions. However, if it continues, it is likely to affect your alertness the next day, and potentially your health. This is, in my mind, what happens to dolphins when humans approach them at inappropriate times. I call it "loving them to death," but acknowledge that "death" might be too strong a term. "Loving them to distraction" may be closer to the truth.

There are bays in the Hawaiian Islands where the above scenario plays out every day that dolphins are present, except for brief respites when it is too windy and tourists stay off and out of the water. One of these bays is Kealakekua on the Kona (or western, leeward) coast of the Big Island. Several colleagues and I studied dolphins there in the 1970s and early 1980s, and we found that, on average, the bay was occupied by a Spinner school on 79% of days. Although there was boating activity – mainly fishing vessels – in Kealakekua Bay back then, there was very little tourism centered on the dolphins. Since the late 1980s, however, humans have interacted heavily with dolphins there (in part owing to people who claim to commune telepathically with the dolphins and who invite others, for a fee, to do likewise). The dolphins now utilize the bay on only 58% of days. This does not, of course, prove that increased human activity is the sole cause of the changed residency patterns of the Spinners, but it is highly likely to have contributed. Other factors could have changed the dolphins' behavior – changing food and predator patterns out at sea, for example. And, anyway, shifts in habitat use are not a definite indicator that dolphins have been disturbed. Instead, we might get at this by asking whether their natural behavior rhythms are changed by human presence; and, in the long-term, whether these changes harm them physiologically or reproductively. For Spinners, we know the answer to the first question: yes, humans affect them near shore, and quite often. A student of mine, Anna Forest-Barber, recently completed a study in the Big Island's Kealakekua Bay that shows that when swimmers, kayaks, or motorboats approach resting dolphins, especially if they do so "aggressively," with many starts, stops, and turns, they almost always disrupt at least a part of the school. Dolphins that were resting may begin leaping. Or they may dash ahead in an explosive rush of whitewater foam. We do not know yet if short-term behavioral changes such as these translate into long-term harm to individuals, or if they compromise dolphin breeding, decreasing their populations.

What is to be done? The USA has strict regulations about not molesting or purposefully changing behavior of marine mammals. These laws are part of the 1972 Marine Mammal Protection Act, and people who wilfully break the rules, even by approaching animals too closely with no intent to harm them, can be prosecuted, fined several thousand dollars, and even sent to jail. In Hawaii, there are some unlicensed organizations and individuals who wilfully break the rules, either because they believe it is their right to approach whales and dolphins in an unrestricted fashion, or because they want to make a tourism dollar, or both. As a researcher who also loves these dolphins, I can only encourage you, the enlightened traveller, to respect these beautiful animals, and view them usually from a distance. If you are on a boat sailing or motoring along shore and a group of dolphins approaches, by all means continue on your course and let them ride your bow and stern waves. Let them leap and cavort and watch them as they vie and jostle for a place in the best "pressure riding" position at the bow. Listen to their whistles, which are so loud that they transmit into the air when the dolphins are near the surface and socially active. Dolphins in these situations are awake and they want to play; you will not harm them by allowing them to do so. But do not cut into their school, or try to entice them to be near you by doubling back and attempting to pick up animals who have lost interest in you. Above all, do not approach Spinner schools within bays – they are likely to be resting. You *will* disturb them; so they are, as the Hawaiians say, Kapu – "off-limits." Watch them from a distance, preferably from a cliff or hillside from shore. Spy on them with your binoculars as they fluidly move just underneath the waves, and "love them" secure in the knowledge that you are not harming them by your presence.

Chapter 10

Underwater Hawaii

by Richard Francis
Department of Ichthyology
California Academy of Sciences

Introduction

Any visitor to Hawaii who does not explore its underwater realm is missing at least half the fun. All that is required, other than a bathing suit, is mask, snorkel and fins. For those who want to explore further and deeper, scuba equipment will be required.

The Hawaiian Islands are part of a vast biogeographic region known as the *Tropical Indo-Pacific*. The boundaries are roughly as follows. The central axis extends from southern Japan to Australia's Great Barrier Reef and includes the Philippines, Indonesia, and New Guinea. The region extends westward into the Indian Ocean, from Madagascar to the Red Sea; and eastward through Melanesia, Micronesia, and Polynesia. The Hawaiian Islands form the easternmost outpost of Polynesia and hence of the entire Tropical Indo-Pacific. The coral reefs of the Indo-Pacific region have a higher level of biodiversity than the reefs found in the other three biogeographic regions that harbor coral reefs (the Tropical Eastern Pacific, the Tropical Eastern Atlantic, and the Tropical Western Atlantic (the Caribbean)).

The Indo-Pacific is the center of distribution and point of evolutionary origin of far more families of marine tropical animals – ranging from corals to fishes – than the other tropical marine provinces. Within this province itself, the major center of distribution for marine families, and hence the region of greatest biodiversity, is the triangle extending roughly from New Guinea to the Philippines and then westward through Indonesia. As you move eastward from there – through Melanesia (New Guinea region, Solomon Islands, etc.), Micronesia (Palau, Guam, Wake Island, etc.) and finally, Polynesia (Tahiti, Samoa, Hawaii, etc.) – you find a decreasing gradient of biodiversity. So, for example, there are fewer species of coral and coral reef fishes in Polynesia than in Micronesia. Within that vast region that is Polynesia, the west to east gradient continues so that, for example, Tahiti has more coral species than Hawaii. Partly because of this west to east gradient, and partly because of the extreme distance between Hawaii and the rest of Polynesia, the islands have fewer types of coral than any other part of the tropical Indo-Pacific. But Hawaii is amply compensated for its relative dearth of species, because a large number of the species found in Hawaii are found nowhere else in the world. In fact, Hawaii contains a greater number of *endemic* tropical marine fishes than any other region except the Red Sea.

Why is this? It seems obvious that the Hawaiian Islands would have a lot of endemic species because, after all, islands are known for this. Because of their isolation from the mainland, any animals that happen to arrive will evolve in isolation (p. 9). Over time, these new independently evolving groups result in new species that never existed on the mainland. But we don't seem to have that sort of isolation in marine environments – there is a vast ocean connecting Hawaii to the rest of the tropical Pacific. This seeming continuity is deceiving, however. The ocean does not provide one homogeneous habitat for marine creatures. And this is particularly true of corals and all the coral reef species that depend on them. For coral, most of that vast expanse of water within the Indo-Pacific is inhospitable, primarily because it is too deep. Coral can grow only at depths to which light penetrates, because the symbiotic algae upon which it depends for nutrients can only photosynthesize in the presence of sunlight. Hence coral growth is confined to continental shelves, islands, and *atolls* (reef islands that enclose shallow lagoons). But coral also requires a hard substrate. Muddy or sandy bottoms won't do. This makes most continental regions unsuitable. Because of the silt and detritus from river runoff, there is often little exposed rock underwater near the mainland or off large islands. Cloudy, silty river runoff precludes coral growth in another way as well: it reduces the amount of light available. Coral can grow only in relatively clear water.

But how does that explain the large number of Hawaiian endemic species? Especially when you consider that coral and other reef species send larvae into the *plankton* (the ocean's huge population of tiny floating organisms) which then drift for long periods of time before settling out. You might think that any given species might be able, therefore, to colonize all of those habitats in the Indo-Pacific that are suitable for coral growth. But the length of time that a larva can spend in the plankton is finite, and, moreover, it can only go where the currents take it. It is certainly true that those species with longer larval stages tend to be more widely distributed than those with shorter larval stages, but even those with the longest larval stages can only colonize incrementally. In the meantime, chances are that in a given location, some other species with similar ecological requirements is already entrenched. So even when the larvae of an immigrant species settle out on a reef, the species must find a niche for itself. Most don't.

For sexually reproducing species there is a further problem. Consider the case of the Imperial Angelfish. This is a species that occurs throughout much of the Indo-Pacific, but not in Hawaii. Every once in a while, however, a diver finds one off the Big Island. But one angelfish does not a successful colonization make. It must find a mate and reproduce, otherwise the colonization ends when that individual dies. And most of the time a lone colonizer remains that way. Such has also been the case with Emperor Angelfish in Hawaii.

Because Hawaii is so distant from any other potential source of immigrants, it has proven particularly difficult to colonize. Hence, colonization is rare enough that those fish that have managed to get there are rarely reinforced by others of their species from the outside. If there is sufficient time between successful colonizations, the initial arrivals may evolve independently for so long that they will no longer interbreed with the next batch of successful colonizers from the same mother species. Instead, the second group of colonizers, in the presence of their already established cousins, will have to start evolving on their own, perhap eventually producing a new line of new species. Conversely, the isolation of the Hawaiian Islands makes it difficult for any species that has evolved in Hawaii to gain a foothold elsewhere. So those species that evolve in Hawaii tend to be confined to Hawaii.

Habitats

It is the *coral reefs* themselves, of course, that are the main underwater attraction. Hawaii is particularly blessed with good snorkeling reefs, particularly on the Big Island and Maui. These are *patch reefs* in relatively calm water, protected from wave action by outer reefs. The conditions are ideal for fish-watching, which, as I hope to show, can be very entertaining. Once you have entered the reef, take some time to take in the big picture, the spectacle of color and motion, the sheer abundance of life. Then, if you are a novice, focus on one of the colorful common species such as the Yellow Tang (Plate 67). Watch their social behavior; look at their mouths; how is their odd mouth related to their feeding habits? Another beauty is the Achilles Tang (Plate 66), much more common here than elsewhere in the Pacific.

You will find aggregations of a number of other fishes swimming above the reef as well. Black Durgeon (Plate 68), Ringtail Surgeons (Plate 66) and Convict Surgeonfish (Plate 67), to name a few. Closer to the coral you will find butterflyfishes (Plates 59, 60) of several kinds, as well as the Moorish Idol (Plate 66), everyone's favorite. Notice that the larger butterflyfish always occur in pairs. Watch how they feed, and watch how they behave when they encounter other butterflyfish, particularly those of their own species.

Parrotfishes (Plate 64) are some of the larger coral reef inhabitants. The characteristic beak for which they are named is put to good use as they munch coral, both living and dead. If you are quiet you can hear the crunching sounds underwater. Every once in a while you may notice a wispy cloud of sandy material in the middle of the water column, which seems to appear out of nowhere. Chances are, however, that there is a parrotfish nearby and it has just unloaded the remains of its lunch, destined, ultimately, for your favorite beach.

Many of the fishes remain in or very near the coral. Several species of damselfishes (Plate 62) stake out feeding territories, and during the breeding season the males become particularly aggressive because they are guarding eggs. They need to be vigilant because there are numerous potential egg eaters on the reef; the wrasses (Plate 63) are especially a threat in this regard. There are many species of wrasses in Hawaii, most of them quite beautiful. They are generally somewhat cigar shaped, and they are quite active. In the nooks and crannies, look for soldierfish and squirrelfish (Plates 54, 55), blennies (Plate 65) and gobies (Plate

65). Only the most careful observers will be able to find the incredible frogfish (Plate 54), though it is not uncommon. Their cryptic coloration effectively makes them look like just another bit of the reef. It is in the crevices at the bottom of the reef that you are most likely to find the morays (Plate 53). They are opening their mouths to breathe, not to impress you with their formidable teeth. Invertebrates such as crabs (Plates 80, 81) and shrimp (Plate 79) can be found scurrying about, as well as the much slower brittle stars.

After exploring the reef patch, move into a sandy area between patches. The fauna changes dramatically. Goatfish (Plate 58) of various sorts are particularly common here, as well as various burrowing species such as the Sand Goby (Plate 65), garden eels (Plate 53), and tilefish (Plate 57). You will notice that there are many fewer species and much less life in general, in the sandy areas. Most inshore species are confined to the reefs themselves.

Sometimes there is a transition zone between the reef and the sand, consisting of *coral rubble*. Some fishes prefer the rubble to either sand or living reefs. There is usually enough structure here for the smaller species to use as shelter.

Now look above, near the surface. You probably did not notice a host of well camouflaged fishes resting just below the surface. They are generally long and thin and silvery gray. Two of the more common species are halfbeaks (Plate 54) and houndfish (Plate 54). Somewhat farther below the surface look for barracuda (Plate 68).

After exploring the inner patch reefs, move to the outer reefs. Here there is more wave action, especially on the seaward side. Some fishes prefer to live near where the waves are crashing. One such species is the Sailfin Tang (Plate 68). Notice how it can seem to double its body size by lifting its dorsal and anal fins. The fish in this area use the ebb and flow of the surge to move quickly from one area to another.

Divers will want to explore the water outside of the outer reef, especially along the reef's outer flank. When the coral growth is dense enough, and the water beyond the reef deep enough, a wall forms. These areas provide some of the richest habitats for coral reef species, vertebrates and invertebrates alike. Look for pelagic fishes beyond the wall. Large predatory jacks (Plate 57) pass swiftly in groups of varying size. Look for hammerheads (Plate 52) and eagle rays (Plate 51) as well.

Invertebrate Life

The majority of reef inhabitants are not fish but an extremely varied assortment of animals that are collectively referred to as *invertebrates*. The *corals* themselves are the most prominent invertebrates and they come in myriad colors and shapes. There are actually several groups of corals of which the hard corals comprise only one. Among the common types of hard coral in Hawaii are the *Pocillopora* (Plate 71) which tend to form dense heads that look like cauliflowers. *Porites* (Plate 71) are encrusting corals that form domes of various sorts. Corals of the genus *Fungia* (Plate 71) are named for their mushroom shapes, and those of the genus *Favia* are often referred to as brain corals. Hawaii has very few branching corals of the sort common throughout the rest of the Indo-Pacific. Members of the genus *Alveopora* are among the most branched, but they are not common here.

There are several species of *fire coral* (Plate 70), so named because of the skin irritation they cause when touched. *Hydroids*, such as *Ralpharia* (Plate 70), are feather-like creatures. The soft corals include the *sea fans*, *gorgonians*, and *leather corals*. The *sea anenomes* (Plate 71) are coral cousins, which along with *jellyfish* (Plate 71) comprise the vast group of invertebrates known as *cnidarians*. All members of this group possess stinging parts known as *nematocysts*.

The so-called *flatworms* (Plate 72) are among the most beautiful reef inhabitants. They are sometimes confused with *nudibranchs* (see below) but they are not at all closely related. The *marine worms*, or *polychaetes*, include the *feather dusters* and *tube worms* (Plate 73).

The *sponges* (Plate 70) are among the most primitive multicellular animals. They add much of the color to the reef environments and provide homes for many reef inhabitants. Some form large barrel-like structures, others form amorphous mats, and still others look like amazingly colorful puffs of velvet.

The *mollusks* comprise perhaps the largest marine group. The *gastropods* (*snails* and *slugs*) are particularly bountiful. These include, in addition to the *cone shells, cowries, olives, tritons, volutes, helmets, turbans, limpets* and *sea hares* (Plates 73 to 77), the incredibly gorgeous *nudibranchs* (Plate 76). The Spanish Dancer (Plate 77) – which comes in varying shades of red – is one of the largest and most commonly seen, because it sometimes swims through the water. It is named for the undulations it makes while doing so. But there are many others as well that would reward a careful search.

The *bivalves* comprise another large group of mollusks. It includes the *cockles, clams, scallops* and *oysters*. Also belonging to the mollusks are the *squids* and *octopi* (Plate 78), collectively referred to as *cephalopods*. These are among the most intelligent creatures in the sea, and well worth looking for.

The phylum Arthropoda includes many inhabitants of both land and sea. The largest group of marine arthropods are called *crustaceans*. This group includes all the *shrimp* and *crabs*, as well as *lobsters* (Plates 79, 80). These can be further subdivided into several families each. Among the more interesting shrimp species are those that serve as *cleaners*, several of which are illustrated in the plates. They set up cleaning stations in much the same way as the cleaner wrasses.

Finally, the *echinoderms* comprise a large and diverse phylum of marine animals, the most famous of which are the *sea stars* (Plates 81, 82). Another large group consists of *sea urchins* (Plates 82, 83) and their relatives the *sand dollars*. The *sea cucumbers* (Plate 83) are perhaps the oddest members of this group. They look like caterpillars on steroids.

Vertebrate Life

Surgeonfishes (Family Acanthuridae)

This family is well represented in Hawaii and includes both the Yellow Tang (Plate 67) and the Achilles Tang (Plate 66). *Surgeonfishes* are so called because of the scalpel-like projections at the base of the tail, which they use to slash at other fishes. They are primarily algae grazers and they have exceptionally long intestines that are essential for the digestion of this food. Members of one subgroup of surgeonfishes are referred to as *unicornfish*. Among the most spectacular reef inhabitants, unicornfish are named for the projections extending forward from their heads. The Bluespine Unicornfish (Plate 67) is typical; the horn-like projections give it a somewhat sinister look.

Yellow Tangs and some other surgeonfishes often form mating aggregations in the late afternoon each day, and engage in mass spawnings on the outer reef sections near dusk. It is worth remaining in the water as the sun begins to sink in order to watch the procession of Yellow Tangs as they migrate toward the outer reefs. They proceed deliberately, almost in single file, never stopping to feed, intent on a more important goal.

The Moorish Idol (Plate 66) is on everyone's list of the top five most gorgeous reef creatures. This species is closely related to surgeonfishes but comprises its own distinct family. You will usually find these yellow, black and white beauties in mated pairs, probing the nooks and crannies with their long snouts. Notice how much more deliberately they move compared with the more skittish surgeonfishes. Their grace only enhances their spectacular physical presence.

Butterflyfishes (Family Chaetodontidae)

Butterflyfishes also typically occur in mated pairs, especially the larger species, such as the Teardrop Butterflyfish (Plate 60), Oval Butterflyfish (Plate 59), Ornate Butterflyfish (Plate 59), Lined Butterflyfish (Plate 59), and the Reticulated Butterflyfish (Plate 60). Interestingly, they pair up as juveniles, long before they become sexually mature. It is suspected that during the early part of the association, if they both happen to be of the same sex, one or the other member of the pair will change sex. Alternatively, both may have the capacity to mature as males or females, so that they must decide this matter between themselves. Once paired, they remain paired for life – which can exceed 20 years – with a degree of fidelity that surpasses that of most birds and mammals.

Many species of butterflyfish feed on live coral polyps, for which activity their narrow snouts are ideally suited. In the Longnose Butterflyfish (Plate 60) the snout extends into needle-shaped pincers, ideal for probing the reef's recesses. This family of fishes is particularly well represented in Hawaii. Among the species found nowhere else are the Bluestripe Butterflyfish (Plate 59) and the Milletseed Butterflyfish (Plate 59).

Angelfishes (Family Pomacanthidae)

These beauties are closely related to the butterflyfishes, but they are not nearly as well represented here as in the rest of the Indo-Pacific. However, four of the six species that do live here are found nowhere else. *Angelfishes* resemble the butterflyfishes in having small mouths and laterally compressed, deep bodies. All angelfish, however, have a prominent cheek spine, a feature that butterflyfishes lack. Their larval and juvenile development is also quite different. Most of the Hawaiian species are algae grazers. Many, and perhaps most, angelfish, undergo sex change. They begin life as females; those that live long enough to attain a large size then change into males. This is referred to as *protogynous (female first) sex change*.

The most beautiful of the bunch is the Flame Angelfish (Plate 61). They are quite shy and not common, but well worth the effort to find. Look for them among areas of dense rubble. The most common member of this family by far is Potter's Angelfish (Plate 61), another endemic species.

Damselfishes (Family Pomacentridae)

These small fishes comprise another important component of the reef community. They are not nearly as colorful as the butterflyfishes or angelfishes, but behaviorally they are among the most interesting reef inhabitants. These species, which feed on benthic algae, are highly territorial. Among the most pugnacious reef inhabitants, they strike out at any fish that dares enter their territories, even much larger surgeonfish and butterflyfish. When in breeding condition they will even attack divers. Two species that fall into this category are the Hawaiian Sergeant (Plate 61) and the Pacific Gregory (Plate 62).

Unlike most reef fishes, *damselfishes* lay their eggs on the substrate (bottom). They then carefully guard them against marauding wrasses and surgeonfishes for one to two weeks, the males tending the eggs. When the larvae hatch, they become planktonic but they enter the plankton at a much more advanced state of development and for a shorter period of time than most reef fishes. Since they have fairly short planktonic stages, damselfishes tend to have smaller geographic ranges than, say, butterflyfishes and surgeonfishes; as you might expect, a large proportion of the Hawaiian species are endemic.

Many damselfishes feed on *zooplankton* (the generally tiny and microscopic animals that drift through the world's oceans), which they pluck from the water one at a time. Some of these species are also territorial and stay close to their refuges. Among them are two of my favorites, the Chocolate-dip Chromis (Plate 61) and the Hawaiian Dascyllus (Plate 61), both endemics. The latter species can be found in branching coral, living in groups consisting of juveniles, several females, and a male. The male started out as a female.

Wrasses (Family Labridae)

This is one of the largest families of fishes. Though they vary greatly in size, shape and habits, they all have a single, continuous dorsal fin and they tend to stay close to the bottom. Many *wrasses* move in a distinctive jerky manner as they explore the substrate for food. They primarily use their pectoral fins to swim, bringing their tails into play only when rapid movement is required.

Many wrasse species undergo dramatic color changes as they mature, and this is often accompanied by a sex change as well. Formerly these different developmental stages were often mistakenly identified as distinct species. A convention has arisen in which the first color pattern in a sexually mature fish is referred to as the *initial phase*, and the second color pattern as the *terminal phase*. Some species, such as the Hawaiian Hogfish (Plate 62), have a distinct *juvenile coloration* as well. Juveniles of the aptly named Rockmover Wrasse (Plate 63) mimic drifting algae and do not remotely resemble the adults in color or shape.

In some species, such as the Hawaiian Cleaner Wrasse (Plate 63) and Elegant Coris (Plate 62), all of the initial phase fish are females. In these species the color change is accompanied by sex change to male (protogynous sex change). In other species, such as the Saddle Wrasse (Plate 63), the initial phase fish may be either male or female. Some initial phase males eventually undergo a color change to become terminal phase males. In addition, some females subsequently undergo both a color change and a sex change to become terminal phase males. The two male types in these *diandric* ("two-male") species exhibit completely different reproductive behavior. The terminal phase males defend a territory to which they attract females with their vigorous courtship displays. The much smaller initial phase males, however, use their female-like appearance to get close to the courting couple, which they then shower with their own sperm, a deceitfully effective way for a small male to compete reproductively with its larger counterparts. In addition, initial phase males sometimes form marauding gangs that overwhelm the territorial defense of the terminal phase males.

All wrasses are fascinating to watch and brimming with personality. One of the most famous is the Cleaner Wrasse (Plate 63). The male stakes out a territory to which he attracts several females to form a harem. His territory attracts more than mates, however; it attracts his clients as well. And his clients include such large predatory fish as jacks and snappers. They come here in order to have their external parasites removed and the Cleaner Wrasses are happy to oblige, systematically probing the surface and, in what looks initially like suicide, inside the mouth as well. The cleaner completely disappears into the maw of larger fish, often emerging through the gills. When one client is done, the next fish, which has been patiently waiting in the queue, steps up for his ministrations. *Cleaning stations*, as these territories are called, can usually be found in fairly prominent locations on the reef such as outcrops. They are well worth seeking out.

Other wrasses of note include the Dr. Seuss-like Bird Wrasse (Plate 63), with its trunk-shaped snout, the endemic Pearl Wrasse (Plate 62), and the beautiful Ornate Wrasse (Plate 63), one of the most common inshore species.

Parrotfishes (Family Scaridae)

Parrotfishes are closely related to the wrasses and share their complex life histories, including the color and sex changes. Parrotfishes are distinguished by their beak, formed by the fusion of several front teeth. Further back in the mouth are powerful molars formed of bony plates, the lower convex, the upper concave. They put both their beak and pharyngeal molars to good use in first removing and then grinding chunks of hard coral in order to extract the algae. The sounds they make in the process are quite audible underwater. Parrotfishes manage to digest their food without the aid of a stomach. Instead they have an exceptionally long intestine. When the coral residue reaches the end of the line (intestine) the parrotfish excretes it in wispy clouds of fine sand, destined some day for one of Hawaii's famous beaches.

The striking Spectacled Parrotfish (Plate 64) is a Hawaiian endemic, named for the pattern of lines on the terminal phase males. As in some other parrotfishes, the initial phase fish form aggregations, which seem to swarm over the reef with the rising tide, taking their bites of coral on the move. The terminal phase males, however, tend to be highly territorial. In general, the terminal phase parrotfishes are brightly colored, while the initial phase fish are some shade of red or brown, and often mottled.

Gobies (Family Gobeidae)

Gobies comprise the largest fish family and they are particularly abundant in tropical marine environments. Relatively few have made it to Hawaii, however, owing to their brief larval stage. By way of compensation, many of the Hawaiian species are endemic. Because of their diminutive size, gobies are generally overlooked by snorkelers and all but the more observant divers. They are, however, fascinating creatures. Like damselfishes they lay eggs on the substrate, which are tended by the father. The Hawaiian Shrimp Goby (Plate 65) has a fascinating relationship with a snapping shrimp. The fish lives in the shrimp's burrow; in return the goby keeps a lookout for predators while the none too keen-sighted shrimp goes about its labors. At the first sign of danger, the goby darts into the burrow along with the alerted shrimp. This is a nice example of between-species symbiosis of the sort known as mutualism (p. 71): "if you scratch my back, I'll scratch yours."

Some of Hawaii's most interesting gobies do not live on the reef, or even in the ocean. They have invaded the rivers and become entirely adapted to freshwater. The most amazing feature of these freshwater gobies is their ability to climb up steep surfaces, including those over which waterfalls cascade. For this, they utilize their pelvic fins, which have been modified to form a sucker-like attachment. They have been able to occupy streams at elevations thousands of feet up the steepest mountains. Keep your eyes out for them when you are hiking.

Blennies (Family Blenniidae)

This is another large family of ground-hugging fishes which, owing to a short larval phase, is not well represented in Hawaii. Like the gobies, *blennies* lay eggs on the substrate, which the male tends. They are highly territorial and pugnacious. The Scarface Blenny (Plate 65) is one of the most common and easily spied, because it likes to perch on a coral projection, keeping its eyes on things. The Zebra Rockskipper (Plate 65) is famous for its ability to jump from tidepool to tidepool.

Triggerfishes (Family Balistidae)

The name *triggerfish* derives from a mechanism by which these species erect the stout first dorsal spine by means of the movement of the second dorsal spine. Presumably this helps discourage any would-be predators. I find triggerfishes mesmerizing, both because they are often quite beautiful, and because of the unique way they move through the water by means of undulations of the dorsal and anal fins. A group of Black Durgeons (Plate 68) is one of the more beautiful spectacles in nature. Most triggerfish species, however, are solitary. They feed on hard-shelled invertebrates such as sea urchins and crabs. Though their mouths are small, their teeth are formidable.

They too lay eggs, but unlike gobies or blennies, it is the mother who cares for them. And it is best to steer clear of her while she is so engaged; more than a few divers have received nasty bites from some of the larger species. Both the Lagoon Triggerfish (Plate 68) and the Reef Triggerfish (sometimes called Picasso Trigger; Plate 68) resemble pieces of abstract art.

Puffers (Family Tetradontidae) and Porcupinefishes (Family Diodontidae)

No survey of reef fishes would be complete without mentioning these two closely related groups of unique reef fishes. *Puffers*, such as the Spotted Puffer (Plate 69) and Crown Toby (Plate 69), are so called because of their ability to inflate themselves in the presence of predators. *Porcupinefishes*, such as the Spiny Balloonfish (Plate 69), add to this defense mechanism an array of spines that are erected in the process. When they are inflated, the pectoral fins, by means of which they propel themselves, become ineffective, so they tend to list and roll in a comical manner. Though comical looking, they should be treated with respect – especially the larger individuals – because they can inflict a nasty bite. Like the triggerfishes, puffers and porcupinefishes feed on invertebrates with hard exoskeletons. The puffers tend to be diurnal, but the larger-eyed porcupinefish are nocturnal.

Sharks and Rays

Several shark species can be found on or around the reefs. Of these, the Blacktip Reef Shark (Plate 52) is most likely to be encountered by day in shallow water. These slender and sleek creatures are quite timid but they have been known to bite the legs of wading humans. The Whitetip Reef Shark (Plate 52) is nocturnal, but can be found in the caves where it rests during the day. The Scalloped Hammerhead (Plate 52) is more of an open water species that you are most likely to see off reef walls. Tiger Sharks are rarely seen but they represent the greatest threat to divers. It is extremely rare, however, for an encounter with a Tiger Shark to result in an attack. The Galápagos Shark is another species that is potentially dangerous.

Of the rays, the Spotted Eagle Ray (Plate 51) is the most commonly seen. These enchanting animals often perform acrobatics underwater and above. I have noticed that they seem particularly prone to leap out of the water when they are in pairs but I don't know what to make of this. Manta Rays (Plate 51) are among the most impressive and awe-inspiring creatures on land or sea. A group of Mantas, gracefully winging their way through the water – with their entourages of remoras (Plate 57b) – is a sight you will remember for the rest of your life.

Chapter 11

Insects and Other Arthropods

by Pete Oboyski
Pacific Island Ecosystems Research Center
US Geological Survey – Biological Resources Division
Hawaii National Park

- Origins
- Biodiversity
- Adaptation
- Charismatic Microfauna (or, Pretty Little Bugs and Spiders)
- Butterflies
- Alien Insects
- Conservation

Origins

Hawaii's insect fauna is special, not only for what is present, but also for what is not present. In fact, only half of the insect orders and less than a quarter of the insect families known throughout the world are native to Hawaii. Conspicuously absent from the native fauna are aquatic insects such as mayflies, stoneflies, and caddisflies and terrestrial groups like ants, cockroaches, and yellowjackets. Much of the reason for this paucity is the great expanse of ocean separating these animals from their mainland origins. It is estimated that the currently known fauna of more than 5000 native insects and spiders is derived from fewer than 400 successful colonizations (p. 100) in the past 70 million years.

How could so few colonizers give rise to so many species? Hawaii's diverse habitats and layout of its islands deserve much of the credit. From shoreline to summit, habitat types range from dry grasslands to wet forests, from fresh lava flows to old-growth forests, and from balmy ocean beaches to frozen alpine deserts. The few fortunate survivors of transoceanic travel found little competition in their new home and made use of ecological niches not before available to them. Through adaptive radiation (p. 54), insects and spiders made the subtle changes necessary to capitalize on the bounty of the islands. Next, geographic isolation and time set species apart.

Geographic isolation occurs at two scales in Hawaii: *pae aina* and *kipuka*. The Hawaiian islands form an archipelago, or pae aina, that extends for 3075 km (1910 miles) from Kure atoll to the Big Island of Hawaii. Movement of insects between islands is infrequent owing to the distance between islands, (the two closest of the main islands, Maui and Kahoolawe (Map 4, p. 41), have a channel of 10 km, 6 miles). Deeply dissected valleys from hundreds of thousands of years of erosion on the older islands have made movement difficult even between sites on the same island.

Isolation also occurs at the level of kipuka, or *islands of soil and vegetation that are older than the surrounding landscape*. Kipuka are formed when flowing lava is diverted around portions of land, resulting in refuges for many plants and animals, isolating them from their neighbors. During periods of isolation, plants and animals diverge to the point where they no longer recognize each other as the same species when they regain contact. This is especially true for insects, which may pass many generations in isolation, giving each population the opportunity to adapt to their habitat in their own unique way.

Biodiversity

Among the great success stories of insect speciation is the Hawaiian fruit fly, *Drosophila* (Plate 84). Over 500 species of flies in the family Drosophilidae have been discovered so far in Hawaii and the counting continues. Many people are familiar with the tiny, drab fruit flies found around rotting fruits and vegetables, including *Drosophila melanogaster*, the mainstay of genetic studies. But in Hawaii, some species of fruit flies, often called *pomace flies* here, grow ten times the size of their mainland relatives and have patterned wings and elegant mating dances. Many of these flies are obligated to feed on particular species of plants. Here they find mates, court, and lay eggs, allowing their larvae to develop on their favorite rotting plant. This specialization is what has allowed this group to form as many different species as it has. Unfortunately, several species have specialized on plants that have become very rare in the wild and so have themselves become rare. Thirteen species of *Drosophila* have been suggested as candidates for the threatened and endangered species list.

Other insect groups in Hawaii have had similar explosions of diversity. Hawaiian relatives of the casebearing clothes moth, *Hyposmocoma* (Plate 84), number more than 350 species. The half-inch case that each species builds and drags around as a caterpillar is unique in its materials and design, some looking like smooth polished wood, others like cemented sand, and still others like flaking leaves. More than 200 species of crickets, twice as many species as over the entire continental United States, are found in Hawaii and nowhere else. Likewise, more than 200 species of ground beetles are native only to Hawaii. Other insect families including seed bugs, planthoppers, sap beetles, woodboring beetles, weevils, and several families of wasps and flies have over 100 endemic species in Hawaii. Many of these groups found success by exploiting the diverse plant life available to them.

Adaptation

The adaptation to new niches did not stop with the exploitation of Hawaii's diverse flora. Several species of Hawaiian insects and spiders have specialized in habitats containing little to no vegetation at all. Native crickets (*Caconemobius fori*; Plate 84) and wolf spiders (*Lycosa* sp.; Plate 84) often colonize new lava flows even before the first plant seeds begin to sprout. Each feeds on the windblown debris (plant seeds and insects) that gets deposited on the lava fields. Wolf spiders have also adapted to living at the highest peaks of Hawaii, enduring subfreezing nightly temperatures and sudden blizzards interspersed with unrelenting sun.

Sharing the extreme summit habitat with the wolf spiders are two species of seed bugs. The "wekiu bug" (*Nysius wekiuicola*; Plate 84) can be found at the summit of Mauna Kea, while its cousin, the "aa bug" (*Nysius aa*; Plate 84) can be found on Mauna Loa. The closest relatives of these summit bugs feed on seeds in the valleys below. The wekiu and aa bugs, however, have adapted to feeding on the remains of insects that are deposited on the mountain tops by the wind. A species of noctuid moth can also be found at the summit of Mauna Kea, feeding on lichens and windblown debris. The spiders and bugs can withstand frigid temperatures but must burrow into the ashy soil beneath the rocks to ride out colder times. The noctuid caterpillar, by contrast, has the capacity to be frozen in an ice cube, be thawed out, then walk away.

As a result of their volcanic origins, the Hawaiian Islands are riddled with lava tubes (p. 27) and caves. Again, insects and spiders are there to take advantage of the unique habitat. Roots of native ohia trees that penetrate the surface layers of lava and hang free within the lava tubes are colonized by pale, blind planthoppers (*Oliarus* spp.; Plate 85). They share their cave habitat with other cave-adapted species including crickets (Plate 85), wolf spiders, thread-legged bugs, millipedes, caterpillars and a water-treader (*Cavaticovelia aaa*; Plate 85) that prefers to tread on cave walls. Though pale against the dark walls of the lava tubes, most of these creatures are difficult to detect because of their small size and ability to retreat into inconspicuous hideouts.

One of Hawaii's most unique insects is the predacious caterpillar, *Eupithecia* (Plate 85). These small (to 2.5 cm, 1 in) inchworm moth caterpillars sit motionless on their perches awaiting unsuspecting flies and other small insect prey. When a potential prey item touches its back end the caterpillar whips around with its grappling-hook-like legs and snatches up its victim. Once in the caterpillar's grasp the struggling prey is consumed head first until all that remains are the discarded wings. All of the 18 *Eupithecia* species discovered so far have been found in wet and humid forests on all of the major islands. Each is camouflaged on its favorite perch, some projecting like sticks from a plant stem, others nestling in notches they chew from leaflets.

Charismatic Microfauna (or, Pretty Little Bugs and Spiders)

Another of Hawaii's small and obscure inhabitants is the happy face spider (*Theridion grallator*; Plate 85). Adults are only 1 cm (a half-inch) long with legs fully extended. Happy face spiders wait on the undersides of leaves where they stalk the silhouette of their prey on the surface above. Once within striking range the spider reaches around to the upper leaf surface and snatches its victim from below. The name "happy face" comes from the patterns found on the top of the spider's abdomen. Black and red spots are arranged on a yellow backdrop to form a smiling clown-like face. Variation in the pattern can be found both within populations of spiders and between spiders from different areas, but all are considered the same species.

Not all of Hawaii's insects are small and obscure. The koa bug (*Coleotichus blackburniae*; Plate 86) sucks the sap of koa and aalii plants by sticking its beak-like proboscis through the plant's outer layers. This 2-cm (1-in) long bug is a relative of stink bugs and has an iridescent green and red shield over its back that is sometimes marked with blue and yellow. The immature bugs are disk-shaped and dark blue-black with red stripes across their abdomens. Koa bugs were once widespread on all the main islands, but have become less abundant in recent decades. This is due, in part, to parasitism by flies introduced to control pest stink bugs. Koa bugs can sometimes be found on ornamental acacia trees planted in parking lots and landscaped yards.

Another brightly colored native insect group is the Hawaiian damselflies. The endemic genus, *Megalagrion*, contains at least 30 species, ranging in color from yellow and black to brilliant reds (Plate 86). Although, like their larger relatives the dragonflies, damselfly nymphs (the youngsters, which are wingless) live in aquatic environments, several have adapted to life out of ponds and streams. Some nymphs are found on wet rocks near streams, others in water that accumulates in the leaf axils of certain native plants, and one lives on moist surfaces in tree canopies. The largest of Hawaii's endemic insects is the giant aeshnid dragonfly, *Anax strenuus*, a close relative to the green darner, *Anax junius*, which is also found in Hawaii. The giant aeshnid has a wingspan of up to 15 cm (6 in) and is much darker in color than the green darner. Only two other dragonflies are native to the islands, the Hawaiian skimmer (*Nesogonia blackburni*) and the globe skimmer (*Pantala flavescens*).

Butterflies

Two butterfly species are endemic to Hawaii, the Kamehameha butterfly (*Vanessa tameamea*; Plate 86) and Blackburn's little blue butterfly (*Udara blackburni*; Plate 86). The Kamehameha butterfly is a showy orange and black "brush-footed" butterfly related to the red admiral and the painted lady, both of which are also found in Hawaii. The Kamehameha butterfly can be identified by its deep red-orange color that is on a greater portion of its wings than those of its relatives. The prickly larvae (caterpillars) feed on the leaves of native nettle plants such as mamaki. Blackburn's little blue butterfly is a small "gossamer-wing" butterfly with iridescent bluish-brown scales on the upperwing surface. It can be distinguished from similar looking alien species by the iridescent green scales on the underside of its wings. The favorite foods of the slug-like looking larvae are native koa and aalii plants.

Three mainland butterflies naturally colonized Hawaii before 1900: the painted lady (*Vanessa cardui*), Hunter's butterfly (or American painted lady;

Vanessa virginiensis), and the monarch butterfly (*Danaus plexippus*; Plate 86). The painted lady, which feeds on thistles, hollyhocks, and other plants, and Hunter's butterfly, which feeds on *Gnaphalium*, were already well established when they were first documented in the late 1800s. The monarch became established following the arrival of its favorite host plant, milkweed, in the 1840s. Monarchs have since become a pest species in Hawaii by feeding on crown flower plants, the flowers of which are important in lei making. Albino monarchs have become common in some areas of Hawaii. The albino form superficially resembles the citrus swallowtail, a butterfly first noted in Honolulu in 1971. The citrus swallowtail is a very adept flier and often eludes would-be bird predators. It is suggested that albino monarchs, despite their lack of warning coloration, are avoided by birds wary of chasing elusive prey; that is, birds may mistake the albino monarch for the swift and elusive swallowtail. Unlike mainland monarchs, the Hawaiian populations are not known to migrate.

Ten other butterflies, for a total of 15, can be seen in Hawaii: the banana skipper (*Erionota thrax*), fiery skipper (*Hylephila phyleus*), citrus swallowtail (*Papilio xuthus*), cabbage butterfly (*Pieris rapae*), gulf fritillary (*Agraulis vanillae*), red admiral (*Vanessa atalanta*), smaller lantana butterfly (*Strymon bazochii*), larger lantana butterfly (*Tmolus echion*), bean butterfly (*Lampides boeticus*), and western pigmy blue (*Brephidium exilis*). Most of these species were first recorded in Hawaii in the 1970s, though they may have been present earlier. The cabbage butterfly arrived in the 1890s with cabbage shipped from California. The smaller and larger lantana butterflies were both introduced from Mexico in 1902 to control the pest plant *Lantana*.

Alien Insects

Many common mainland insects never reached Hawaii on their own. The first human arrivals must have found Hawaii to be a true paradise, with no mosquitos, no ants, no roaches, and no yellowjacket wasps. Today all of these are household pests and many are negatively affecting the environment. Over 40 species of ants have infiltrated both urban and natural areas from sea level to 2830 m (9300 ft). Ants can be effective scavengers as well as voracious predators, decimating native insects, attacking hatchlings of turtles and birds, and competing with native predators for rapidly disappearing native prey. The western yellowjacket (*Vespula pensylvanica*) was first discovered on Kauai in 1919, but has since become well established on all the major islands. This ground-nesting wasp is a common "picnic pest" that scavenges on garbage and food remains. However, yellowjackets have become established in natural areas up to 2900 m (9500 ft) elevation, where they presumably prey on other insects and spiders.

Mosquitos are another pest of humans in Hawaii, but have had a greater impact on native birds. Avian malaria carried by the southern house mosquito (*Culex quinquefasciatus*) is a major culprit in the demise of many endemic honeycreepers (pp. 58; 151). Mosquitos breed in standing water associated with human habitat such as buckets, old tires, and cesspools. They also colonize forests by using as breeding pools gouged-out tree fern trunks filled with water. The ferns are knocked down by feral pigs, which dig out the starchy centers to eat. Southern house mosquitos emerge from these cavities and bite birds, injecting their saliva, sucking the bird's blood, and transmitting the malaria from one bird to the next. The southern house mosquito also carries dog heartworm and is one of five mosquito species that bite humans in Hawaii. This mosquito and the night-biting floodwater mosquito (*Aedes nocturnus*) generally feed at dusk or after dark, while the forest day (or Asian tiger) mosquito (*Aedes albopictus*) bites during the day. Bites from the other two species, the yellowfever (*Aedes aegypti*) and bromeliad (*Wyeomyia mitchelli*) mosquitos, are much less common. In an effort to control these pests, three species of "cannibal" mosquitos (*Toxorhynchites* spp.) were introduced to attack larvae of the human biting mosquitos. *Anopheles* mosquitos, which carry human malaria, are NOT established in Hawaii.

Other pests of humans include the large centipede, lesser brown scorpion, and widow spiders. The large centipede (*Scolopendra subspinipes*) reaches lengths up to 18 cm (7 in) and packs a mean bite. It and the lesser brown scorpion (*Isometrus maculatus*) live in dry lowland areas under rocks and logs, and sometimes hide out in houses. The brown widow (*Latrodectus geometricus*) appears similar to the southern black widow (*Latrodectus mactans*). Both may have an hourglass-shaped red patch on the underside of the abdomen, but the brown widow is often lighter and more mottled in color. Both prefer dark secluded hideouts. Black widows are more common in mid- to upper-elevation dry areas, whereas brown widows are discovered more often around human habitation. The brown violin spider (*Loxosceles rufescens*), a relative of the brown recluse, has also been reported from the islands of Kauai, Oahu and Maui. A bite or sting from any of these venomous animals should be treated by a physician.

Conservation

The greatest impacts to Hawaii's native arthropods have come from introduced species and habitat loss. Introduced plants promote populations of introduced insects while out-competing native plants that native insects depend on. Birds, reptiles and amphibians, along with predatory and parasitic insects, continue to be introduced to the islands, assaulting an already stressed ecosystem. Native forests continue to be fragmented by agriculture and development. Native trees, harvested for their high-quality wood, are replaced by tree farms of alien species. Rare plants, once depended on by many native birds and insects, are relegated to steep cliffs and ravines inaccessible to introduced sheep, goats and pigs. But despite how rare many Hawaiian arthropods have become, including many facing extinction, none are yet recognized as endangered species.

The best hope for Hawaii's native insects and spiders lies in academic interests, natural reserves, and public awareness. Because Hawaii's flora and fauna are so unique, they have attracted the attention of many scholars and scientists. Universities, government agencies and private research groups find Hawaii an ideal laboratory for studying the processes of evolution and adaptation. Nature preserves and forests are managed in many different habitats to promote the survival of native species. Schools and public interest groups increasingly are teaching the general public about conservation. Though Hawaii has lost many species to extinction, and will lose many more, it continues to be a treasure trove of insects and spiders to those who take the time to appreciate them.

References and Additional Reading

Below is a list of the most comprehensive and authoritative references for Hawaiian natural history – the main sources I used to compile the information in this book, and the sources you should seek out and read if you desire further information on the subjects they treat.

Berger A. J. (1981) *Hawaiian Birdlife, 2nd ed*. University of Hawaii Press, Honolulu, HI.

Carlquist S. (1980) *Hawaii: A Natural History*. Pacific Botanical Tropical Garden, Lawai, Kauai, HI.

Culliney J. L. (1988) *Islands in a Far Sea*. Sierra Club Books, San Francisco.

Eldredge L. G. and S. E. Miller (1995) How many species are there in Hawaii? *Bishop Museum Occasional Papers* 41: 1–16.

Fielding A. and E. Robinson (1987) *An Underwater Guide to Hawaii*. University of Hawaii Press, Honolulu, HI.

Gosliner T. M., D. W. Behrens and G. C. Williams (1996) *Coral Reef Animals of the Indo-Pacific*. Sea Challengers, Monterey, CA.

Harrison C. S. (1990) *Seabirds of Hawaii: Natural History and Conservation*. Cornell University Press, Ithaca, NY.

Hawaiian Audubon Society (1996) *Hawaii's Birds*. Hawaiian Audubon Society, Honolulu, HI.

Hazlett R. W. and D. W. Hyndman (1996) *Roadside Geology of Hawaii*. Mountain Press Publishing, Missoula, MT.

Hitch T. K. (1992) *Islands in Transition: The Past, Present, and Future of Hawaii's Economy*. First National Bank, Honolulu, HI.

Hoover J. P. (1993) *Hawaii's Fishes: A Guide for Snorkelers, Divers, and Aquarists*. Mutual Publishing, Honolulu, HI.

Howarth F. G. and W. P. Mull (1992) *Hawaiian Insects and Their Kin*. University of Hawaii Press, Honolulu, HI.

Kaufman G. D. and P. H. Forestell (1986) *Hawaii's Humpback Whales*. Pacific Whale Foundation Press and Island Heritage Publishing, Aiea, HI.

Kear J. and A. J. Berger (1980) *The Hawaiian Goose*. Buteo Books, Vermillion, SD.

Lever C. (1994) *Naturalized Animals: The Ecology of Successfully Introduced Species*. T. and A. D. Poyser, London.

McKeown S. (1996) *A Field Guide to Reptiles and Amphibians in the Hawaiian Islands*. Diamond Head Publishing, Los Osos, CA.

McMahon R. (1996) *Adventuring in Hawaii*. Sierra Club Books, San Francisco, CA.

Meffe G. K. and C. R. Carroll (1997) *Principles of Conservation Biology, 2nd. ed*. Sinauer, Sunderland, MA.

Munro G. C. (1944) *Birds of Hawaii*. Bridgeway Press, Rutland, Vermont.

Perrins C. M. and A. L. A. Middleton (1985) *The Encyclopedia of Birds*. Facts on File Publications, New York.

Pratt H. D. (1993) *Enjoying Birds in Hawaii: A Birdfinding Guide to the Fiftieth State*. Mutual Publishing, Honolulu, HI.

Pratt H. D. (1996) *A Pocket Guide to Hawaii's Birds*. Mutual Publishing, Honolulu, HI.

Pratt H. D., P. L. Bruner and D. G. Berrett (1987) *A Field Guide to the Birds of Hawaii and the Tropical Pacific*. Princeton University Press, Princeton, NJ.

Randall J. E. (1996) *Shore Fishes of Hawaii*. Nature World Press, Vida, OR.

Scott J. M., S. Mountainspring, F. L. Ramsey and C. B. Kepler (1986) *Forest Bird Communities of the Hawaiian Islands: Their Dynamics, Ecology, and Conservation*. Studies in Avian Biology 9: 1–431.

Scott S. (1988) *Oceanwatcher*. Green Turtle Press, Honolulu, HI.

Scott S. (1991) *Plants and Animals of Hawaii*. Bess Press, Honolulu, HI.

Stone C. P. and L. W. Pratt (1994) *Hawaii's Plants and Animals: Biological Sketches of Hawaii Volcanoes National Park*. Hawaii Natural History Association, National Park Service, and Cooperative National Park Resources Studies Unit, Honolulu, HI.

Stone C. P. and J. M. Scott (1985) *Hawaii's Terrestrial Ecosystems: Preservation and Management*. Cooperative National Park Resources Studies Unit, University of Hawaii, Honolulu, HI. (A multi-authored edited work.)

VanderWerf E. A. (1998) Elepaio (*Chasiempis sandwichensis*). In *The Birds of North America*, No. 344. A. Poole and F. Gill, eds. The Birds of North America, Inc., Philadelphia, PA.

van Riper S. G. and C. van Riper (1982) *A Field Guide to the Mammals in Hawaii*. Oriental Publishing, Honolulu, HI.

Waller G., M. Dando and M. Burchett (1996) *SeaLife: A Complete Guide to the Marine Environment*. Smithsonian Institution Press, Washington, DC.

Habitat Photos

1 Dry forest and scrub area at Mauna Kea Forest Reserve, off Saddle Road, on the Big Island. Mauna Loa is in the background.

2 Trail at Upper Waiakea Forest Reserve, off Saddle Road, on the Big Island. Mauna Kea in the background. Hiking is over old, rough lava flows.

3 Nature trail at Kipuka Puaulu (also called Bird Park), off Mauna Loa Strip Road, Hawaii Volcanoes National Park, the Big Island.

4 Shaded portion of Crater Rim Trail, through native Hawaiian rainforest, Hawaii Volcanoes National Park, the Big Island.

5 Spectacular coastal view from Waipio Valley Lookout in the northern part of the Big Island.

6 Nualolo Trail on a misty, dripping day at Kauai's Kokee State Park.

7 A stream crosses the Alakai Swamp Trail, Kokee State Park, Kauai.

8 Afternoon mist at Kokee State Park, near Puu O Kila Lookout, Kauai.

9 Surf, beach, and greenery along the Na Pali Coast, as seen from the Na Pali Coast (Kalalau) Trail, near the trailhead, Kauai.

10 Moving along the narrow boardwalks through beautiful short-stature native forests, Pihea Trail, Kokee State Park, Kauai.

11 Kauai's Waimea Canyon – "the Grand Canyon of the Pacific" – from a lookout at Kokee State Park.

12 Native Hawaiian cloud forest habitat at The Nature Conservancy's Waikamoi Preserve, Maui.

13 Coastal vegetation and black lava boulders jutting out into the Pacific, Waianapanapa State Park, just north of Hana, Maui.

14 Trail through mixed native and alien forest in The Nature Conservancy's Waikamoi Preserve, near its boundary with Haleakala National Park, Maui.

15 Bridge spanning a gulch in low elevation wet forest along the trail to Waimoku Falls, in the southeast section of Haleakala National Park, Maui.

16 Molokai - stunning view from a trail as it reaches a high lookout at The Nature Conservancy's Kamakou Preserve.

17 Pepeopae Trail through middle elevation wet forest and bog at The Nature Conservancy's Kamakou Preserve, Molokai.

18 Hauula Loop Trail, inland from Hauula Beach County Park, northeastern coast of Oahu.

19 View from Nuuanu Valley Lookout at the end of the Pauoa Flats Trail/Puu Ohia Trail, just north of Honolulu, Oahu.

Identification Plates

Plates 1–86

Abbreviations on the Identification Plates are as follows:

M; male
F; female
IM; immature

The species pictured on any one plate are not necessarily to scale.

Text by Sarah Reichard

Plate 1a

White Shrimp Plant
Justicia betonica (Non-native)
The clusters of flowers at the tip of the plants do look something like white shrimps. There is a large, persistent, greenish outer structure (calyx) around the white to lavender flowers. This species forms large clumps in disturbed areas on most of the islands.

Plate 1b

Mango
Mangifera indica (Non-native)
A tree of lowland wet forest, this species fruits heavily at times and not at all at others. The fruits of the wild trees are not as flavorful as those of commercial trees, but may be eaten. The leaves are large and evergreen and the flowers small and greenish.

Plate 1c

Christmasberry
Brazilian Pepper
Schinus terebinthifolius (Non-native)
This aggressive weed has become a serious problem in open dry areas on all the islands. The first common name of this species, used only in Hawaii, stems from its tendency to produce bright red fruit at Christmas time. Although poisonous when consumed in large amounts, the dried fruits have a peppery taste and are the pink peppercorns of the gourmet trade – leading to the second common name. The flowers are small and white.

Plate 1d

Maile
Alyxia oliviformis (Native)
One of the most prized species for leis, maile is found in nearly all habitat types. The plants are woody vines with variable leaf shapes. Leaves are fragrant when crushed.

Plate 1e

Hawaiian Holly
Kawaʻu
Ilex anomala (Native)
This holly relative has thick leathery leaves, small white flowers, and purplish-black fruits. It is common in wet forests and bogs on all the islands.

Plate 1f

Octopus Tree
Schefflera actinophylla (Non-native)
Yes, this is the same plant found in bank lobbies around the world. A popular ornamental, *Scheffleras* have attractive large glossy leaves and interesting spikes of flowers and fruits (the fruits are increasingly finding their way into leis). It is spreading rapidly in wet forests on all the islands.

a White Shrimp Plant

b Mango

c Christmasberry

d Maile

e Hawaiian Holly

f Octopus Tree

Plate 2a

‘Ohe

Tetraplasandra hawaiensis (Native)

This small tree has large leaves with dense yellowish hairs on the underside. The small fruits are reddish and are found at the ends of branches. This species is found in low wet forests.

Plate 2b

Silversword

‘Ahinahina

Argyroxiphium sandwicense (Native)

Despite the complicated genus name (pronounced something like a sneeze), this is one of the most spectacular species anywhere. The sword-like leaves are covered with silver-colored hairs, leading to the common name. The plant remains as a basal rosette for many years, finally shooting up a tall spike of pink daisy-like flowers. The plant dies after flowering. There are two subspecies, one found only in the crater of Haleakala on Maui and the other in a single population on Mauna Kea on the Big Island.

Plate 2c

Kupaoa

Dubautia scabra (Native)

The leaves on this weakly stemmed shrub are usually rough when rubbed between the fingers. There are few leaves towards the ends of the branches, where the flowers are found. The daisy-like flowers are small and the outer ray petals are purplish. It is found in wet forests and open areas on several of the islands.

Plate 2d

Na‘ena‘e

Dubautia menziesii (Native)

These shrubs may grow to 2 m (6 ft) tall and the branches are often covered with white hairs. The leaves have a light fragrance when crushed and the daisy-like flowers have orangish outer ray petals. They are found on Maui in high elevation open areas.

Plate 2e

Sourbush

Pluchea symphytifolia (Non-native)

A scruffy, mostly nondescript shrub, this species is now common along disturbed roadsides on all the islands. It tolerates salt and is often found in areas that are periodically inundated with sea water. The leaves are densely hairy underneath and the unimpressive daisy-like flowers have pinkish outer ray petals.

Plate 2f

Wedelia

Wedelia trilobata (Non-native)

Wedelia is a small mat-forming groundcover with bright yellow daisy-like flowers. Although it is a popular ornamental in Hawaiian landscapes, it has escaped cultivation.

a ʻOhe

b Silversword

c Kupaoa

d Naʻenaʻe

e Sourbush

f Wedelia

Plate 3a

Jacaranda
Jacaranda mimosifolia (Non-native)

Jacaranda is a street tree found on many of the islands but it has escaped and established itself in lowland dry forests. The leaves are fern-like and the flowers are a lovely purple color in inflorescences (multi-flower structures) at the end of the branches.

Plate 3b

African Tulip Tree
Spathodea campanulata (Non-native)

Large red flowers that attract birds are the obvious feature of this pest tree of wet lowland forests. The leaves are compound, with hairs on the underside of the leaves. The seeds have long papery wings that catch the breeze and disperse the seeds throughout the forest.

Plate 3c

Tree Heliotrope
Tournfortia argentea (Non-native)

This small seaside tree is actually a relative of the lowly forget-me-not. It is characterized by large leaves and small white flowers in stiff spikes on the ends of the branches. It is found on all main islands.

Plate 3d

Clermontia
Clermontia clermontiodes (Native)

A small shrub/tree whose key feature (and of other members of the genus) is the interesting flowers found in the axils of leaves. The flowers have curved tubes that are several inches long. The birds that pollinate these flowers have beaks that are curved to match. In this species the tube is greenish to purple on the outside, creamy white on the inside. You may find it in mid-elevation wet forests on Oahu and Kauai.

Plate 3e

Haha
Cyanea angustifolia (Native)

If you are hiking in the wet forests of the Koolau Mountains on Oahu you may encounter this interesting species. This large shrub has long narrow leaves with long tapering tips. The white to purple flowers are in clusters in the axils of the leaves and are tubular, with lobes that spread open at the end.

Plate 3f

Koli'i
Trematolobelia kauaiensis (Native)

This is a small shrub with long narrow leaves. The flowers, in clusters at the ends of the branches, are bright red and have a curved tubular shape from which the staminal column protrudes. You will find this species in wet forests of Kauai, but related species may be found on other islands.

a Jacaranda

b African Tulip Tree

c Tree Heliotrope

d Clermontia

e Haha

f Koli'i

Plate 4a

Papaya
Carica papaya (Non-native)
Papayas are an important fruit in Hawaii and some seem to have wandered away from backyard gardens to establish in low wet forests. The fruits are found only on female plants, so not all plants encountered will have fruits, but all are recognizable by the large, deeply lobed leaves.

Plate 4b

Ironwood
Australian Pine
Casuarina equisetifolia (Non-native)
This species is not a pine at all, but a flowering plant from Australia. It has been planted and used for a number of purposes and has spread from those plantings. It is now very common on the dry side of all the islands. The long pendulous branches have small scale-like leaves and the flowers are in dense clusters resembling cones.

Plate 4c

Olomea
Perrottetia sandwicensis (Native)
The shiny leaves of this species are distinctive, with reddish veins and stems. The margins of the leaves are somewhat toothed. The flower clusters are in the axils of the leaves and are greenish to red. Fruits are bright red. This is a common tree in mid-elevation wet forests on all main islands.

Plate 4d

Alexandrian Laurel
Kamani
Calophyllum inophyllum (Non-native)
The Polynesians introduced this tree both for medicine and wood. Leaves are large and leathery and the veins are closely spaced and parallel to each other. The flowers are notable for the numerous stamens. Kamani may be found in coastal areas on most of the islands.

Plate 4e

Indian Almond
Terminalia catappa (Non-native)
The greenish red fruits on this coastal tree do somewhat resemble almonds and they are edible. The dark green, glossy leaves are wider towards the far end and they are crowded at the ends of the branches, where there are also spikes of small white flowers. It may be found on all the major islands.

Plate 4f

Woodrose
Merremia tuberosa (Non-native)
The common name, woodrose, derives from the capsular fruits which do look remarkably like wooden roses. This vine covers disturbed areas and may be found on the main islands along roadsides and growing up telephone poles.

a Papaya

b Ironwood

c Olomea

d Alexandrian Laurel

e Indian Almond

f Woodrose

Plate 5a

Morning Glory
Koali ʻAwa
Ipomoea indica (Native)

This morning glory species is common in disturbed dry areas on all the main islands. The leaves are broadly heart-shaped and the flower is blue to purple (sometimes white).

Plate 5b

Hawaiian Dodder
Kaunaʻoa
Cuscuta sandwichiana (Native)

This species resembles golden orange wires twisting over other plants – this is because it is a parasite and does not need green chlorophyll for photosynthesis. Instead, it gains its carbohydrates from connecting to a host plant. It uses many host plants and occurs mostly in coastal areas.

Plate 5c

Lama
Diospyros sandwicensis (Native)

This persimmon relative is a small tree found in moist low to mid-elevation forests on the major islands. The leaves are dark and leathery and the upper surface may have an indented network of veins. The flowers and fruits are supported by leathery bracts. Fruits are orange.

Plate 5d

Pukiawe
Styphelia tameiameiae (Native)

This attractive species is very variable in form and it occurs in a number of different habitats. It is found in low to high elevation forests, usually in wetter areas but also in dry. The plant is usually a low shrub with small, narrow, leathery leaves. The small flowers are white and the fruits are reddish or pink.

Plate 5e

ʻŌhelo
Vaccinium reticulatum (Native)

This shrub resembles huckleberry because it is a close relative. Look for ʻŌhelo berry jam at groceries – it is every bit as delicious as huckleberry. The urn-shaped flowers may be red, yellow, or yellow with red, and the fruits are usually a bluish-black.

Plate 5f

Candlenut Tree
Kukui
Aleurites moluccana (Non-native)

The Polynesians brought this species with them when they colonized the islands. It is now one of the dominant trees in lowland wet forests, recognizable from a distance by the silvery sheen of the tri-lobed leaves. Kukui is the official state tree. Polynesians used this species for many purposes, including burning the oils from the fruits for a light source. Candlenut leis are often found in souvenir shops.

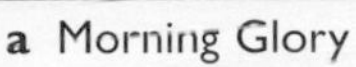

a Morning Glory

b Hawaiian Dodder

c Lama

d Pukiawe

e ʻŌhelo

f Candlenut Tree

Plate 6a

Akoko
Chamaesyce celastroides (Native)

This variable species is a generally fairly nondescript shrub. The leaves vary in size and shape but are generally 15 cm (6 in) long and 7 cm (3 in) wide. The flowers and fruits are in the axils of the leaves and the fruits are upright capsules.

Plate 6b

Castor Bean
Pa'aila
Ricinus communis (Non-native)

This is the species from which the dreaded but medicinal castor oil is derived. It is a fairly attractive shrub commonly found in disturbed areas, open habitats, and forest edges. The leaves are deeply lobed and the rounded reddish brown fruits are covered with soft spines.

Plate 6c

Black Wattle
Acacia mearnsii (Non-native)

This tree has compound fern-like leaves and the new branches and leaves are covered with tiny gray hairs. Flowers are yellow puffs at the end of the branches and fruits are pods that are constricted between each seed. The plants are found in dry forests and pastures on the main islands.

Plate 6d

Koa
Acacia koa (Native)

Koa is a beautiful tree found in middle elevation forests. It was one of the most prized trees in the islands for its strong and lovely wood – so prized that few really large trees exist anymore. Koa wood jewelry and other trinkets are popular souvenirs. The "leaves" (they are really modified stems) are long, narrow, sickle-shaped structures, the flowers are small cream-colored puffs at the ends of branches, and the fruits are flattened pods.

Plate 6e

Moluccan Albizia
Paraserianthes falcataria (Non-native)

These tall trees have feathery compound leaves with a large disk-like nectary at the base of the central stem and smaller ones farther up along it. The pods have papery wings. This species was planted to reforest part of Hawaii and has spread from these plantings.

Plate 6f

Wiliwili
Erythrina sandwicensis (Native)

The wiliwili tree is a small but picturesque tree found in low areas on the dry sides of all the islands. The tree is deciduous (has no leaves) during the dry season but produces spectacular yellow to red (usually orange) flowers on the bare branches. The leaves are compound, with three leaflets that are broader than they are long. The pods generally have only a few seeds in them.

a Akoko

d Koa

b Castor Bean

e Moluccan Albizia

c Black Wattle

f Wiliwili

Plate 7a

Koa Haole
Leucaena leucocephala (Non-native)

The common name for this species means "white person's koa tree" but the species really bears little resemblance to the native Koa. It is a small tree with compound leaves and long dry pods with numerous glossy brown seeds. It has been planted for a number of uses and is one of the most common naturalized trees in tropical regions.

Plate 7b

Sea Bean
Mucuna gigantea (Native)

This woody high-climbing vine may be passed by, unobserved, but those who do spot it have a treat. The creamy colored flowers resemble pea flowers (a close relative) and are found in clusters. The pods are covered with stiff orange-brown hairs. Look for it on the hike above the pools near Hana on Maui.

Plate 7c

Kiawe
Prosopis pallida (Non-native)

One would think that Kiawe is a native species because it is so common along seashores and has many local uses (especially for grilling meats) but, in fact, all the plants are believed to be derived from one individual plant brought to Oahu by Father Bachelot in 1823. The original plant still may be seen at the Catholic Mission on Fort Street in Honolulu. Look for Kiawe along coasts and in low-elevation, dry, disturbed land; and be careful if walking barefoot – the spines on fallen parts of the plant can be wicked.

Plate 7d

Monkeypod
Samanea saman (Non-native)

The monkeypod tree, with a very broad crown, is a popular ornamental species that is now naturalized in disturbed areas around the islands. Guest houses have been built in the canopies of some. The pink flowers are in clusters and the pods are long and thin, and may be twisted when young.

Plate 7e

ʻIlihia
Cyrtandra platyphylla (Native)

"Platyphylla" means "wide leaf" and this species does have wide leaves for a *Cyrtandra.* The leaves are found at the ends of the branches and are opposite each other. The flower is whitish and tubular, flaring at the end of the tube. It is found in mid-elevations in wet forests on Maui and the Big Island.

Plate 7f

Haʻiwale
Cyrtandra longifolia (Native)

Cyrtandra is a large genus and, while the number of actual species in Hawaii is in question, there are certainly plenty of them. Part of the problem is that a single species can be variable in form, making classification difficult. This species is no exception, but it has long, narrow, and leathery leaves. The white axillary flowers resemble its relative, the African Violet, and the berries are white. It is found on Kauai.

a Koa Haole

d Monkeypod

b Sea Bean

e ʻIlihia

c Kiawe

f Haʻiwale

Plate 8a

Naupaka
Scaevola sericea (Native)

This species is found along seashores throughout the South Pacific and on all the islands of Hawaii. It is easily recognizable by the odd, asymmetrically split white flowers, often with purple streaks. The leaves are widest at the far end of the leaf. The fruit is a white berry. There are other species in the genus found in forests on the islands – you may recognize them by the unusual flower.

Plate 8b

Hibiscus
Kokiʻo Keʻokeʻo
Hibiscus arnottianus (Native)

Although the gaudy Chinese hibiscus plants are grown as ornamentals in Hawaii, this species far surpasses them in beauty. Large white flowers (sometimes reddish) with a protruding "staminal column," containing both male and female parts (typical of the hibiscus family), mark this species. It is easily spotted along the mid-elevation trails on Oahu and Molokai.

Plate 8c

Beach Hibiscus
Hau
Hibiscus tiliaceus (Native)

This species is found in salty soils throughout the South Pacific islands. Like *Hibiscus arnottianus*, it has the distinctively shaped flowers of the family, but in this species the flowers are yellow, fading almost to red as they age. The leaves are heart-shaped.

Plate 8d

Milo
Thespesia populnea (Native)

This shrubby tree has shiny rounded leaves with long tips. The flowers are yellow, with a red spot at the base. Fruits are large, flattened, round brown capsules.

Plate 8e

Koster's Curse
Clidemia hirta (Non-native)

The common name should indicate that this is not a beloved species. This low shrub has invaded acres of natural areas in the islands and is the scourge of land managers. The branches and leaves have bristly hairs. The small white flowers are in the leaf axils.

Plate 8f

Lasiandra
Glorybush
Princess Flower
Tibouchina urvilleana (Non-native)

This is a lovely shrub with velvety leaves and bright purple flowers, but in fact it is an introduced species that has become a pest in some areas – notably around the village of Volcano on the Big Island and near Kokee State Park on Kauai. It rarely fruits and spreads mostly from underground stems.

Plate 8g

Miconia
Triana
Miconia calvescens (Non-native)

This species has already destroyed 75% of the forests on Tahiti and is now moving in on the Hawaiian Islands. Land managers on Maui are aggressively pursuing its control and those on the Big Island and Oahu are rising to the challenge as well. A spindly but tall tree, the leaves are very large, have three prominent veins, and are bright purple, especially on the underside. Its many fruits are spread by birds.

a Naupaka

b Hibiscus

c Beach Hibiscus

e Koster's Curse

d Milo

f Lasiandra

g Miconia

Plate 9a

Wauke
Paper Mulberry
Broussonetia papyrifera (Non-native)
Wauke was introduced by the Polynesians, who made "kapa," or barkcloth, from it. It persists from plantings and occasionally spreads into moist areas on all the islands. The leaves are rounded and softly hairy on the undersides. The female flowers are in unshowy rounded clusters in the axils of the leaves.

Plate 9b

Naio
Myoporum sandwicense (Native)
The shrubby trees of naio are variable in form, but the leaves are long and thin, fleshy, and may be sticky when young. The small white flowers are in the axils of the leaves. It may be found at almost all elevations, from dry to wet, but is most common in subalpine forests on all main islands.

Plate 9c

Firetree, Firebush
Faya Tree
Myrica faya (Non-native)
This species was introduced around the middle of the 20th Century and has spread rapidly, especially around Hawaii Volcanoes National Park. This shrubby tree has leathery evergreen leaves and inconspicuous flowers toward the ends of branches. The purplish-black berries are eaten by birds, especially non-native birds.

Plate 9d

Kolea Lau Nui
Myrsine lessertiana (Native)
The leaves of this small tree are leathery, widest at the far end, and are crowded at the ends of branches. Young leaves may be reddish. The flowers are small and purplish and the fruits are black berries. It is found in wet open areas, bogs, and subalpine areas on all the islands.

Plate 9e

Swamp Mahogany
Eucalyptus robusta (Non-native)
The thick, reddish, spongy bark is easily recognizable in the species. Leaves are grayish and fragrant, as are most *Eucalyptus*. The fruit is a small, woody, bluish capsule. This is one of the most commonly planted forestry trees in the islands.

Plate 9f

ʻŌhiʻa Lehua
Metrosideros polymorpha (Native)
One of the most beautiful trees in the world, with a gnarled, picturesque silouette and vivid red (sometimes yellow) shaving-brush-like flowers. This is a very variable species but it is easily recognizable by the shape, flowers, and small rounded evergreen leaves. ʻŌhiʻa is found in a number of habitat types.

Plate 9g

ʻŌhiʻa Lehua
Metrosideros polymorpha (Native)
ʻŌhiʻa flowers (see above), which provide nectar for many Hawaiian birds, bees, and butterflies.

a Wauke

g ʻŌhiʻa Lehua

b Naio

c Firetree

d Kolea Lau Nui

e Swamp Mahogany

f ʻŌhiʻa Lehua

Plate 10a

Strawberry Guava
Psidium cattleianum (Non-native)
Hundreds of acres are covered with this small alien tree with shiny, leathery evergreen leaves, yellow inflorescences, and a peeling bark. The tangy red (sometimes yellow) fruits are sometimes used in jams and drinks.

Plate 10b

Common Guava
Psidium guajava (Non-native)
Guava drinks are popular in Hawaii and this non-native invasive species is the source of the fruit. The large leaves are leathery and have heavy veins. The oversweet large yellow fruits are often encountered on trails, covered in fruit flies. This weedy shrub may be found on all the islands.

Plate 10c

Mountain Apple
Syzygium malaccense (Non-native)
This species was a Polynesian introduction for its tasty, watery fruit. It is common in lowland wet forests and is characterized by pink apple-like fruit streaked with red, and flowers of mostly brilliant pink/red reproductive parts. It is found in wet areas at low elevations.

Plate 10d

'Aulu
Pisonia sandwichensis (Native)
A small tree, 'Aulu has thick leathery leaves with conspicuous veins on the leaves. The petioles are thick. Brownish flowers are in the axils of the leaves. It may be found in dry sites at low to mid-elevations on the main islands.

Plate 10e

Olopua
Hawaiian Olive
Nestegis sandwicensis (Native)
This plant is in the olive family and it does resemble olives somewhat, with leathery glandular leaves and greenish olive-like fruits that become brown at maturity. It is found in dry sites, low to mid-elevations, on all the islands.

Plate 10f

Passionfruit
Liliko'i
Passiflora edulis (Non-native)
These woody vines are notable for their extraordinary fringed lavender flowers. The shiny leaves have three lobes. Fruits are variable in color and are very sweet and juicy. It is found at low to mid-elevations on the main islands.

Plate 10g

Banana Poka
Passiflora mollisima (Non-native)
Legend has it that this species was introduced to cover an outhouse earlier in the 20th Century. Regardless of introduction, this vine now covers many acres of trees and other landscapes. With large pendant pink flowers and banana-like fruit, this is a beautiful ornamental species with serious consequences to the environment. It covers trees in the same way that the infamous Kudzu does in the southeastern USA.

a Strawberry Guava

b Common Guava

g Banana Poka

c Mountain Apple

d 'Aulu

e Olopua

f Passionfruit

Plate 11a

Pyracantha
Pyracantha angustifolia (Non-native)
This thorny shrub is notable for its bright red-orange fruits. It is familiar to gardeners from temperate zone regions and appears to be spreading rapidly in the areas around Kokee State Park and Hawaii Volcanoes National Park.

Plate 11b

Prickly Florida Blackberry
Rubus argutus (Non-native)
This is a serious weed in Hawaii, notable for its vigorous, thick, and very spiny stems and large blackberry fruits. The leaves are compound with three to five leaflets that have small spines on the underside. It is found in wet forests and disturbed sites.

Plate 11c

Hawaiian Raspberry
ʻAkala
Rubus hawaiensis (Native)
This plant's large blackberry fruits tempt, but beware – the red or yellow fruits are bitter. The stems are mostly spineless and the leaves each have three leaflets. Flowers are pinkish-purple. It is found in moist forests at mid to high elevations on the main islands (except Oahu).

Plate 11d

Thimbleberry
Rubus rosifolius (Non-native)
Common in wet forest understories, this low shrub spreads by rhizomes. The compound leaflets are soft and the stems are not thorny. Flowers are white and fruits are small and red. It is common in low wet forests on all the islands.

Plate 11e

Kukaenene
Coprosma ernodeoides (Native)
If you hike on the higher lava fields of Maui or the Big Island you may encounter this trailing shrub. Leaves are tiny and closely crowded, flowers are small and tubular, and fruits are glossy and black.

Plate 11f

Manono
Hedyotis terminalis (Native)
This is one of the most variable species in the Hawaiian flora, perhaps because of frequent hybridization between forms. The leaves are variable, with veins that curve toward the leaf tip. Flowers and fruits are at the ends of branches. Manono occurs on all the main islands, mostly in wetter locations.

Plate 11g

Indian Mulberry
Noni
Morinda citrifolia (Non-native)
Morinda was introduced by the ancient Polynesians who used the large, warty, compound fruit to make a tonic. Modern herbalists now use the fruit to make expensive elixirs said to cure a variety of ills. The shrub has very large glossy leaves and the plants are found in dry coastal areas, especially around traditional Hawaiian encampments.

a Pyracantha

b Prickly Florida Blackberry

c Hawaiian Raspberry

d Thimbleberry

e Kukaenene

f Manono

g Indian Mulberry

Plate 12a

Kopiko
Psychotria hexandra (Native)
The genus *Psychotria* is one of the largest in the world, with several species in Hawaii. This species has thick leathery leaves that often have a very long pointed tip. The white tubular flowers are in clusters at the ends of branches. It is found in wet forests on Oahu and Kauai.

Plate 12b

ʻIliahialoʻe
ʻIliʻahi
Coast Sandlewood
Santalum ellipticum (Native)
The sandlewoods were heavily exported for their fragrant woods and have greatly decreased in number. Leaves on small trees are leathery and grayish. The green-orange flowers are fragrant and are a tube that ends in four lobes. This species is found in dry forests and shrublands on all the islands.

Plate 12c

ʻAʻaliʻi
Dodonaea viscosa (Native)
The key feature on this shrub is the inflated, winged bladder-like fruits, which are reddish purple and quite striking. Leaves are papery and often glandular. ʻAʻaliʻi is especially found in dry open areas from low to mid elevations on all main islands.

Plate 12d

Soapberry
Aʻe
Sapindus saponaria (Native)
This tree has large deciduous compound leaves. The bark is scaly on mature trees. Flowers are found in loose clusters at the ends of branches. Called Soapberry because a soap substitute can be made from the fruits. Seeds are strung into leis. Aʻe is found at mid-elevations on the Big Island's volcanoes.

Plate 12e

ʻAlaʻa
Pouteria sandwicensis (Native)
The sticky milky sap of this species was used by the Polynesians to catch small birds. The trees have thick, leathery leaves with a prominent center vein and numerous lateral veins. Flowers and fruits are in the leaf axils and the fruits are round and orange, black at maturity. ʻAlaʻa occurs in dry mid-elevation forests on the main islands.

Plate 12f

ʻAkia
Wikstroemia uva-ursi (Native)
This is a sprawling shrub with small, sometimes hairy, leaves. The small flowers are yellow and the fleshy fruit is red. It is often found in ornamental plantings and also in low-elevation dry sites on all main islands.

Plate 12g

Lantana
Lantana camara (Non-native)
Walking through a thicket of this aggressive weed will leave your legs shredded by the numerous thick spines. Lantana has tight clusters of small flowers that range in color from yellow to orange within the same cluster. Leaves are usually well chewed by some unseen insect – this is the work of a beetle that was introduced to combat this nasty weed. Lantana is found in dry areas on all the islands.

a Kopiko

b 'Iliahialo'e

c 'A'ali'i

d Soapberry

e 'Ala'a

f 'Akia

g Lantana

Plate 13a

Ti

Cordyline fruiticosa (Non-native)

The Polynesians, who introduced this species, found numerous uses for the large strap-like leaves of the ti plant. They were used as roofing thatch, wrappers, and for clothing. The roots were used for food and beverages. Leaves are generally green but may be edged with red or cream. The leaves are usually at the end of the single stem and the stem has circular markings that show the locations of previous leaves. Ti is found in low sites on all the islands.

Plate 13b

Taro

Colocasia esculenta (Non-native)

Visitors to Hawaii should always try *poi*, the Polynesian staple that is made from the taro root. Even those who do not like poi may like the delicious (but blue-purple) breads that are made from the starchy root. The plant resembles the ornamental "elephant ears" with all the large heart-shaped leaves arising from the base of the plant. Flowers are crowded on a stem that is surrounded by a green tubular leaf. Taro is grown in swampy patches throughout the islands.

Plate 13c

Golden Pothos

Epipremnum pinnatum (Non-native)

Low to middle elevation forests are the place to spot this attractive but weedy vine. The yellow and green leaves cover ground and tree trunks in many areas such as the Koolau Mountains on Oahu and the road to Hana on Maui. If you think you have seen this species hanging in restaurants, you are right – it is a popular indoor plant.

Plate 13d

Coconut Palm

Niu

Cocos nucifera (Native)

Coconut palms line coastal areas around the South Pacific – the large fruits float in the ocean and wash up on beaches throughout the region. The long graceful palm fronds were used by the Polynesians for a number of fiber needs (you can buy woven coconut frond bowls and hats at tourist shops and from roadside vendors at many popular tourist stops) and the fruits, of course, have always been popular. Be careful walking under the trees – the huge fruits do fall on heads!

Plate 13e

Banana

Musa x paradisiaca (Non-native)

The Polynesians brought banana with them and planted small groves throughout the hills. The patches are persistent and one of the great joys of hiking in Hawaii is coming across a small grove in a remote valley and snacking on the fruit (much smaller and more flavorful than the commercial variety). Leaves are quite large and often tattered from the wind.

Plate 13f

Bamboo Orchid

Arundina graminifolia (Non-native)

Fields of orchids are beautiful, even if the orchids are weedy non-natives. The bright pink flowers are at the ends of tall grass-like stems. Look for this species especially around Hilo and Puna on the Big Island, and also on Kauai and Oahu.

Plate 13g

ʻIeʻie

Freycinetia arborea (Native)

This woody vine with long strap-shaped leaves has unusual flowers. The leaf-like bracts under the inflorescence (multiple-flower structures) are fleshy (often chewed by rats) and are orange and pink. The flowers themselves are yellowish. It is common in wet mid-elevation forests on all the islands.

a Ti

b Taro

c Golden Pothos

d Coconut Palm

e Banana

f Bamboo Orchid

g ‘Ie‘ie

Plate 14a

Hala
Pandanus tectorius (Native)

Large fruits are the distinctive trait of this species – they resemble pineapples stuck up in the trees. Leaves are long and strap-like and have sharp spines on the margins. The plants generally have exposed roots propping up the stem. Hala is found along the coast on all the islands. There is a particularly nice patch outside of Hana, on Maui.

Plate 14b

Job's Tears
Coix lachryma-jobi (Non-native)

This is a tall annual grass that has become naturalized on most of the islands. The key feature, and source of the common name, is large shiny dark fruits that hang from the ends of the stems.

Plate 14c

Fountain Grass
Pennisetum setaceum (Non-native)

This is one of the most destructive of the many introduced species in the islands. Fountain Grass is found in dry areas on the islands, especially on the Big Island. As the plants dry they become tinder for a spark or tossed cigarette, the large biomass causing fires that devastate native species not adapted to fire. This species is recognizable by the flowering spikes that superficially resemble those of wheat.

Plate 14d

White Ginger
Hedychium coronarium (Non-native)

This herbaceous plant can be 2 m (6 ft) or more tall. Flowers are white (often with a bit of yellow) and very fragrant. Fruits are capsules with bright red seeds. White Ginger is found in wet forests on all the main islands.

Plate 14e

Kahili Ginger
Hedychium gardnerianum (Non-native)

The inflorescences on this plant are amazing – cylindrical affairs with bright yellow and red flowers that have protruding red stamens. It is common in the wet forests of Hawaii Volcanoes National Park as well as in Kokee State Park and areas near Hana on Maui.

Plate 14f

Pili Grass
Heteropogon contortus (Native)

Pili Grass is widely distributed in coastal areas on all the islands. The leaves are in a basal rosette and the inflorescences resemble wheat. Pili grass in the past was used to thatch houses.

Plate 14g

Māmane
Sophora chrysophylla (Native)

This is a shrub or small tree in dry areas on all the main islands. It has compound leaflets and the leaves and young branches are covered with golden hairs. The pea-shaped flowers are yellow. The pods are short and strongly constricted around the seeds.

a Hala

b Job's Tears

c Fountain Grass

d White Ginger

e Kahili Ginger

f Pili Grass

g Māmane

Plate 15a

‘Ōlapa
Cheirodendron sp. (Native)
Species of *Cheirodendron* are found in wet forests on all the islands. The compound leaves are opposite each other. Flowers and fruits are purple.

Plate 15b

‘Ilima
Sida fallax (Native)
‘Ilima flowers are commonly used in leis and are considered to be an emblem of the island of Oahu. The flowers are in the axils of the leaves and are yellow to orange, sometimes with red at the base. ‘Ilima is found in dry areas on all the islands.

Plate 15c

Mamaki
Waimea
Pipturus albidus (Native)
The leaves of this shrub are papery and covered with hairs. The compressed heads of flowers are found in the axils of the leaves. Mamaki is found in wet forests from low to mid-elevations on all the islands.

Plate 15d

False Staghorn Fern
‘Uluhe
Dicranopteris linearis (Native)
This fern is found in disturbed areas on all the main islands. Stems are quite long and the fronds fork at the ends. Making your way through a dense patch of ‘uluhe is nearly impossible.

Plate 15e

Hapu‘u Pulu
Cibotium glaucum (Native)
Tree ferns need a moist environment so this species occurs in low and mid-elevation wet forests, especially around the Thurston Lava Tubes on the Big Island. The new fronds are densely covered with reddish brown hairs.

Plate 15f

‘Ama‘u
Sadleria cyantheoides (Native)
A beautiful fern on a short trunk, ‘Ama‘u has 1 m-long (3 ft) fronds that are bronze/red when young and dark glossy green when mature. ‘Ama‘u is found on all the islands on young lava flows and in moist forests.

Plate 15g

Hawaiian Greenbrier
Hoi Kuahiwi
Smilax melastomifolia (Native)
The leaves of this woody vine are leathery, glossy, and have conspicuous veins. There may be small spines on the veins. The small flowers are found in the axils of the leaves. This species may be found in wet forests at low to mid-elevations on all main islands.

a ‘Ōlapa

b ‘Ilima

c Mamaki

d False Staghorn Fern

e Hapu‘u Pulu

f ‘Ama‘u

g Hawaiian Greenbrier

Plate 16a

'Ulei
Osteomeles anthyllidifolia (Native)
This is a low shrub of dry areas on all the islands. The leaflets of the compound leaves are small, glossy on the upper side and usually hairy on the underside. Flowers are white, as are the fruits.

Plate 16b

Beach Pea
Mauna Loa
Canavalia carthartica (Non-native)
The flowers on this herbaceous vine are rose-pink. The pods are not constricted between the seeds. Mauna Loa vine is found in low to mid-elevation dry areas, including coastal areas, on most of the islands.

Plate 16c

Kanawao
Brousaissia arguta (Native)
These shrubs superficially resemble their relatives, the ornamental hydrangeas. Leaves are leathery, opposite each other, and leave obvious scars on the stems when removed. Petals vary in color from blue to pink to white and are flexed back, exposing the reproductive parts.

Plate 16d

Rose Apple
Syzygium jambos (Non-native)
These are tall trees with leathery, long, narrow leaves. Creamy stamens on the flowers extend far beyond the petals, resembling a powder puff. Fruits are pinkish white. Rose Apple is found in low-elevation wet forests on all the islands.

Plate 16e

Kukaemoa
Alani
Melicope clusifolia (Pelea clusifolia) (Native)
This is a shrub with whorls of leaves along the stem. The leaves are leathery and generally widest at the far end of the leaf. The veins are prominent. The capsules split open into four parts to reveal the seeds. Kukaemoa is found in wet forests on all the islands.

Plate 16f

Breadfruit
'Ulu
Artocarpus altilis (Non-native)
Breadfruit was introduced by the Polynesians, who used the fruit. The leaves of this tree are large and glossy. It is commonly planted and may be sparingly naturalized.

Plate 16g

Ma'o
Hawaiian Cotton
Gossypium tomentosum (Native)
This small shrub has lobed leaves and bright yellow flowers. The fruits are capsules with seeds that are covered with brownish hairs. It is found in low-elevation dry areas on all the islands.

a 'Ulei

b Beach Pea

c Kanawao

d Rose Apple

e Kukaemoa

f Breadfruit

g Ma'o

Plate 17a

Giant Toad (also called Marine Toad, Cane Toad)
Bufo marinus
Poloka

ID: A large, ugly, warted toad; large, triangular glands on each side of the head behind eyes; males brown, females often combinations of dusky brown, tan, and chocolate; to 19 cm (7.5 in).

HABITAT: Low and middle elevation open and semi-open sites, agricultural areas, suburban gardens; found on the ground; dusk- and night-active.

ISLANDS: BIG, KAU, LAN, MAUI, MOLO, OAHU

ORIGIN: Central and South America; brought to Hawaii in 1932.

Plate 17b

Bullfrog
Rana catesbeiana
Poloka Lana

ID: Large green frog, often green with brown mottling; light belly with darker spots; large tympanum (eardrum) behind eye; webbed toes; 10 to 17 cm (4 to 6.5 in).

HABITAT: Found in or around most freshwater sites at low and middle elevations, including lakes, ponds, marshes, rivers, and streams; day-active; also night-active when breeding.

ISLANDS: BIG, KAU, LAN, MAUI, MOLO, OAHU

ORIGIN: North America; brought to Hawaii during late 1800s.

Plate 17c

Wrinkled Frog
Rana rugosa

ID: Small gray, brownish, or grayish brown frog with smooth skin but lengthwise ridges ("wrinkles") on back and limbs; webbed toes; to about 4 cm (1.5 in).

HABITAT: Found in and around low and middle elevation rivers and streams, generally in shady, cooler sites; often basks on rocks in mid-stream or on bank; day-active.

ISLANDS: BIG, KAU, MAUI, OAHU

ORIGIN: Japan and Korea; brought to Hawaii during 1890s.

Plate 17d

Green and Black Poison-dart Frog (also called Green Poison-arrow Frog)
Dendrobates auratus

ID: Small bright green, turquoise, or dark green frog with brownish or black splotches; 2.5 to 4 cm (1 to 1.5 in).

HABITAT: Low and middle elevation wet forested valleys; found in moist areas on the forest floor or on low vegetation.

ISLANDS: (BIG?) OAHU

ORIGIN: Central America; brought to Hawaii in the 1930s.

Plate 17e

Jackson's Chameleon
Chamaeleo jacksonii

ID: Largish, odd-looking lizard, often green or yellowish brown, but color highly variable; body side-to-side flattened; small eyes set in prominent "turrets," which swivel; males with three "horns" on snout (some females with small horns); prehensile tail often curled; males larger than females, to 15 cm (6 in), plus long tail.

HABITAT: Low and middle elevation wetter forest areas, but also some drier woodlands, backyards, gardens; found in trees, shrubs, hedges; day-active.

ISLANDS: BIG, KAU, LAN, MAUI, OAHU

ORIGIN: Central eastern Africa; released in Hawaii in 1970s.

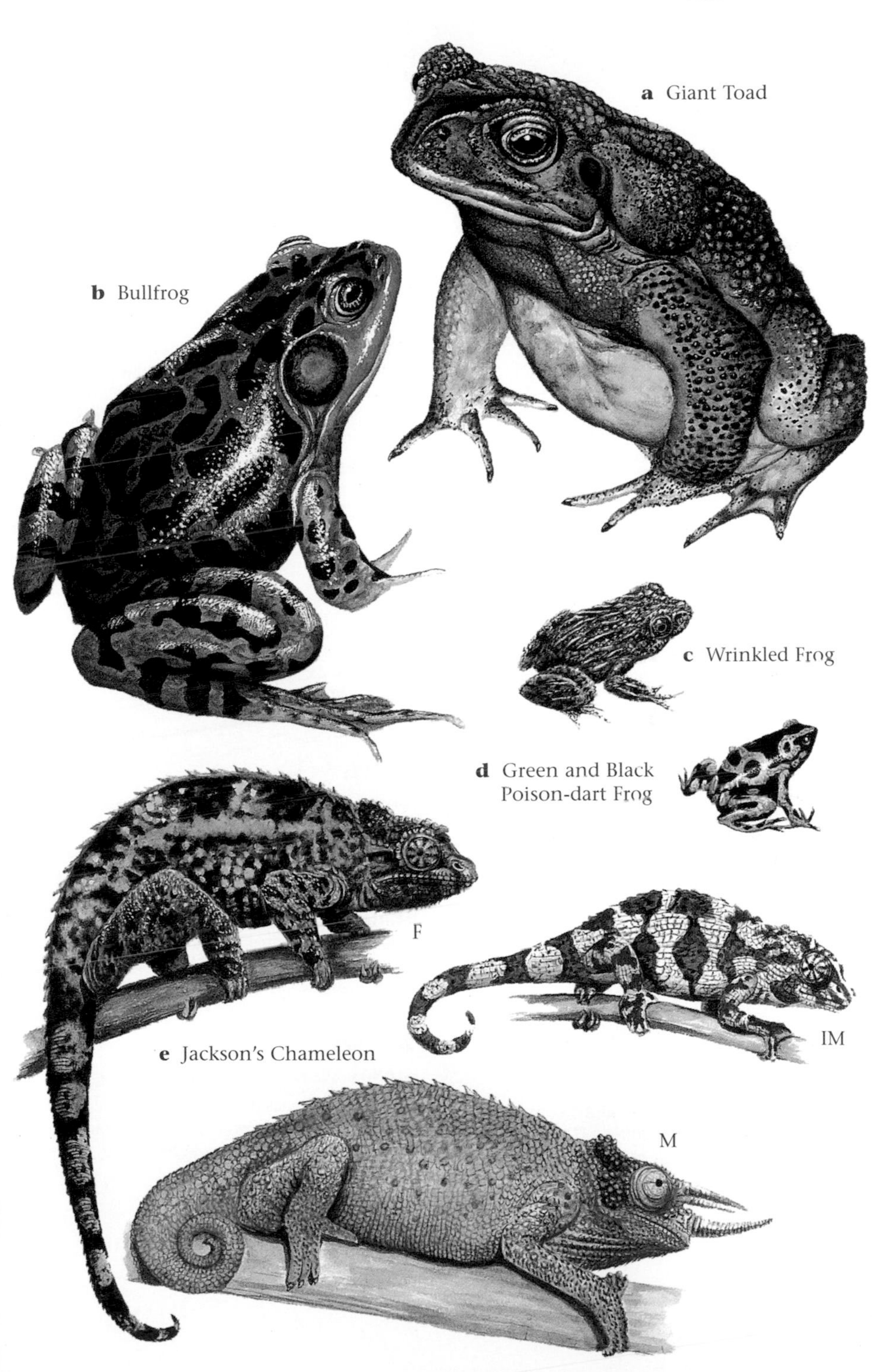
a Giant Toad
b Bullfrog
c Wrinkled Frog
d Green and Black Poison-dart Frog
F
IM
e Jackson's Chameleon
M

Plate 18a

Green Anole
Anolis carolinensis
Moʻo

ID: Small slender lizard with pointed snout; can be various shades of brown or green, including bright green, or grayish; male with pinkish throat flap (dewlap; only sometimes visible); to 7.5 cm (3 in), plus long tail.

HABITAT: Found in trees and shrubs in open and semi-open areas, usually in or near human settlements – on fences, in gardens and agricultural areas, but also in adjacent woodlands; day-active.

ISLANDS: BIG, KAU, MAUI, MOLO, OAHU

ORIGIN: Southeastern USA; established in Hawaii in 1950s.

Plate 18b

Brown Anole
Anolis sagrei
Moʻo

ID: Small light brown, dark brown, or blackish, plump-bodied lizard; male with crest on head and back, and orangish or reddish throat flap (dewlap; only sometimes visible); female often with light band or row of blotches along back; to 6.5 cm (2.5 in), plus long tail.

HABITAT: Found on the ground and in trees and shrubs in open and semi-open areas in and around human settlements – on fences, in gardens and backyards; day-active.

ISLANDS: OAHU

ORIGIN: Cuba and the Caribbean region; established in Hawaii in 1980s.

Plate 18c

Metallic Skink (also called Delicate Skink, Rainbow Skink)
Lampropholis delicata

ID: Small, slender, shiny lizard, brown or grayish brown usually with small dark spots; usually with dark side stripe; grayish tail often with bluish gloss; light belly; to 5 cm (2 in), plus tail.

HABITAT: Low and middle elevations, generally in or near human settlements; found on ground in open and semi-open sites such as gardens, parks, along roads and trails; day-active.

ISLANDS: BIG, KAU, MAUI, MOLO, OAHU

ORIGIN: Australia; in Hawaii since early 1900s.

Plate 18d

Snake-eyed Skink
Cryptoblepharus poecilopleurus

ID: Small brown or grayish shiny lizard with two thin gold lengthwise stripes on back; dark sides with light spots; yellowish belly; fairly long legs (for a skink); to 4.5 cm (1.8 in), plus tail.

HABITAT: Low and some middle elevation open sites, among rocks, leaf litter; most often seen in rocky areas along shore, near beach, especially on old lava flows; day- and dusk-active.

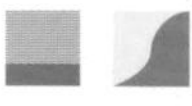

ISLANDS: BIG, KAU, LAN, MAUI, MOLO, OAHU

ORIGIN: This species is broadly distributed on Pacific Islands; may have reached Hawaii on its own, or with early Polynesian settlers.

Plate 18e

Moth Skink
Lipinia noctua

ID: Small brownish lizard with darker sides; yellowish spot on head continuous with light back stripe; to 5 cm (2 in), plus tail.

HABITAT: Usually found in low-elevation sites in more open areas, often around human settlements, in gardens; found on the ground, in leaf litter, and on rock walls; day- and dusk-active.

ISLANDS: BIG, KAU, MAUI, MOLO, OAHU

ORIGIN: This species is broadly distributed on Pacific Islands; may have reached Hawaii on its own, or with early Polynesian settlers.

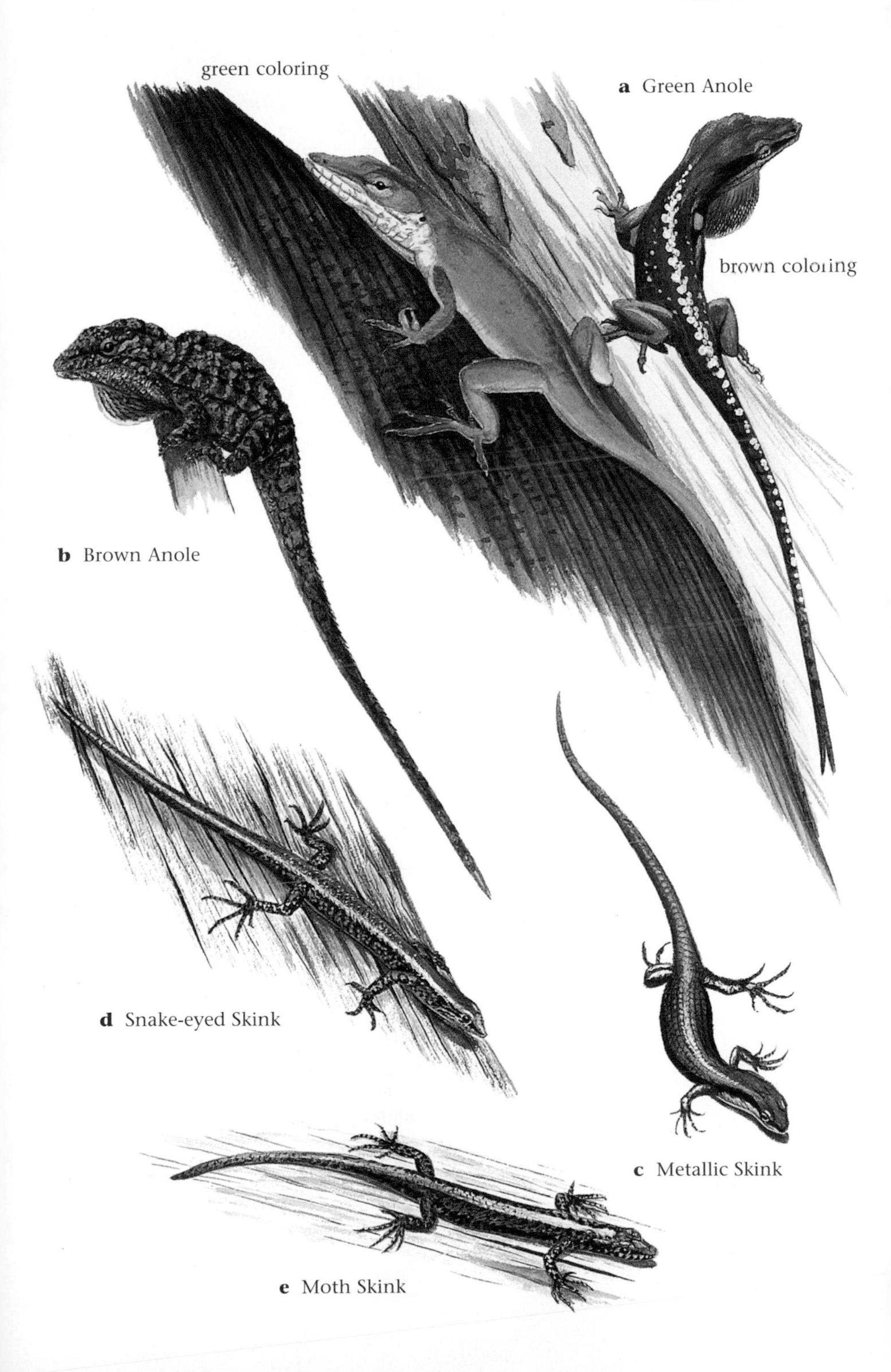
green coloring
a Green Anole
brown coloring
b Brown Anole
d Snake-eyed Skink
c Metallic Skink
e Moth Skink

Plate 19a

Mourning Gecko (also called Scaly-toed Gecko)
Lepidodactylus lugubris
Mo'o Ala

ID: Very small brown gecko with dark eye-line; light and dark bands on back; slightly flattened tail with tooth-like scales on sides; to 4 cm (1.5 in), plus tail.

HABITAT: Found in and around human settlements, on and in buildings; also found on ground, on trees, in gardens; nocturnal.

ISLANDS: BIG, KAU, LAN, MAUI, MOLO, OAHU

ORIGIN: Asia and Indo-Pacific region.

Plate 19b

Stump-toed Gecko (also called Stump-toed Dtella)
Gehyra mutilata
Mo'o Ala

ID: Gray, grayish brown, or blackish lizard with light and dark spots; usually lighter, even whitish, at night; often a thin dark eye-line; to 6.5 cm (2.5 in), plus tail.

HABITAT: Found in and around human settlements, on and in buildings; also found on ground, on trees, in gardens; nocturnal.

ISLANDS: BIG, KAU, LAN, MAUI, MOLO, OAHU

ORIGIN: Native to many South Pacific islands.

Plate 19c

Tree Gecko (also called Waif Gecko)
Hemiphyllodactylus typus
Mo'o Ala

ID: Very small, slender, gray, brown, or blackish lizard (lighter at night) with narrow tail; dark stripe from eye to shoulder; two dark spots at base of tail; light belly; to 4.5 cm (1.8 in), plus tail.

HABITAT: Found in forested areas, on trees, among rocks, or on ground; nocturnal.

ISLANDS: BIG, KAU, LAN, MAUI, MOLO, OAHU

ORIGIN: Native to many South Pacific islands.

Plate 19d

Indo-Pacific Gecko (also called Fox Gecko)
Hemidactylus garnotii
Mo'o Ala

ID: Gray, light brown, blackish, or whitish lizard, often with light or dark flecks; tail slightly flattened, orangish or pinkish underneath; to 5.5 cm (2.2 in), plus tail.

HABITAT: Found at low and middle elevations in and around human settlements, on and in buildings; also found on ground, on trees, in gardens; nocturnal.

ISLANDS: BIG, KAU, LAN, MAUI, MOLO, OAHU

ORIGIN: Perhaps India, southeast Asia; occurs naturally on many Pacific Islands.

Plate 19e

House Gecko (also called Leaf-toed Gecko)
Hemidactylus frenatus
Mo'o Ala

ID: Small brown, beige, grayish or whitish gecko with tiny spots or flecks; some have vague, wavy stripes running from eye to tail; light belly; 5 cm (2 in), plus tail.

HABITAT: Found at low and middle elevations, in and around human settlements, on and in buildings; also on ground and fences, in trees; nocturnal.

ISLANDS: BIG, KAU, LAN, MAUI, MOLO, OAHU

ORIGIN: Asia, but has now spread to many parts of the world through its association with people.

Plate 19f

Gold Dust Day Gecko
Phelsuma laticauda
Mo'o Ala

ID: Green or yellowish green lizard with bright yellow spots on upper back; reddish or brownish bars on back and, often, on head; to 7 cm (2.75 in), plus tail.

HABITAT: Found in and around human settlements, in buildings; also in trees, shrubs, gardens; day-active.

ISLANDS: BIG, MAUI, OAHU

ORIGIN: Native to Madagascar and other islands off African coast.

a Mourning Gecko
b Stump-toed Gecko
c Tree Gecko
d Indo-Pacific Gecko
e House Gecko
f Gold Dust Day Gecko

Plate 20a

Laysan Albatross (also called White Gooney Bird)
Phoebastria immutabilis
Moli

ID: Very large white bird with long narrow wings dark above, dark and light below; dark or gray eye patch; small dark tail; light bill with dark, down-curved tip; to 81 cm (32 in); wingspan to 2 m (6.5 ft).

HABITAT: Pelagic seabird; rarely near land except to nest on oceanic islands with low vegetation.

ISLANDS: BIG, KAU, OAHU, MID

ORIGIN: Native seabird.

Plate 20b

Black-footed Albatross (also called Black Albatross, Brown Gooney Bird)
Phoebastria nigripes

ID: Very large brown or brownish gray seabird with long narrow wings; whitish at base of bill and behind eye; dark bill with down-curved tip; dark legs; to 74 cm (29 in); wingspan to 2.1 m (6.8 ft).

HABITAT: Pelagic seabird; rarely near land except to nest on oceanic islands on beaches or slopes with little vegetation.

ISLANDS: KAU, MID

ORIGIN: Native seabird.

Plate 20c

Short-tailed Albatross (also called Golden Gooney Bird)
Phoebastria albatrus

ID: Very large white seabird with brownish upper wing and tail; back of neck yellowish-buffy; pink down-curved bill with dark tip; pink legs; to 94 cm (37 in); wingspan to 2.2 m (7.2 ft).

HABITAT: Pelagic seabird; rarely near land except to nest on a few small islands.

ISLANDS: MID

ORIGIN: Native seabird.

Note: Endangered, CITES Appendix I and USA ESA listed.

Plate 20d

Great Frigatebird
Fregata minor
ʻIwa

ID: Large black seabird with long, narrow, pointed wings; long forked tail; long gray bill with down-curved tip; male with reddish throat patch (seen only during breeding, so never seen on main islands); female with gray throat and whitish chest; immature bird with white head and chest/belly; to 92 cm (3 ft); wingspan to 2 m (6.5 ft).

HABITAT: Found around seashores; often seen soaring high over coastal areas.

ISLANDS: BIG, KAU, LAN, MAUI, MOLO, OAHU, MID

ORIGIN: Native seabird.

a Laysan Albatross
b Black-footed Albatross
c Short-tailed Albatross
F
IM
M
d Great Frigatebird

Plate 21a

Wedge-tailed Shearwater
Puffinus pacificus

ʻUaʻu-kani

ID: Mid-sized seabird with wedge-shaped tail, light or flesh-colored legs, and grayish bill with down-curved tip; dark form dark above, grayish below with lighter throat and chest; light form dark above, light below; to 43 cm (17 in); wingspan to 1.1 m (42 in).

HABITAT: Pelagic seabird; rarely near land except to nest in burrows on small islands and some seacoasts.

ISLANDS: KAU, LAN, MAUI, MOLO, OAHU, MID

ORIGIN: Native seabird.

Plate 21b

Christmas Shearwater
Puffinus nativitatis

ʻUaʻu-kani

ID: Mid-sized all dark brown seabird with rounded tail; dark bill with down-curved tip; to 38 cm (15 in); wingspan to 80 cm (31 in).

HABITAT: Pelagic seabird; rarely near land except to breed on oceanic islands.

ISLANDS: MOLO, OAHU, MID

ORIGIN: Native seabird.

Plate 21c

Sooty Shearwater
Puffinus griseus

ʻUaʻukani

ID: Mid-sized seabird, sooty brown above, lighter areas below; dark bill with down-curved tip; to 50 cm (20 in); wingspan to 1.1 m (42 in).

HABITAT: Pelagic seabird; rarely near land except to breed on islands outside of Hawaii region on slopes covered with dense vegetation.

ISLANDS: BIG, KAU, OAHU, MID

ORIGIN: Native seabird.

Plate 21d

Newell's Shearwater
Puffinus newelli

ʻAʻo or ʻUaʻu-kani

ID: Mid-sized seabird, dark above, white below, with white extending up onto sides of rump; all black face; black bill with down-curved tip; black legs; to 35 cm (14 in); wingspan to 88 cm (35 in). (Sometimes considered the same species as Townsend's Shearwater, *Puffinus auricularis*, which breeds on islands off western Mexico.)

HABITAT: Pelagic seabird; rarely near land except when nesting in thickly vegetated high mountain sites.

ISLANDS: BIG, KAU, MOLO, OAHU

ORIGIN: Native seabird; endemic to Hawaii.

Note: Threatened, USA ESA listed.

dark form
a Wedge-tailed Shearwater
light form
b Christmas Shearwater
c Sooty Shearwater
d Newell's Shearwater

Plate 22a

Bulwer's Petrel
Bulweria bulwerii

'Ou

ID: Small dark brown seabird with light band on upperwing, longish pointed tail; small dark bill with down-curved tip; to 28 cm (11 in); wingspan to 72 cm (28 in).

HABITAT: Pelagic seabird; rarely near land except when nesting on barren offshore islands.

ISLANDS: KAU, LAN, MAUI, OAHU, MID

ORIGIN: Native seabird.

Plate 22b

Hawaiian Petrel (formerly called Dark-rumped Petrel)
Pterodroma sandwichensis

'Ua'u

ID: Mid-sized seabird, dark above, white below; white forehead; black stripe or patch on underside of each wing; wedge-shaped tail; dark bill with down-curved tip; 43 cm (17 in); wingspan 91 cm (3 ft).

HABITAT: Pelagic seabird; rarely near land except when nesting in rocky areas in high mountain sites.

ISLANDS: BIG, KAU, LAN, MAUI

ORIGIN: Native seabird.

Note: Endangered, USA ESA listed.

Plate 22c

Bonin Petrel
Pterodroma hypoleuca

ID: Small seabird, gray above with darker markings, whitish below with dark wing bars; dark head top; whitish forehead; small dark bill with down-curved tip; 30 cm (12 in); wingspan to 70 cm (27.5 in).

HABITAT: Pelagic seabird; rarely near land except when nesting in burrows on oceanic islands.

ISLANDS: MID

ORIGIN: Native seabird.

Plate 22d

Black-winged Petrel
Pterodroma nigripennis

ID: Small seabird mostly very dark above but with bluish gray head top and back of neck; light below with dark wingbars; whitish forehead; black eye patch; dark-tipped grayish tail; small dark bill with down-curved tip; to 30 cm (12 in); wingspan to 70 cm (27.5 in).

HABITAT: Pelagic seabird; rarely near land except when nesting in burrows and crevices on oceanic islands outside of Hawaii region.

ISLANDS: OAHU (but probably occurs throughout the region at sea)

ORIGIN: Native seabird.

Plate 22e

Mottled Petrel
Pterodroma inexpectata

ID: Mid-sized seabird, gray above with darker markings; white below but with gray belly and dark wingbars; black eye patch; small black bill with down-curved tip; to 35 cm (14 in); wingspan to 80 cm (31.5 in).

HABITAT: Pelagic seabird; rarely near land except when nesting in burrows and crevices on islands outside of Hawaii region.

ISLANDS: KAU, MOLO, OAHU (but probably occurs throughout the region at sea)

ORIGIN: Native seabird.

c Bonin Petrel
a Bulwer's Petrel
b Hawaiian Petrel
d Black-winged Petrel
e Mottled Petrel

Plate 23a

Masked Booby
Sula dactylatra

ʻA

ID: Large white seabird with black skin around face, black and white wings, and silvery or darkish tail; yellow or greenish bill and feet; youngsters with brownish, mottled plumage, with mostly white chest/belly; to 81 cm (32 in); wingspan to 1.6 m (5.1 ft).

HABITAT: Coastal and offshore waters; seen flying low over the ocean; breeds on atolls, small islands.

ISLANDS: OAHU, MID

ORIGIN: Native seabird.

Plate 23b

Brown Booby
Sula leucogaster

ʻA

ID: A mid-sized seabird with brown back, brown neck, white belly, and greenish or yellowish, cone-shaped, sharply pointed bill; greenish or yellowish feet; pointed wings; youngsters with pale brownish belly; to 76 cm (30 in); wingspan to 1.4 m (4.5 ft).

HABITAT: Coastal; found around seashores and islands; breeds on small islands.

ISLANDS: BIG, KAU, LAN, MAUI, MOLO, OAHU

ORIGIN: Native seabird.

Plate 23c

Red-footed Booby
Sula sula

ʻA

ID: Largish white seabird with black on wings, pink skin around face, and blue bill; red legs and feet; youngsters brownish with dark legs and bill; 71 cm (28 in); wingspan to 1 m (40 in).

HABITAT: Coastal; found around seashores and islands; nests in trees on islands, including Oahu and Kauai.

ISLANDS: BIG, KAU, LAN, MAUI, MOLO, OAHU, MID

ORIGIN: Native seabird.

a Masked Booby
IM
IM
b Brown Booby
c Red-footed Booby
IM

Plate 24a

White-tailed Tropicbird
Phaethon lepturus
Koa'e Kea

ID: Largish slender white seabird (sometimes with a pinkish hue) with black markings on wings, black eyestripe, yellowish bill, and very long, thin, white tail streamers; to 80 cm (32 in), including tail; wingspan to 96 cm (38 in).

HABITAT: Soars over ocean, shoreline, and through inland valleys; nests in crevices or on ground.

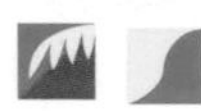

ISLANDS: BIG, KAU, LAN, MAUI, MOLO, OAHU, MID

ORIGIN: Native seabird.

Plate 24b

Red-tailed Tropicbird
Phaethon rubricauda
Koa'e 'ula

ID: Largish chunky white seabird (sometimes with a pinkish hue) with black eyestripe, red bill, and very long, thin, red tail streamers; to 106 cm (42 in), including tail; wingspan to 1.1 m (44 in).

HABITAT: Soars over ocean and shoreline; nests in crevices or on ground.

ISLANDS: KAU, LAN, OAHU, MID

ORIGIN: Native seabird.

Plate 24c

Common Fairy-Tern (also called White Tern or Common White-Tern)
Gygis alba
Manu-o-ku

ID: Small all white seabird with black eyering; forked tail; pointed bluish bill; to 31 cm (12 in); wingspan to 80 cm (31 in).

HABITAT: Flies and soars over shorelines, coastal areas, Honolulu; nests, roosts in trees.

ISLANDS: OAHU, MID

ORIGIN: Native seabird.

Note: Oahu population considered threatened by State of Hawaii.

Plate 24d

Ring-billed Gull
Larus delawarensis

ID: Mid-sized seabird; white with gray back and tops of wings, black wingtips, brown streaks on head, yellowish legs, yellowish bill with dark ring; juveniles are streaked brownish; to 50 cm (20 in); wingspan to 1.2 m (4 ft).

HABITAT: Coastal, but also inland at reservoirs, garbage dumps, along rivers, estuaries.

ISLANDS: BIG, KAU, MAUI, MOLO, OAHU

ORIGIN: North American seabird; visits Hawaii.

Plate 24e

Laughing Gull
Larus atricilla

ID: Mid-sized seabird; white with dark gray back and tops of wings, black wing tips, dark legs; dark bill (during summer breeding, black head and reddish bill); juveniles are brownish or brownish and gray; to 44 cm (17 in); wingspan to 1 m (40 in).

HABITAT: Coastal.

ISLANDS: BIG, KAU, MAUI, OAHU, MID

ORIGIN: North and South American seabird; visits Hawaii.

c Common Fairy-Tern
a White-tailed Tropicbird
b Red-tailed Tropicbird
d Ring-billed Gull
e Laughing Gull

Plate 25a

Gray-backed Tern (also called Spectacled Tern)
Sterna lunata
Pakalakala

ID: Mid-sized seabird, dark above, light below, with deeply forked tail; white forehead and stripe above eye, black line through eye; straight pointed black bill; juvenile with dark scaled pattern above, light below; 38 cm (15 in); wingspan to 76 cm (30 in).

HABITAT: Pelagic seabird; rarely near land except when breeding on cliffs, beaches, bare ground on oceanic islands.

ISLANDS: OAHU, MID

ORIGIN: Native seabird.

Plate 25b

Sooty Tern
Sterna fuscata
ʻEwaʻewa

ID: Mid-sized seabird dark/blackish above, light below, with forked tail; white forehead, black stripe through eye; straight pointed black bill; juvenile all dark except grayish belly and under wings; to 44 cm (17 in); wingspan to 94 cm (37 in).

HABITAT: Pelagic seabird; rarely near land except when breeding on flat, vegetated areas of oceanic islands.

ISLANDS: OAHU, MID

ORIGIN: Native seabird.

Plate 25c

Brown Noddy (also called Common Noddy)
Anous stolidus
Noio koha

ID: Mid-sized brown seabird with dark brown wings, long narrow dark tail; light head cap; long, straight black bill; black legs; 44 cm (17 in); wingspan to 85 cm (33 in).

HABITAT: Pelagic seabird; nests on cliffs, in trees, or on ground on islands.

ISLANDS: BIG, KAU, LAN, MAUI, MOLO, OAHU, MID

ORIGIN: Native seabird.

Plate 25d

Black Noddy (also called White-capped Noddy, Hawaiian Noddy)
Anous minutus
Noio

ID: Mid-sized sooty black seabird with long narrow grayish tail; light head cap; slender, straight, black bill; legs vary in color, usually yellowish orange around main islands, black at MID; to 38 cm (15 in); wingspan to 71 cm (28 in).

HABITAT: Shorelines of off-shore and oceanic islands; nests in cliffs on islands.

ISLANDS: BIG, KAU, LAN, MAUI, MOLO, OAHU, MID

ORIGIN: Native seabird.

Plate 25e

Pomarine Jaeger
Stercorarius pomarinus

ID: Mid-sized heavy-bodied brown seabird with a few extra-long tail feathers (streamers); dark form all dark with white wing patches below; light form with black cap, white throat and belly, dark chest band, yellow neck patches; juveniles brown barred and mottled; to 51 cm (20 in); wingspan to 1.3 m (4.2 ft).

HABITAT: Coastal.

ISLANDS: OAHU

ORIGIN: Breeds on North American and Eurasian arctic tundra; winters in Hawaii, other sites.

a Gray-backed Tern
IM
IM
b Sooty Tern
c Brown Noddy
light form
IM
d Black Noddy
dark form
IM
e Pomarine Jaeger

Plate 26a

Common Fairy-Tern (also called White Tern or Common White-Tern)
Gygis alba
Manu-o-ku

ID: Small all-white seabird with black eyering; forked tail; pointed bluish bill; to 31 cm (12 in); wingspan to 80 cm (31 in).

HABITAT: Flies and soars over shorelines, coastal areas, Honolulu; nests, roosts in trees.

ISLANDS: OAHU, MID

ORIGIN: Native seabird.

Note: Oahu population considered threatened by State of Hawaii.

Plate 26b

Gray-backed Tern (also called Spectacled Tern)
Sterna lunata
Pakalakala

ID: Mid-sized seabird, dark above, light below, with deeply forked tail; white forehead and stripe above eye, black line through eye; straight pointed black bill; juvenile with dark scaled pattern above, light below; 38 cm (15 in); wingspan to 76 cm (30 in).

HABITAT: Pelagic seabird; rarely near land except when breeding on cliffs, beaches, bare ground on oceanic islands.

ISLANDS: OAHU, MID

ORIGIN: Native seabird.

Plate 26c

Sooty Tern
Sterna fuscata
ʻEwaʻewa

ID: Mid-sized seabird dark/blackish above, light below, with forked tail; white forehead, black stripe through eye; straight pointed black bill; juvenile all dark except grayish under wings and belly; to 44 cm (17 in); wingspan to 94 cm (37 in).

HABITAT: Pelagic seabird; rarely near land except when breeding on flat, vegetated areas of oceanic islands.

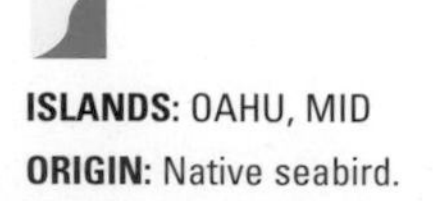

ISLANDS: OAHU, MID

ORIGIN: Native seabird.

Plate 26d

Brown Noddy (also called Common Noddy)
Anous stolidus
Noio koha

ID: Mid-sized brown seabird with dark brown wings, long narrow dark tail; light head cap; long, straight black bill; black legs; 44 cm (17 in); wingspan to 85 cm (33 in).

HABITAT: Pelagic seabird; nests on cliffs, in trees, or on ground on islands.

ISLANDS: BIG, KAU, LAN, MAUI, MOLO, OAHU, MID

ORIGIN: Native seabird.

Plate 26e

Black Noddy (also called White-capped Noddy, Hawaiian Noddy)
Anous minutus
Noio

ID: Mid-sized sooty black seabird with long narrow grayish tail; light head cap; slender, straight, black bill; yellowish legs during breeding; to 38 cm (15 in); wingspan to 71 cm (28 in).

HABITAT: Shorelines, oceanic and off-shore islands; nests on cliffs on islands.

ISLANDS: BIG, KAU, LAN, MAUI, MOLO, OAHU, MID

ORIGIN: Native seabird.

b Gray-backed Tern
c Sooty Tern
d Brown Noddy
a Common Fairy-Tern
e Black Noddy

Plate 27a

Cattle Egret
Bubulcus ibis

ID: Large white bird with thickish neck, yellow bill, and dark legs; during breeding, head, chest, and back with yellowish buff color, and bill and legs reddish; immature bird is white with yellowish bill; 50 cm (20 in); wingspan to 91 cm (3 ft).

HABITAT: Low and middle elevation agricultural areas; found foraging in fields or following tractors; also marshes.

ISLANDS: BIG, KAU, LAN, MAUI, MOLO, OAHU

ORIGIN: Old World; recently spread naturally to South America, then to North America; introduced to Hawaii late 1950s.

Plate 27b

Black-crowned Night-Heron
Nycticorax nycticorax
ʻAukuʻu

ID: Large, stocky grayish bird with black back and head top; whitish throat, belly; short, yellowish legs; immature bird is streaked brown; 64 cm (25 in); wingspan to 1.1 m (3.6 ft).

HABITAT: Wetlands – ponds, marshes, streams, rivers; roosts in trees, nests in trees, shrubs.

ISLANDS: BIG, KAU, LAN, MAUI, MOLO, OAHU

ORIGIN: Hawaiian native.

Plate 27c

Hawaiian Stilt
Himantopus knudseni
Aeʻo

ID: Mid-sized, slender white and dark marsh bird with long reddish legs and long, thin, straight black bill; male with black back, female with brown back; to 40 cm (16 in).

HABITAT: Low elevation aquatic sites; found in ponds, lagoons, mudflats, agricultural ponds.

ISLANDS: BIG, KAU, LAN, MAUI, MOLO, OAHU

ORIGIN: Hawaiian native; endemic.

Note: Endangered, USA ESA listed.

Plate 27d

Nene (also called Hawaiian Goose)
Branta sandvicensis
Nēnē

ID: Mid-sized brown or brownish gray goose; black head; light cheek; characteristic buffy or whitish "furrowed" neck; to 66 cm (26 in).

HABITAT: Sparsely vegetated lava flows on volcanic slopes, at 1500 to 2400 m (5000 to 8000 ft); also dry scrublands, grasslands, golf courses; open lowlands of Kauai.

ISLANDS: BIG, KAU, MAUI

ORIGIN: Hawaiian native; endemic.

Note: Endangered, CITES Appendix I and USA ESA listed.

Plate 27e

"Cackling" Canada Goose
Branta canadensis minima

ID: A smallish brown goose with black head and neck, broad white throat band; 64 to 89 cm (25 to 35 in); a small subspecies of the widely distributed Canada Goose; occasionally other subspecies, a bit larger, also visit Hawaii.

HABITAT: Occurs in a large variety of habitats, but almost always near water.

ISLANDS: BIG, KAU, MAUI, MOLO, OAHU, MID

ORIGIN: Native mostly to North American mainland; visits Hawaii in winter; Cackling variety breeds in western Alaska.

a Cattle Egret
b Black-crowned Night-Heron
breeding
IM
non-
breeding
F
M
c Hawaiian Stilt
d Nene
e "Cackling" Canada Goose

Plate 28a

Common Moorhen (also called Common Gallinule, Hawaiian Common Moorhen)
Gallinula chloropus
'Alae 'ula

ID: Mid-sized blackish water bird with yellow-tipped red bill and red forehead; yellowish legs; feet not webbed; to 35 cm (14 in).

HABITAT: Lakes, ponds, reservoirs, marshes, irrigation ditches; walks in shoreline vegetation and swims.

ISLANDS: KAU, MOLO, OAHU

ORIGIN: Native Hawaiian.

Note: The Hawaiian subspecies of this broadly distributed bird is endangered, USA ESA listed.

Plate 28b

Hawaiian Coot
Fulica alai
Alae Ke'oke'o

ID: Mid-sized dark gray duck-like waterbird with whitish pointed bill and forehead (some have red forehead); grayish or yellowish legs; feet lobed, not webbed; 39 cm (15 in).

HABITAT: Lakes, ponds, marshes, estuaries.

ISLANDS: BIG, KAU, LAN, MAUI, MOLO, OAHU

ORIGIN: Native Hawaiian.

Note: The Hawaiian subspecies of this broadly distributed bird is endangered, USA ESA listed.

Plate 28c

Pied-billed Grebe
Podilymbus podiceps

ID: Smallish duck-like brown bird with small, stout yellowish bill, light throat; during breeding season, bill gray with black bar, throat black; 34 cm (13.5 in).

HABITAT: Freshwater ponds, lakes, marshes.

ISLANDS: BIG, KAU, OAHU

ORIGIN: Hawaiian visitor, native to North and South America; a small, very recently established natural breeding population on the Big Island (at Aimakapa Pond), if it persists, will transform this species into a Hawaiian native.

Plate 28d

Hawaiian Duck
Anas wyvilliana
Koloa

ID: Mid-sized mottled brown duck; both sexes resemble female Mallard, the male being a bit darker; male bill greenish, female bill orangish or grayish; orange legs; male to 51 cm (20 in); female smaller.

HABITAT: Ponds, marshes, wet pastures, reservoirs, drainage ditches, streams, river valleys.

ISLANDS: BIG, KAU, OAHU

ORIGIN: Hawaiian native.

Note: Endangered, USA ESA listed.

Plate 28e

Mallard
Anas platyrhynchos

ID: Largish duck with orange legs; male grayish with green head, reddish brown chest, white band across chest, yellowish bill; female mottled brown with orangish bill; to 60 cm (2 ft).

HABITAT: Ponds, marshes, wet pastures, reservoirs, drainage ditches, streams, river valleys.

ISLANDS: BIG, KAU, LAN, MAUI, MOLO, OAHU

ORIGIN: Hawaiian visitor; breeds over North America, Eurasia.

Note: Most Mallards in Hawaii are feral domestic birds, as shown. Mallards are probably the most numerous ducks on Earth.

a Common Moorhen
red-shield form
b Hawaiian Coot
typical form
M
d Hawaiian Duck
F
F
e Mallard
M
c Pied-billed Grebe

Plate 29a

Northern Pintail
Anas acuta
Koloa maoli

ID: Large duck with long tail, grayish bill; male grayish with brown head, white chest; female mottled brown; to 66 cm (26 in).

HABITAT: Shallow freshwater ponds, marshes, rivers; coastal lagoons.

ISLANDS: BIG, KAU, LAN, MAUI, MOLO, OAHU

ORIGIN: Hawaiian winter resident; breeds over northern North America, Eurasia.

Plate 29b

Northern Shoveler
Anas clypeata
Koloa mapu

ID: Mid-sized duck with large, spatula-like bill longer than head, blue and green wing patches, and orange legs; male with green head, white chest, brownish belly, blackish tail/rear; female mottled brown; to 51 cm (20 in).

HABITAT: Shallow freshwater ponds, marshes, rivers; coastal lagoons, tidal mudflats.

ISLANDS: BIG, KAU, LAN, MAUI, MOLO, OAHU

ORIGIN: Hawaiian winter resident; breeds over northern North America, Eurasia.

Plate 29c

Green-winged Teal (also called Common Teal)
Anas crecca

ID: Small duck with green wing patch (seen mainly in flight), gray bill and legs; male grayish and brown with reddish brown head with broad green eyestripe, light patch near tail outlined with black; female mottled brown with dark eyeline; 38 cm (15 in).

HABITAT: Ponds, small lakes, marshes.

ISLANDS: BIG, KAU, LAN, MAUI, MOLO, OAHU

ORIGIN: Hawaiian visitor; breeds over northern North America, Eurasia.

Plate 29d

Blue-winged Teal
Anas discors

ID: Smallish brown duck with conspicuous blue wing patch (seen mainly in flight); male with gray head, white crescent in front of eye, white patch near tail; female mottled with dark eyeline; to 40 cm (16 in).

HABITAT: Fresh and brackish water; lakes, ponds, marshes.

ISLANDS: BIG, KAU, MAUI, MOLO, OAHU

ORIGIN: Hawaiian visitor; breeds over North America.

Note: Very occasionally breeds in Hawaii.

Plate 29e

Garganey
Anas querquedula

ID: Smallish brown duck with bluish-gray and green wing patches (seen mainly in flight), gray bill; breeding male with reddish brown head, broad white eyestripe; female mottled brown with fairly distinct light and dark facial streaks; nonbreeding male resembles female; to 40 cm (15.5 in).

HABITAT: Coastal marshes, ponds, lagoons.

ISLANDS: BIG, MOLO, OAHU, MID

ORIGIN: Hawaiian visitor; breeds over Eurasia.

F
M
a Northern Pintail
F
M
b Northern Shoveler
F
M
c Green-winged Teal
F
M
d Blue-winged Teal
F
M
e Garganey

Plate 30a

American Wigeon
Anas americana

ID: Mid-sized brown or reddish brown duck with white wing patch, dark-tipped blue bill; male with grayish head, white forehead and crown, green eye streak; female brownish with gray head; 48 cm (19 in).

HABITAT: Ponds and reservoirs.

ISLANDS: BIG, KAU, LAN, MAUI, MOLO, OAHU, MID

ORIGIN: Hawaiian winter resident; breeds in North America.

Plate 30b

Eurasian Wigeon
Anas penelope

ID: Mid-sized duck with dark-tipped blue bill; male gray with reddish brown head, yellowish/tawny forehead and crown, white wing patch; female dark brown with grayish or reddish brown head; to 51 cm (20 in).

HABITAT: Ponds and reservoirs.

ISLANDS: BIG, KAU, LAN, MAUI, MOLO, OAHU, MID

ORIGIN: Hawaiian winter visitor; breeds in northern Europe, Asia.

Plate 30c

Lesser Scaup
Aythya affinis

ID: Mid-sized dark duck with short white wing stripe; male gray with black head (with purplish gloss), chest, tail, and bluish bill and feet; female dark brown with white face patch, dark bill; to 42 cm (16.5 in).

HABITAT: Ponds and reservoirs.

ISLANDS: BIG, KAU, LAN, MAUI, MOLO, OAHU, MID

ORIGIN: Hawaiian winter resident; breeds in North America.

Plate 30d

Greater Scaup
Aythya marila

ID: Mid-sized dark duck with longer white wing stripe, slightly larger than Lesser Scaup, with a more smoothly rounded head; male gray with black head (with greenish gloss), chest, and tail, and bluish bill and feet; female very similar to female Lesser Scaup but with different head shape; to 46 cm (18 in).

HABITAT: Ponds and reservoirs.

ISLANDS: BIG, KAU, LAN, MAUI, MOLO, OAHU, MID

ORIGIN: Hawaiian winter visitor; breeds in North America, Europe, Asia.

Plate 30e

Tufted Duck
Aythya fuligula

ID: Mid-sized dark duck with bold white wing stripe, dark head crest (not always seen), dark-tipped blue bill; male black with purplish glossed head and white sides/belly; female blackish brown sometimes with small white area on face near bill; to 45 cm (17.5 in).

HABITAT: Ponds and reservoirs.

ISLANDS: BIG, KAU, LAN, MAUI, MOLO, OAHU, MID

ORIGIN: Hawaiian winter visitor; breeds in Europe, Asia.

Plate 30f

Ring-necked Duck
Aythya collaris

ID: Mid-sized dark duck with slightly peaked head, bluish gray dark-tipped bill with white ring, white belly; male black with purplish glossed head, gray sides, narrow white ring at base of bill; female dark brown with lighter face, white eyering; 43 cm (17 in).

HABITAT: Ponds, lakes, marshes, reservoirs, estuaries, coastal lagoons.

ISLANDS: BIG, KAU, LAN, MAUI, MOLO, OAHU, MID

ORIGIN: Hawaiian winter visitor; breeds in North America.

F
a American Wigeon
M
F
b Eurasian Wigeon
M
F
c Lesser Scaup
M
d Greater Scaup
F
M
e Tufted Duck
M
F
f Ringed-necked Duck
F
M

Plate 31a

Osprey
Pandion haliaetus

ID: Large brownish bird with white head; dark stripe through eye; gray legs; wing in flight has backward "bend;" underside of wing white with darker stripes and markings; to 60 cm (2 ft); wingspan to 1.7 m (5.5 ft).

HABITAT: Low, middle and some higher elevations; seen flying or perched in trees near water – ocean shores, ponds, reservoirs.

ISLANDS: BIG, KAU, MAUI, MOLO, OAHU

ORIGIN: Hawaiian winter visitor; breeds over many parts of the world.

Plate 31b

Hawaiian Hawk
Buteo solitarius

'Io

ID: Mid-sized bird-of-prey with yellow skin at base of bill, yellowish feet, and grayish tail with brown bars; dark form all brown; light form brown with whitish chest/belly often with dark streaks; light form immature bird is pale buffy/yellowish on head and below, with dark eyeline; to 46 cm (18 in).

HABITAT: Low, middle, and higher elevations; forest, woodlands, forest edges, and more open sites such as some agricultural areas, lava flows.

ISLANDS: BIG (reports of occasional visits to MAUI, OAHU)

ORIGIN: Hawaiian native; endemic.

Note: Threatened, USA ESA and CITES Appendix II listed.

Plate 31c

Peregrine Falcon
Falco peregrinus

ID: Mid-sized bird-of-prey, dark gray above, light below; chest, belly with bars or streaks; yellow skin at base of bill; yellow feet; female often with chest tinged brownish; immature bird with brown back, brown streaked chest; to 50 cm (19.5 in); wingspan to 1.2 m (4 ft).

HABITAT: Found almost anywhere; often near cliffs.

ISLANDS: BIG, KAU, LAN, MAUI, MOLO, OAHU, MID

ORIGIN: Hawaiian visitor; breeds over many parts of the world.

Note: Listed as endangered (CITES Appendix I, USA ESA), but not really globally threatened.

Plate 31d

Barn Owl
Tyto alba

ID: Mid-sized tawny/brownish streaked owl with heart-shaped face; dark eyes; pale bill; male with white chest/belly, female with yellowish/tawny chest/belly; 41 cm (16 in).

HABITAT: Semi-open and open sites such as open woodlands, forest edges, parklands, agricultural sites, urban areas, from low to high elevations; nocturnal, but sometimes active during the day.

ISLANDS: BIG, KAU, LAN, MAUI, MOLO, OAHU (maybe MID)

ORIGIN: Native to North America and several other continents; introduced to Hawaii from North America in late 1950s.

Plate 31e

Short-eared Owl (also called Hawaiian Owl)
Asio flammeus

Pueo

ID: Mid-sized brown and white owl, heavily streaked; yellow eyes; dark bill; 38 cm (15 in).

HABITAT: Occurs in a variety of habitats, from wet and dry forests to open fields and pastures, at low to high elevations; mostly seen in open areas; day- and night-active.

ISLANDS: BIG, KAU, LAN, MAUI, MOLO, OAHU, MID

ORIGIN: Hawaiian native subspecies of an owl broadly distributed in Eurasia, North Africa, and the Americas.

Note: Considered endangered on Oahu (State of Hawaii listed); but the species is abundant in many mainland regions, and not threatened.

a Osprey
b Hawaiian Hawk
dark
form
light
form
IM
light form
M
c Peregrine
Falcon
F
F
M
d Barn Owl
e Short-eared Owl

Plate 32a

Black Francolin

Francolinus francolinus

ID: Mid-sized chicken-like bird with orange legs; male black with streaks and spots, reddish brown neck, white cheek patch; female mottled/streaked brownish with reddish brown neck patch and whitish throat; to 36 cm (14 in).

HABITAT: Open areas such as grasslands, pastures, roadsides.

ISLANDS: BIG, KAU, MAUI, MOLO

ORIGIN: India and Middle East; introduced to Hawaii in late 1950s.

Plate 32b

Gray Francolin

Francolinus pondicerianus

ID: Mid-sized, light brownish, barred, chicken-like bird with orangish throat; gray, brown, or reddish legs; to 35 cm (14 in).

HABITAT: Open areas such as grasslands, dry scrublands, golf courses, roadsides.

ISLANDS: BIG, KAU, LAN, MAUI, MAUI, MOLO, OAHU

ORIGIN: India and Middle East; introduced to Hawaii in late 1950s.

Plate 32c

Erckel's Francolin

Francolinus erckelii

ID: Largish chicken-like brown bird, heavily streaked above and below; reddish brown cap on head; to 43 cm (17 in).

HABITAT: Mostly in open and semi-open sites such as open woodlands, forest edges, grasslands, roadsides, but also some forests.

ISLANDS: BIG, KAU, LAN, MAUI, MOLO, OAHU

ORIGIN: Africa; introduced to Hawaii in late 1950s.

Plate 32d

Japanese Quail

Coturnix japonica

ID: Small, plump, brown streaked ground bird with light belly, light stripe over eye, short tail; male with orangish and/or blackish throat, female with whitish throat; to 19 cm (7.5 in).

HABITAT: Open areas such as dry grasslands, pastures.

ISLANDS: BIG, KAU, LAN, MAUI, MOLO

ORIGIN: Asia; introduced to Hawaii in early 1920s.

Plate 32e

California Quail

Callipepla californica

ID: Mid-sized plump gray or blackish ground bird with lighter belly with black scaling pattern and black head plume; male with black face with white markings; to 27 cm (10.5 in).

HABITAT: Open and semi-open areas such as grassland, scrubland, agricultural sites, and forest edges.

ISLANDS: BIG, LAN

ORIGIN: West Coast of North America; introduced to Hawaii during mid-1800s.

Plate 32f

Chestnut-bellied Sandgrouse

Pterocles exustus

ID: Mid-sized pigeon-like ground bird with long tapering tail; male light brown with dark chest bar, dark belly; female brownish with dark streaks; to 33 cm (13 in).

HABITAT: Very open dry sites such as grasslands, pastures, fields.

ISLANDS: BIG

ORIGIN: Asia and Africa; introduced to Hawaii in early 1960s.

F
b Gray Francolin
M
d Japanese Quail
a Black
Francolin
c Erckel's
Francolin
e California
Quail
M
F
f Chestnut-bellied
Sandgrouse
M
F

Plate 33a

Chukar
Alectoris chukar

ID: Mid-sized grayish ground bird with black face stripe; light and dark barred sides and belly; reddish bill and legs; to 37 cm (14.5 in).

HABITAT: Bare stony slopes with grassy or shrubby vegetation in middle and higher elevation dry areas.

ISLANDS: BIG, KAU, LAN, MAUI, MOLO

ORIGIN: Europe and Asia; introduced to Hawaii during 1920s.

Plate 33b

Kalij Pheasant
Lophura leucomelana

ID: Large chicken-like ground bird with bare-skin red face, dark or light crest; male is dark bluish or blackish with grayish lower chest and light barring on lower back; female is brownish with lighter streaks, bars; to 74 cm (29 in).

HABITAT: Middle and higher elevation wet and dry forests, forest edges, thickets, roadsides.

ISLANDS: BIG

ORIGIN: Asia; introduced to Hawaii in early 1960s.

Plate 33c

Red Junglefowl
Gallus gallus
Moa

ID: Large ground bird resembling a chicken, which it actually is; male reddish brown with red comb and wattle, dark belly and tail; female smaller than male, drab brownish; to 75 cm (29 in).

HABITAT: Flat or gently sloping areas in forests at low and middle elevations.

ISLANDS: KAU, LAN, OAHU

ORIGIN: Asia; introduced to Hawaii by early Polynesians.

Note: Only survives on mongoose-free islands; Oahu population reintroduced, but probably not self-sustaining.

Plate 33d

Ring-necked Pheasant (also called Common Pheasant)
Phasianus colchicus

ID: Large ground bird with long pointed tail; male reddish brown with greenish/dark head, white neck band, red wattles; female drab brown, streaked; to 89 cm (35 in). A green form (either a subspecies or separate species, *Phasianus versicolor*) occurs on slopes of Big Island's Mauna Loa; male with greenish body.

HABITAT: Open and semi-open sites including open woodlands, forest edges, grasslands, savannah, fields, pastures.

ISLANDS: BIG, KAU, LAN, MAUI, MOLO, OAHU

ORIGIN: Asia; introduced to Hawaii during late 1800s.

Plate 33e

Common Peafowl
Pavo cristatus

ID: Large, unmistakable chicken-like bird with conspicuous crest; male bluish/greenish with huge, colorful tail; female smaller, brownish with white throat, colorful neck; to 1 m (40 in), plus very long tail in male. Many albinos in Hawaii.

HABITAT: Dense forests and some semi-open sites – woodlands, forest edges; on ground or in trees.

ISLANDS: BIG, KAU, MAUI, OAHU

ORIGIN: Asia; introduced to Hawaii in early 1860s.

Plate 33f

Wild Turkey
Meleagris gallopavo

ID: Huge dark/brownish ground bird with bare-skin head and neck; male larger, brighter, iridescent, with reddish/pinkish head; female smaller, duller; to 1 m (40 in).

HABITAT: Forests, open woodlands, forest edges and adjacent open areas such as pastures, fields.

ISLANDS: BIG, LAN, MAUI, MOLO

ORIGIN: North America; introduced to Hawaii in late 1780s.

a Chukar
b Kalij Pheasant
M
F
c Red Junglefowl
M
F
d Ring-necked Pheasant
M
F
f Wild Turkey
M
F
e Common Peafowl
M
F

Plate 34a

Pacific Golden-Plover
Pluvialis fulva
Kolea

ID: Mid-sized, slender, long-legged shorebird; mottled dark brown and golden above, lighter below; indistinct light eyestripe; to 26 cm (10 in).

HABITAT: Low, middle, and higher elevations; fields, agricultural areas, lawns, parks, mudflats, mountainous grassy slopes and meadows.

ISLANDS: BIG, KAU, LAN, MAUI, MOLO, OAHU, MID

ORIGIN: Native migrant; breeds in arctic Alaska, Asia.

Plate 34b

Black-bellied Plover (also called Gray Plover)
Pluvialis squatarola

ID: Mid-sized shorebird with largish head and stout black bill; brownish gray above, lighter below; white rump and belly; black and white barred tail; to 29 cm (11.5 in).

HABITAT: Shorelines, mudflats.

ISLANDS: BIG, KAU, LAN, MAUI, MOLO, OAHU, MID

ORIGIN: Native migrant; breeds in arctic North America.

Plate 34c

Semipalmated Plover
Charadrius semipalmatus

ID: Smallish brown shorebird with white forehead, eyeline, neckband, and underparts; brown chest bar; yellowish legs; during breeding, orangish bill with dark tip; to 19 cm (7.5 in).

HABITAT: Shorelines, mudflats, sand dunes; occasionally lake, pond margins.

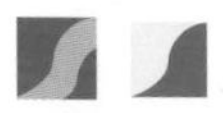

ISLANDS: BIG, KAU, MAUI, MOLO, OAHU, MID

ORIGIN: Native migrant; breeds in northern North America.

Plate 34d

Ruddy Turnstone
Arenaria interpres
‘Akekeke

ID: Small, robust, dark brown shorebird with white throat and belly, orange legs; dark U-shaped chest markings in breeding bird, blotchy brown non-breeding; smallish, pointed black bill; to 25 cm (9.5 in).

HABITAT: Shorelines, sand and pebble beaches, grassy fields, mudflats. On MID only, on ironwood forest floor.

ISLANDS: BIG, KAU, LAN, MAUI, MOLO, OAHU, MID

ORIGIN: Native migrant; breeds in arctic North America, Eurasia.

Plate 34e

Wandering Tattler
Heteroscelus incanus
‘Ūlili

ID: Mid-sized dark gray shorebird with short yellowish legs; whitish belly and eyeline; long straight dark bill; to 29 cm (11 in).

HABITAT: Rocky shorelines, beaches, mudflats, coral reefs; also rocky streams; sometimes perches in trees.

ISLANDS: BIG, KAU, LAN, MAUI, MOLO, OAHU, MID

ORIGIN: Native migrant; breeds in Siberia, Alaska, northwest Canada.

b Black-bellied Plover
breeding
a Pacific Golden-Plover
c Semipalmated Plover
non-breeding
e Wandering Tattler
breeding
non-breeding
d Ruddy Turnstone

Plate 35a

Sanderling
Calidris alba
Hunakai

ID: Small light gray shorebird with darker shoulder area and white head, chest and belly; straight black bill; black legs; to 20 cm (8 in).

HABITAT: Seashores; rocky and sandy shorelines, mudflats.

ISLANDS: BIG, KAU, LAN, MAUI, MOLO, OAHU, MID

ORIGIN: Native migrant; breeds in arctic North America, Eurasia.

Plate 35b

Pectoral Sandpiper
Calidris melanotos

ID: Mid-sized shorebird with streaked brown back, chest; white throat, belly; yellowish legs; yellowish bill with dark tip; to 23 cm (9 in).

HABITAT: Freshwater and brackish wetlands, ponds, mudflats.

ISLANDS: BIG, MAUI, OAHU, MID

ORIGIN: Native migrant; breeds in northern North America, Eurasia.

Plate 35c

Sharp-tailed Sandpiper
Calidris acuminata

ID: Mid-sized brownish shorebird with white throat, belly; whitish eyeline; reddish brown head top; yellowish to greenish brown legs; to 22 cm (8.5 in).

HABITAT: Freshwater and brackish wetlands, ponds, mudflats.

ISLANDS: BIG, KAU, LAN, MAUI, MOLO, OAHU, MID

ORIGIN: Native migrant; breeds in Siberia and Alaska.

Plate 35d

Long-billed Dowitcher
Limnodromus scolopaceus

ID: Mid-sized plump gray shorebird with white belly and whitish eyeline; very long, straight, dark bill; yellowish legs; to 30 cm (12 in).

HABITAT: Ponds, mudflats, coastal tidal flats.

ISLANDS: BIG, KAU, MAUI, OAHU, MID

ORIGIN: Native migrant; breeds in northern North America, Siberia.

Plate 35e

Least Sandpiper
Calidris minutilla

ID: Small brownish shorebird with dark chest, white belly, light eyeline; smallish, straight, pointed black bill; yellowish legs; to 15 cm (6 in).

HABITAT: Freshwater and brackish wetlands, ponds, mudflats.

ISLANDS: BIG, KAU, MAUI, OAHU

ORIGIN: Native migrant; breeds in northern North America.

Plate 35f

Lesser Yellowlegs
Tringa flavipes

ID: Mid-sized grayish or brownish shorebird with white throat, belly; light eyeline; long, straight black bill; white and black barred tail; yellow legs; to 26 cm (10 in).

HABITAT: Inland and coastal wetlands, ponds, mudflats.

ISLANDS: BIG, KAU, LAN, MAUI, MOLO, OAHU, MID

ORIGIN: Native migrant; breeds in northern North America.

b Pectoral Sandpiper
c Sharp-tailed Sandpiper
a Sanderling
f Lesser Yellowlegs
d Long-billed Dowitcher
e Least Sandpiper

Plate 36a

Bristle-thighed Curlew
Numenius tahitiensis
Kioea

ID: Large brownish shorebird with tawny/yellowish streaked neck, chest; whitish belly; buffy-orange rump, tail; light eyeline; long, down-curved bill; to 44 cm (17 in).

HABITAT: Grassy fields at all elevations; sand bars, beaches, mudflats. On MID only, on ironwood forest floor.

ISLANDS: BIG, KAU, LAN, MAUI, MOLO, OAHU, MID

ORIGIN: Native migrant; breeds in Alaska.

Note: Endangered, USA ESA listed.

Plate 36b

Whimbrel
Numenius phaeopus

ID: Large brownish shorebird with streaked neck, chest; light belly; rump sometimes white; brown and light barred tail; bold light and dark head stripes; long down-curved bill; to 46 cm (18 in).

HABITAT: Muddy, rocky, or sandy beaches, mudflats, wet grasslands.

ISLANDS: OAHU

ORIGIN: Native migrant; breeds in northern North America, northern Eurasia.

Plate 36c

Spotted Dove (also called Spotted-necked Dove; Lace-necked Dove, Chinese Dove)
Streptopelia chinensis

ID: Mid-sized grayish/brownish dove with rosy chest, belly; black patch on neck with white spots; to 31 cm (12 in).

HABITAT: On the ground or in trees in open and semi-open areas, low and middle elevations; forest clearings and edges, cultivated areas, parks, gardens, urban areas.

ISLANDS: BIG, KAU, LAN, MAUI, MOLO, OAHU

ORIGIN: Southern Asia; introduced to Hawaii during 1800s.

Plate 36d

Zebra Dove (also called Barred Dove)
Geopelia striata

ID: Small brownish gray dove with dark barring; rosy belly; bluish face, bluish skin around eye; reddish legs; to 21 cm (8 in).

HABITAT: On the ground in low and middle elevation open, dry areas including lightly forested sites, dry scrubland, agricultural sites, and other peopled areas – town parks, gardens, city streets.

ISLANDS: BIG, KAU, LAN, MAUI, MOLO, OAHU

ORIGIN: Southeast Asia; introduced to Hawaii during 1920s.

Plate 36e

Rose-ringed Parakeet
Psittacula krameri

ID: Mid-sized light green parrot with dark red bill and long, narrow, pointed bluish or greenish tail; male with narrow black line on throat, pinkish neck band; to 43 cm (17 in).

HABITAT: Plantations and other agricultural areas, suburban gardens.

ISLANDS: BIG, KAU, OAHU

ORIGIN: Africa, Asia; introduced to Hawaii probably during 1960s.

a Bristlc-thighed Curlew
b Whimbrel
c Spotted Dove
d Zebra Dove
e Rose-ringed Parakeet

Plate 37a

Hawaiian Crow
Corvus hawaiiensis
ʻAlalā

ID: Large dull blackish or dark brown bird with large, stout, pointed bill; black legs; to 50 cm (20 in).

HABITAT: Middle and higher-elevation wet forests and forest edges in central Kona (western) part of the Big Island (on private property, the McCandless Ranch; may eventually become a national wildlife refuge).

ISLANDS: BIG

ORIGIN: Hawaiian native; endemic.

Note: Critically endangered, USA ESA listed.

Plate 37b

Omao (also called Hawaiian Thrush)
Myadestes obscurus
ʻOmaʻo

ID: Mid-sized grayish brown bird with gray chest/belly; small dark bill; black legs; immature grayish with brown scalloped pattern on chest/belly; 18 cm (7 in).

HABITAT: Higher elevation wet forests (native ohia forests; usually above 1200 m, 4000 ft) and high elevation savannah and lava flow scrub areas; windward (eastern) side of Big Island only.

ISLANDS: BIG

ORIGIN: Hawaiian native; endemic.

Plate 37c

Puaiohi (also called Small Kauai Thrush)
Myadestes palmeri

ID: Smallish brown or brownish gray bird with gray chest/belly; light eyering; some with whitish patch above or below eye; small, dark, slender bill; pinkish legs; immature grayish with brown scalloped pattern on chest/belly; 17 cm (6.5 in).

HABITAT: Dense undergrowth near streams in higher elevation wet forests (native ohia forest) in Alakai Swamp region.

ISLANDS: KAU

ORIGIN: Hawaiian native; endemic.

Note: Endangered, USA ESA listed.

Plate 37d

Kauai Elepaio
Chasiempis sclateri
Kauaʻi ʻElepaio

ID: Small gray or grayish brown bird with white throat, belly, rump, and wing markings; orangish/brown chest band; white-tipped tail often held vertically; immature bird is reddish brown/tawny with white belly; 14 cm (5.5 in).

HABITAT: Higher elevation wet forests, especially dense, wet ohia forests.

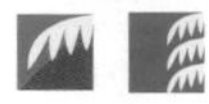

ISLANDS: KAU

ORIGIN: Hawaiian native; endemic.

Plate 37e

Oahu Elepaio
Chasiempis ibidis
Oʻahu ʻElepaio

ID: Small reddish brown bird with white belly, rump, and wing markings; blackish throat/upper chest with white markings; white-tipped tail often held vertically; 14 cm (5.5 in).

HABITAT: Native and introduced drier and seasonally wet forests at lower and middle elevations.

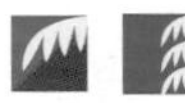

ISLANDS: OAHU

ORIGIN: Hawaiian native; endemic.

Note: Proposed endangered species, USA ESA.

Plate 37f

Hawaii Elepaio
Chasiempis sandwichensis
Hawaiʻi ʻElepaio

ID: Small brown bird with white belly, rump, and wing markings; reddish brown chest with streaks to sides; dark throat with light markings (male) or white throat (female); white-tipped tail often held vertically; 14 cm (5.5 in).

HABITAT: Wet forests at middle and high elevations (darker form birds); some leeward-side dry forests on Mauna Kea (light form).

ISLANDS: BIG

ORIGIN: Hawaiian native; endemic.

IM
b Omao
IM
d Kauai Elepaio
a Hawaiian Crow
e Oahu Elepaio
IM
IM
c Puaiohi
dark form
M
f Hawaii Elepaio
light form
M

Plate 38a

White-rumped Shama (also called Shama Thrush)

Copsychus malabaricus

ID: Mid-sized black (male) or gray (female) bird with rusty or reddish brown belly; white rump; long dark tail with white sides; smallish black bill; pinkish legs; to 27 cm (10.5 in).

HABITAT: Low elevation forests and forest edges, thickets, gardens.

ISLANDS: KAU, OAHU

ORIGIN: Asia; introduced to Hawaii in 1940.

Plate 38b

Northern Mockingbird

Mimus polyglottos

ID: Mid-sized dull gray or brownish gray bird with lighter chest/belly and darker wings, tail; white patches on wings and tail seen in flight; dark bill and legs; 25 cm (10 in).

HABITAT: Dry forests and woodlands, forest edges, parklands, dry scrub areas, at low, middle, and some higher elevations.

ISLANDS: BIG, KAU, LAN, MAUI, MOLO, OAHU

ORIGIN: North America; introduced to Hawaii in early 1930s.

Plate 38c

Common Myna (also called Indian Myna)

Acridotheres tristis

ID: Mid-sized plump chocolate-brown bird with black head; white wing patches seen in flight; black tail with white tip; yellow bill, legs, and patch behind eye; to 25 cm (10 in).

HABITAT: Low and middle elevation dry areas; woodlands, open forest, forest edges; settled areas; agricultural sites, urban areas.

ISLANDS: BIG, KAU, LAN, MAUI, MOLO, OAHU, MID

ORIGIN: India; introduced to Hawaii in 1865, Midway in 1974.

Plate 38d

Western Meadowlark

Sturnella neglecta

ID: Mid-sized chunky bird, brown streaked above, bright yellow below with broad V-shaped chest band; white tail edges seen in flight; white stripe above eye; long dark bill; flesh-colored legs; to 25 cm (10 in).

HABITAT: Low-elevation open areas such as grassland, pastures, agricultural fields, golf courses.

ISLANDS: KAU

ORIGIN: North America; introduced to Kauai in 1931.

Plate 38e

Eurasian Skylark

Alauda arvensis

ID: Mid-sized brownish streaked bird with light chest with brown streaks, whitish belly; usually inconspicuous crest; white tail edges seen in flight; smallish pointed light bill; to 19 cm (7.5 in).

HABITAT: Low, middle and some higher elevation grasslands, roadsides, dry scrub, woodlands, woodland edges.

ISLANDS: BIG, LAN, MAUI, MOLO, OAHU

ORIGIN: Europe; introduced to Hawaii in 1865.

a White-rumped Shama
F
M
b Northern Mockingbird
d Western Meadowlark
c Common Myna
e Eurasian Skylark

Plate 39a

Greater Necklaced Laughing-thrush (also called Black-gorgeted Laughing-thrush)
Garrulax pectoralis

ID: Largish brown or reddish brown bird, white below with bold black chest band; long tail with light tip; whitish face with dark streaks; 33 cm (13 in).

HABITAT: Low and middle elevation wet forests, stream valleys; in trees and low in undergrowth.

ISLANDS: KAU

ORIGIN: Southeast Asia; introduced to Hawaii in early 1900s.

Plate 39b

Melodious Laughing-thrush (also called Hwamei, Chinese Thrush)
Garrulax canorus

ID: Mid-sized brown bird with white "spectacles"; yellowish bill and legs; to 25 cm (10 in).

HABITAT: Low, middle, and high-elevation wet and dry forests, scrub areas, thickets, parks, gardens.

ISLANDS: BIG, KAU, MAUI, MOLO, OAHU

ORIGIN: China; introduced to Hawaii in early 1900s.

Plate 39c

Red-billed Leiothrix (also called Pekin Robin, Hillrobin)
Leiothrix lutea

ID: Beautifully marked small greenish or greenish gray bird with yellow throat, orangish or yellow chest, light belly; red and yellow patch seen on folded wing; pale yellow around eye; reddish bill; forked tail; 14 cm (5.5 in).

HABITAT: Wet forests at various elevations (but prefers middle and higher elevation sites); dense woodlands; tree plantations; shrubby areas.

ISLANDS: BIG, MAUI, MOLO, OAHU

ORIGIN: Asia; introduced to Hawaii in early 1900s.

Plate 39d

Japanese Bush-Warbler (also called Uguisu)
Cettia diphone

ID: Small bird, olive-brown or olive-gray above, lighter below; light eyestripe; pale bill and legs; 14 cm (5.5 in).

HABITAT: Middle and some higher elevation wet forests, mostly in dense undergrowth, brushy areas, thickets.

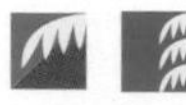

ISLANDS: BIG, KAU, LAN, MAUI, MOLO, OAHU

ORIGIN: Japan; introduced to Hawaii in 1930s.

Plate 39e

Japanese White-eye (also called Chinese White-eye)
Zosterops japonicus

ID: Small greenish bird with yellow throat/chest, white belly, white eyering; small pointed bill; to 11.5 cm (4.5 in).

HABITAT: Almost all terrestrial habitats at low to high elevations, from wet and dry forests to agricultural areas, parks and gardens.

ISLANDS: BIG, KAU, LAN, MAUI, MOLO, OAHU

ORIGIN: Southeast Asia, China, Japan; introduced to Hawaii in 1930s.

a Greater Necklaced Laughing-thrush
b Melodious Laughing-thrush
d Japanese Bush-Warbler
c Red-billed Leiothrix
e Japanese White-eye

Plate 40a

Northern Cardinal
Cardinalis cardinalis

ID: Mid-sized, with red crest and thick reddish orange bill; male red with black surrounding bill; female brownish with reddish tinge on wings and tail; to 23 cm (9 in).

HABITAT: Low, middle, and some higher elevation forests with dense undergrowth; brushy fields, thickets, forest edges, suburban areas, parks and gardens.

ISLANDS: BIG, KAU, LAN, MAUI, MOLO, OAHU

ORIGIN: North America; introduced to Hawaii in late 1920s.

Plate 40b

Red-crested Cardinal (also called Brazilian Cardinal)
Paroaria coronata

ID: Mid-sized blue-gray bird with red head, crest, throat, upper chest; white lower chest/belly; thick grayish bill; dark legs; 19 cm (7.5 in).

HABITAT: Scrubby and brushy areas mostly in drier lowlands; parks, gardens, lawns; on the ground or in bushes, trees.

ISLANDS: KAU, LAN, MAUI, MOLO, OAHU

ORIGIN: South America; introduced to Hawaii in late 1920s.

Note: Near-threatened, CITES APPENDIX II listed.

Plate 40c

Yellow-billed Cardinal
Paroaria capitata

ID: Mid-sized bluish black bird with black throat, white chest/belly, red head; thick bill and legs orangish or yellowish; 18 cm (7 in).

HABITAT: Dry scrub, brushy areas, mostly in drier lowlands; residential areas; on the ground or in trees, bushes; also around open-air restaurants.

ISLANDS: BIG

ORIGIN: South America; introduced to Hawaii perhaps during late 1920s.

Note: Near-threatened, CITES APPENDIX II listed.

Plate 40d

Red-vented Bulbul
Pycnonotus cafer

ID: Mid-sized blackish bird with lighter belly; black crest; whitish rump seen in flight; white-tipped black tail; crimson under tail; 22 cm (8.5 in).

HABITAT: Found in settled areas (parks, gardens, suburban and urban residential areas) but also agricultural sites and some forests.

ISLANDS: OAHU

ORIGIN: India; introduced to Hawaii during the 1950s.

Plate 40e

Red-whiskered Bulbul
Pycnonotus jocosus

ID: Mid-sized dark brown bird with white chest/belly; red patch behind eye; red under tail; 18 cm (7 in).

HABITAT: Found mostly in settled areas: parks, gardens, suburban and urban residential areas; some agricultural sites, forests.

ISLANDS: OAHU

ORIGIN: India; introduced to Hawaii during the mid-1960s.

F
M
a Northern Cardinal
e Red-whiskered Bulbul
c Yellow-billed Cardinal
d Red-vented Bulbul
b Red-crested Cardinal

Plate 41a

Saffron Finch
Sicalis flaveola

ID: Adult is smallish yellow bird with darker, streaked back and orange-tinged head; female a bit duller than male; immature duller with streaked back, chest; to 17 cm (6.5 in).

HABITAT: Dry forests, forest edges, grasslands, roadsides, settled areas, parks, lawns.

ISLANDS: BIG, OAHU

ORIGIN: South America; introduced to Hawaii in 1960s.

Plate 41b

House Sparrow (also called English Sparrow)
Passer domesticus

ID: Male is a small brown bird with gray on top of head, whitish side of head, black throat/chest, black streaks on back, gray belly; female is duller brown above, grayish below, with light brown eyestripe; 14 cm (5.5 in).

HABITAT: Usually associated with people – found in villages, towns, cities, parks.

ISLANDS: BIG, KAU, LAN, MAUI, MOLO, OAHU

ORIGIN: Europe, Asia, North Africa; introduced to Hawaii in late 1800s.

Plate 41c

House Finch
Carpodacus mexicanus
Papaya-bird

ID: Small streaked bird with dark bill and legs; male brown streaked with reddish or yellowish face, throat/chest, rump; female brownish, streaked above and below; 14 cm (5.5 in).

HABITAT: A wide range of habitats at low, middle, and higher elevations, from woodlands (mostly dry but some wetter areas), ranches, and agricultural sites to parks, towns, and even cities.

ISLANDS: BIG, KAU, LAN, MAUI, MOLO, OAHU

ORIGIN: North America; introduced to Hawaii in mid-19th Century.

Plate 41d

Yellow-fronted Canary
Serinus mozambicus

ID: Very small dark greenish/olive bird; bright yellow below, yellow rump and facial stripes; gray on top of head, neck; brownish wings, bill and legs; 11 cm (4.5 in).

HABITAT: Semi-open and open woodlands, forest edges, parkland, city parks.

ISLANDS: BIG, OAHU

ORIGIN: Africa; introduced to Hawaii in mid-1960s.

Plate 41e

Island Canary (also called Common Canary)
Serinus canaria

ID: Small pale yellow bird; whitish wings, tail; pinkish bill, legs; to 15 cm (6 in).

HABITAT: Suburban areas, ironwood groves.

ISLANDS: MID

ORIGIN: North Atlantic's Canary and Azores Islands; first introduced to Midway in early 1900s.

a Saffron Finch
IM
b House Sparrow
F
M
F
c House Finch
M
yellow form
red form
M
d Yellow-fronted Canary
e Island Canary

Plate 42a

Common Amakihi (also called Hawaii Amakihi)
Hemignathus virens
Hawai'i 'Amakihi

ID: Small yellowish green bird with small, short, gray down-curved bill; yellow chest/belly; dark line through eye, pale line above eye; dark legs; female duller; 11 cm (4.5 in).

HABITAT: Middle and higher elevations, in leafy branches of trees (often in flowers feeding on nectar) of drier native forests and woodlands, some wetter forests; also subalpine scrub areas; occasionally to low elevations.

ISLANDS: BIG, MAUI, MOLO

ORIGIN: Hawaiian native; endemic.

Plate 42b

Kauai Amakihi
Hemignathus kauaiensis
Kaua'i 'Amakihi

ID: Small olive or yellowish green bird with pinkish gray down-curved bill much larger than that of Common Amakihi; dull yellowish green chest/belly; dark eyeline; dark legs; female a bit duller; 11 cm (4.5 in).

HABITAT: Higher elevation native wet forests; often creeps on smaller tree trunks or larger branches.

ISLANDS: KAU

ORIGIN: Hawaiian native; endemic.

Plate 42c

Oahu Amakihi
Hemignathus flavus
O'ahu 'Amakihi

ID: Small greenish gray bird with small, short, gray down-curved bill; bright yellow from chin to undertail; dark legs; female duller with two light wing bars; 11 cm (4.5 in).

HABITAT: Higher elevation wet forests and some native drier forests; takes nectar from flowers and gleans leaves and branches for insects.

ISLANDS: OAHU

ORIGIN: Hawaiian native; endemic.

Plate 42d

Anianiau
Hemignathus parvus
'Anianiau

ID: Very small yellow or yellowish green bird with no black in lores (small feathered area between bill and eye); small, narrow, only slightly down-curved pale bill; pale legs; female duller; 10 cm (4 in).

HABITAT: Middle and higher elevation native wet forests; found in trees and shrubs.

ISLANDS: KAU

ORIGIN: Hawaiian native; endemic.

Plate 42e

Akekee (formerly called Kauai Akepa)
Loxops caeruleirostris
'Akeke'e

ID: Small yellow bird greenish above with yellow cap and rump, and long, dark, notched tail; black face; short bluish, cone-shaped bill; female duller; 11 cm (4.5 in).

HABITAT: Middle and higher elevation native wet forests; found (almost exclusively) in ohia tree canopy.

ISLANDS: KAU

ORIGIN: Hawaiian native; endemic.

a Common Amakihi
F
M
b Kauai Amakihi
F
c Oahu Amakihi
M
d Anianiau
F
M
e Akekee

Plate 43a

Akiapolaau
Hemignathus munroi
ʻAkiapōlāʻau

ID: Small, stocky olive or greenish bird with yellow head, chest/belly; dark patch in front of eye; bill black, upper part very long, thin, down-curved, lower part shorter, thicker, straight; black legs; short tail; female duller; to 14 cm (5.5 in).

HABITAT: Higher elevation native wet forests and some drier woodlands; found on tree trunks and branches, often pecking vigorously with lower bill only.

ISLANDS: BIG

ORIGIN: Hawaiian native; endemic.

Note: Endangered, USA ESA listed.

Plate 43b

Maui Parrotbill
Pseudonestor xanthophrys

ID: Small olive or greenish bird, yellow below, yellow stripe above eye; large, stout bill, upper part larger, darker, down-curved, lower part straight; black legs; short tail; female duller; to 14 cm (5.5 in).

HABITAT: High-elevation native wet forests, East Maui; found on trunks and branches of trees and shrubs.

ISLANDS: MAUI

ORIGIN: Hawaiian native; endemic.

Note: Endangered, USA ESA listed.

Plate 43c

Maui Alauahio (formerly called Maui Creeper)
Paroreomyza montana
Maui ʻAlauahio

ID: Small greenish bird; yellow below and on forehead; small straight bill, dark above, light below; female duller; 11 cm (4.5 in).

HABITAT: High-elevation wet forests and some drier subalpine scrub areas; found usually in trees.

ISLANDS: MAUI

ORIGIN: Hawaiian native; endemic.

Plate 43d

Akikiki (formerly called Kauai Creeper)
Oreomystis bairdi
ʻAkikiki

ID: Small gray or brownish gray bird with lighter or whitish chest/belly; short pinkish bill, slightly down-curved; short tail; pale legs; immature bird with white eyering; 13 cm (5 in).

HABITAT: High-elevation native wet forests; found on tree trunks.

ISLANDS: KAU

ORIGIN: Hawaiian native; endemic.

Plate 43e

Hawaii Creeper
Oreomystis mana

ID: Small olive-greenish or olive-grayish bird with lighter chest/belly; whitish throat; dark eye patch; small, pale, slightly down-curved bill; dark legs; to 13 cm (5 in).

HABITAT: Middle and high-elevation native wet forests; found on tree trunks.

ISLANDS: BIG

ORIGIN: Hawaiian native; endemic.

Note: Endangered, USA ESA listed.

a Akiapolaau
b Maui Parrotbill
c Maui Alauahio
F
M
d Akikiki
e Hawaii Creeper

Plate 44a

Palila
Loxioides bailleui

ID: Mid-sized bird with gray back, yellow head/throat/chest; pale gray belly; yellowish/olive wings; black eye patch; short, stout, black bill; female a bit duller; to 18 cm (7 in).

HABITAT: Higher-elevation mamane-naio forests, slopes of Mauna Kea; found in trees.

ISLANDS: BIG

ORIGIN: Hawaiian native; endemic.

Note: Endangered, USA ESA listed.

Plate 44b

Akepa
Loxops coccineus
ʻĀkepa

ID: Very small bird with small yellowish or grayish bill, dark wings, and long dark, notched tail; male reddish orange (BIG) (as shown) or orangish/yellowish (MAUI); female greenish or gray above, lighter below, with yellow chest; 10 cm (4 in).

HABITAT: Higher-elevation native forests; found in tree canopy.

ISLANDS: BIG, MAUI

ORIGIN: Hawaiian native; endemic.

Note: The smallest living native Hawaiian bird; endangered, USA ESA listed.

Plate 44c

Apapane
Himatione sanguinea
ʻApapane

ID: Small deep red bird with dark wings and tail; white lower belly and under tail; narrow, slightly down-curved black bill; black legs; 13 cm (5 in).

HABITAT: Middle and higher-elevation forests and some brushy areas with scattered trees; found in flowering trees.

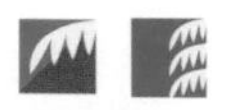

ISLANDS: BIG, KAU, LAN, MAUI, MOLO, OAHU

ORIGIN: Hawaiian native; endemic.

Plate 44d (and book cover)

Iiwi
Vestiaria coccinea
ʻIʻiwi

ID: Smallish bright red bird with black wings and tail; long, strongly down-curved orangish bill; legs same color as bill; 15 cm (6 in).

HABITAT: Middle and high-elevation forests; found in flowering trees and shrubs.

ISLANDS: BIG, KAU, MAUI, MOLO, OAHU

ORIGIN: Hawaiian native; endemic.

Plate 44e

Akohekohe (formerly called Crested Honeycreeper)
Palmeria dolei
ʻĀkohekohe

ID: Mid-sized bird, dark/blackish with gray and reddish streaks/spots; red on back of neck; light/whitish bushy crest; black tail with white tip; black bill and legs; 18 cm (7 in).

HABITAT: High-elevation native wet forests, East Maui; found in flowering ohia trees.

ISLANDS: MAUI

ORIGIN: Hawaiian native; endemic.

Note: Endangered, USA ESA listed.

a Palila
F
b Akepa
M
c Apapane
d Iiwi
e Akohekohe

Plate 45a

Red-cheeked Cordonbleu
Uraeginthus bengalus

ID: Small brown bird with blue throat, chest, sides; light brown belly; male with red face patch; bill with pinkish tinge; to 13 cm (5 in).

HABITAT: Low and middle elevation open dry grassy areas with scattered trees; feeds on ground, flies up to trees when alarmed.

ISLANDS: BIG

ORIGIN: Africa; introduced to Hawaii in mid-1960s or a bit earlier.

Plate 45b

Lavender Waxbill
Estrilda caerulescens

ID: Very small grayish or bluish gray bird with crimson/deep red rump, tail; darker sides with white spots; bill with crimson tinge; to 11 cm (4.5 in).

HABITAT: Low and middle elevation open dry scrubby areas, edges of cultivated sites, roadsides, vacant lots; edges of dry woodlands; on the ground and in shrubs.

ISLANDS: BIG

ORIGIN: Africa; introduced to Hawaii in early 1960s or a bit earlier.

Plate 45c

Orange-cheeked Waxbill
Estrilda melpoda

ID: Very small brownish bird with gray chest/belly; gray head with orange face patch; orangish bill; reddish rump; to 10.5 cm (4 in).

HABITAT: Grassy clearings near golf courses, canefields, weedy patches, roadsides; on the ground and in shrubs.

ISLANDS: MAUI, OAHU

ORIGIN: Africa; introduced to Hawaii during 1960s.

Plate 45d

Common Waxbill
Estrilda astrild

ID: Very small finely barred brown or grayish brown bird with lighter chest/belly; red eyeline; reddish bill; longish tail brown above, blackish below; to 10.5 cm (4 in).

HABITAT: Grasslands, tall grass fields, thickets, weedy patches, marsh areas; on the ground and in shrubs.

ISLANDS: OAHU

ORIGIN: Africa; introduced to Hawaii during 1960s.

Plate 45e

Black-rumped Waxbill
Estrilda troglodytes

ID: Very small, finely barred brown or grayish brown bird with lighter chest/belly, sometimes with pinkish tinge; red eyeline; reddish bill; square-tipped black tail; to 10.5 cm (4 in).

HABITAT: Grasslands, grassy fields, thickets, scrubby/weedy areas; on the ground and in shrubs.

ISLANDS: BIG

ORIGIN: Africa; introduced to Hawaii during 1950s.

a Red-cheeked Cordonbleu

b Lavender Waxbill

c Orange-cheeked Waxbill

d Common Waxbill

e Black-rumped Waxbill

Plate 46a

Red Avadavat (also called Strawberry Finch, Red Munia)
Amandava amandava

ID: Very small with red bill, red rump, dark wings, pale legs; male is red with white spots; female and non-breeding male are grayish brown or yellowish brown; to 11 cm (4.5 in).

HABITAT: Grasslands, tall grass fields, scrubby/weedy areas, reeds/marshes, edges of cultivated sites, fallow cropfields; on the ground and in shrubs.

ISLANDS: BIG, KAU, OAHU

ORIGIN: Asia; introduced to Hawaii in early 1900s.

Plate 46b

Warbling Silverbill (also called African Silverbill)
Lonchura malabarica

ID: Very small sandy-colored bird with whitish chest/belly; black rump and longish black tail; bluish gray bill; to 11 cm (4.5 in).

HABITAT: Dry open sites such as grasslands, thickets, scrub areas, edges of cultivated lands; on the ground and in shrubs.

ISLANDS: BIG, KAU, LAN, MAUI, MOLO, OAHU

ORIGIN: Africa; introduced to Hawaii probably in mid-20th Century.

Plate 46c

Nutmeg Mannikin (also called Spotted Munia, Ricebird, Spice Finch)
Lonchura punctulata

ID: Very small brown bird with dark brown face; chest with black and white scalloped pattern; light belly; blackish bill and legs; immature bird is plain brown with black bill; 10 cm (4 in).

HABITAT: Open and semi-open sites such as grasslands, fields, parks, forest edges and clearings, at low, middle, and some higher elevations; on the ground and in shrubs.

ISLANDS: BIG, KAU, LAN, MAUI, MOLO, OAHU

ORIGIN: Southern Asia; introduced to Hawaii in mid-1860s.

Plate 46d

Chestnut Mannikin (also called Chestnut Munia, Black-headed Munia)
Lonchura malacca

ID: Very small reddish brown bird with black head and belly; largish silvery or bluish bill; dark legs; immature bird is dull brown with reddish brown tail; 11 cm (4.5 in).

HABITAT: Open areas such as grasslands, roadsides, scrub areas, edges of marshland and edges of cultivated areas, lawns; on the ground and in shrubs.

ISLANDS: KAU, OAHU

ORIGIN: Southern Asia; introduced to Hawaii in 1960s.

Plate 46e

Java Sparrow
Padda oryzivora

ID: Small gray bird with black head, large white face patch; black tail; light or brownish belly; large pink bill; to 16 cm (6 in).

HABITAT: Low-elevation open and semi-open sites such as grasslands, gardens, lawns, parks, edges of open woodlands, edges of cultivated areas; on the ground or in trees and shrubs.

ISLANDS: BIG, KAU, MAUI, MOLO, OAHU

ORIGIN: Indonesia; introduced to Hawaii in late 1960s.

a Red Avadavat
F
M
b Warbling Silverbill
d Chestnut Mannikin
IM
IM
c Nutmeg Mannikin
e Java Sparrow

Plate 47a

Hawaiian Hoary Bat
Lasiurus cinereus

ID: Mid-sized grayish or reddish gray densely furred bat; long tail contained within furred tail membrane that fills area between legs; head and body length 5 to 9 cm (2 to 3.5 in); wingspan 27 to 35 cm (10.5 to 14 in); the only bat in Hawaii.

HABITAT: Prefers drier areas below 1200 m (4000 ft), but also seen at higher elevations, such as high on Mauna Kea (BIG) and at Haleakala (MAUI); roosts in trees by day, forages in late afternoon and evenings in open areas, forest clearings, and over water.

ISLANDS: BIG, KAU, LAN (?), MAUI, MOLO, OAHU

ORIGIN: Native to Hawaii; ancestors probably colonized when blown off course during migration in the Americas.

Note: Endangered, USA ESA listed (although may be more common than once thought).

Plate 47b

Norway Rat (also called Brown Rat, Laboratory Rat)
Rattus norvegicus

ID: Gray, brown, or grayish brown rodent with smallish ears; 18 to 23 cm (7 to 9 in), plus tail that is shorter than head and body.

HABITAT: Usually found living in association with people in towns, homes, farms and agricultural fields, most often at lower elevations; found on and under the ground; day- and night-active.

ISLANDS: BIG, KAU, LAN, MAUI, MOLO, OAHU

ORIGIN: Asia; introduced to Hawaii during 1800s.

Plate 47c

Black Rat (also called Roof Rat, House Rat)
Rattus rattus

ID: Brown or blackish rat with whitish, gray, or blackish belly; largish ears; usually 18 to 20 cm (7 to 8 in), plus tail that is longer than head and body.

HABITAT: Found living in close association with people in towns, homes, farms and agricultural fields, at low and middle elevations, but also common in scrub areas and wet and dry forests; found on and under the ground; also climbs trees; day- and night-active.

ISLANDS: BIG, KAU, LAN, MAUI, MOLO, OAHU
ORIGIN: Asia; introduced to Hawaii late 1800s.

Plate 47d

Pacific Rat (also called Hawaiian Rat, Polynesian Rat)
Rattus exulans

ID: Small blackish or brownish rat with medium-sized ears, and tail with conspicuous scaly rings; usually about 13 cm (5 in), plus tail that is about same length as head and body.

HABITAT: Low and middle elevation drier sites, agricultural areas and fields, sugarcane and pineapple plantations, brushy gulches; occasionally seen at higher elevations; found on and under the ground; day- and night-active.

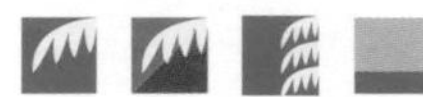

ISLANDS: BIG, KAU, LAN, MAUI, MOLO, OAHU

ORIGIN: Southeast Asia; introduced to Hawaii by early Polynesian settlers.

Plate 47e

House Mouse
Mus musculus

ID: Very small tawny to dark brown rodent, often with grayish sides; small feet and long slender tail; usually about 8 to 8.5 cm (3 to 3.5 in), plus tail about as long as head and body.

HABITAT: Occurs over a wide range of habitats, in association with people and in wild areas, from sea level to 3900 m (12,800 ft); prefers drier sites such as beaches, grasslands, scrub areas and drier woodlands.

ISLANDS: BIG, KAU, LAN, MAUI, MOLO, OAHU

ORIGIN: Western Europe; introduced to Hawaii in late 1700s or early 1800s.

Plate 47f

Mongoose (also called Small Indian Mongoose)
Herpestes auropunctatus

ID: Slender, weasel-like brown mammal with light/yellowish speckles or tinge; underside generally lighter; front claws sharp and curved; hind foot naked to heel; fluffy tail nearly as long as body; 25 to 30 cm (10 to 12 in), plus tail.

HABITAT: Various habitats in low, middle and some higher-elevation sites, but generally below 2400 m (8000 ft); prefers lower-elevation drier areas; found on the ground and in trees; day-active.

ISLANDS: BIG, MAUI, MOLO, OAHU

ORIGIN: Middle East, India, Central Asia; introduced to Hawaii in late 1800s.

a Hawaiian Hoary Bat
b Norway Rat
two-tone form
c Black Rat
black form
brown form
d Pacific Rat
e House Mouse
f Mongoose

Plate 48a

Brush-tailed Rock Wallaby
Petrogale penicillata
Kanakalu

ID: Grayish brown kangaroo-like mammal with large ears, long, dense fur, and long, cylindrical, bushy tail; to about 60 cm (2 ft) tall; adults weigh 4 to 8 kg (9 to 18 lb); you will know it when you see it.

HABITAT: Rocky slopes with crevices and caves in the Kalihi Valley and environs, just north of Honolulu, along Route 63, the Likelike Highway; usually at between 100 and 400 m (330 and 1300 ft) elevation; somewhat nocturnal, but also day-active.

ISLANDS: OAHU

ORIGIN: Australia; brought to Hawaii in early 1900s.

Plate 48b

Feral Pig
Sus scrofa
Pua'a

ID: A dark pig, usually blackish, but sometimes with brown and/or tawny patches, or with spots or white feet; youngsters blackish or brownish, often with lighter horizontal stripes; adults 50 to 90+ kg (110 to 200+ lb); females smaller than males.

HABITAT: Occurs in forest, forest edge, and more open areas such as pastures and grasslands, at a range of elevations, but prefers middle elevations; day- and night-active.

ISLANDS: BIG, KAU, MAUI, MOLO, OAHU

ORIGIN: Eurasia; brought to Hawaii by Polynesians and Europeans.

Plate 48c

Feral Goat
Capra hircus
Kao or Kunana

ID: Short-haired goat, generally black and/or brown, sometimes with lighter spots or markings; some with beard and shaggy mane; some with curved horns; 30 to 70 kg (65 to 150 lb); females smaller than males.

HABITAT: Middle and higher-elevation forests and more open areas, including open ranges and rocky uplands and slopes; may also be at some lower elevations on the Big Island; day-active.

ISLANDS: BIG, KAU, MAUI, MOLO, OAHU

ORIGIN: Asia; brought to Hawaii by Europeans by late 1700s.

Plate 48d

Axis Deer (also called Spotted Deer)
Axis axis
Kia

ID: Largish brown, reddish brown, or yellowish brown coarse-haired deer with white spots (at least part of the year); lighter underneath; usually 1 to 1.5 m (3 to 5 ft) long, plus tail; 75 to 90 cm (30 to 36 in) high at shoulder; females to 55 kg (120 lb), males to 90 kg (200 lb). (Note: Another species, the Black-tailed, or Mule, Deer, *Odocoileus hemionus*, occurs in highlands of western Kauai, but is rarely seen.)

HABITAT: Found in lower-elevation drier forests and mid-elevation, wetter forests, including rainforests; also sometimes in semi-open and more open sites; day- and night-active; moves around mostly in early morning and late afternoon.

ISLANDS: LAN, MAUI, MOLO

ORIGIN: India; brought to Hawaii by diplomats, as a gift, in the 1860s.

Plate 48e

Feral Donkey (domestic form of the African Wild Ass)
Equus asinus
Kekake

ID: Brownish or grayish horse-like mammal; to 2 m (6.5 ft) long; to 1.3 m (4.3 ft) high at shoulder; weight to 250 kg (550 lb).

HABITAT: Often seen in semi-open and open sites over parts of the leeward side of the Big Island, including the South Kohala area.

ISLANDS: BIG

ORIGIN: North Africa and Middle East; brought to Hawaii by Europeans in early 1800s.

a Brush-tailed
Rock Wallaby
IM
b Feral Pig
d Axis Deer
c Feral Goat
e Feral Donkey

Plate 49a

Hawaiian Monk Seal
Monachus schauinslandi
'Ilio-holo-i-ka-uaua

ID: A large seal, slate-gray or brownish above, sometimes tinged greenish with algae; light gray or yellowish brown below; to 2.3 m (7.5 ft).

HABITAT: Secluded sandy beaches and shallow lagoons, mostly at eight small islets and atolls in the Northwestern Hawaiian Islands, but occasionally at main islands.

ISLANDS: KAU, MAUI, MOLO, OAHU, MID

ORIGIN: Hawaiian native.

Note: Endangered, CITES Appendix I and USA ESA listed.

Plate 49b

Pacific Bottle-nosed Dolphin
Tursiops truncatus gillii
Nai'a

ID: Light gray, dark gray, or slate-blue dolphin with pronounced beak; flippers and flukes often darker; belly often lighter, whitish or pinkish; curved line of mouth resembles a smile; dorsal fin about 25 cm (10 in) high; 2 to 3.5 m (6.5 to 11.5 ft).

HABITAT: Coastal bays and lagoons, off-shore areas; often leaps out of water; in small to medium-sized schools; sometimes mixes with Humpback Whales.

ISLANDS: BIG, KAU, LAN, MAUI, MOLO, OAHU

Plate 49c

Spinner Dolphin
Stenella longirostris
Nai'a

ID: Sleek gray dolphin, often with dark gray back, lighter gray sides, and whitish belly; pronounced slender beak; no spots on body; gray flippers and flukes; often a dark eyestripe; dorsal fin, sometimes curved forward instead of rearward, about 25 cm (10 in) high; 1.7 to 2.1 m (5.5 to 7 ft).

HABITAT: Coastal bays and lagoons, and off-shore areas; in small to medium-sized schools; leaps from water and spins.

ISLANDS: BIG, KAU, LAN, MAUI, MOLO, OAHU, MID

Plate 49d

Spotted Dolphin
Stenella attenuata
Nai'a

ID: Dark gray or blackish dolphin often with light spots (most Hawaiian specimens are nearly spotless); belly whitish with or without gray spots; tip of snout often white; to 2.5 m (8 ft).

HABITAT: Coastal waters and off-shore areas; often seen in channels between islands.

ISLANDS: BIG, KAU, LAN, MAUI, MOLO, OAHU

Plate 49e

Rough-toothed Dolphin
Steno bredanensis
Nai'a

ID: Dark gray, bluish- or purplish gray, or blackish dolphin, sometimes with a few light spots or streaks; whitish belly separated from darker sides by irregular line; pointed beak; to 2.4 m (8 ft).

HABITAT: Coastal waters and off-shore areas; may prefer deeper waters; usually seen in small groups; seen less often than other dolphins perhaps because it spends more time submerged.

ISLANDS: BIG, KAU, LAN, MAUI, MOLO, OAHU

a Hawaiian Monk Seal

b Pacific Bottle-nosed Dolphin

c Spinner Dolphin

d Spotted Dolphin

e Rough-toothed Dolphin

Plate 50a

Humpback Whale
Megaptera novaeangliae
Kohola

ID: Huge dark/blackish whale with lighter or whitish belly; long whitish flippers with scalloped or knobby edges, to a third of body length; small dorsal fin; wart-like knobs on head; lower jaw with large lump; to 16 m (52 ft).

HABITAT: Coastal and off-shore waters; often seen in small groups (two to eight whales); often breaches (jumps from water) and spyhops (pushes head out of water and rotates it, looking around).

ISLANDS: BIG, KAU, LAN, MAUI, MOLO, OAHU

Note: Endangered, CITES Appendix I and USA ESA listed.

Plate 50b

False Killer Whale
Pseudorca crassidens
Kohola

ID: Mid-sized black whale with rounded head; gray patch on belly between flippers; largish swept-back dorsal fin at mid-back; owing to curve of mouth-line, often appears to be smiling; to 5.9 m (19 ft).

HABITAT: Off-shore and coastal waters, although not usually very close to shore; often seen in largish groups (to 100) that may hunt together; rides bow waves.

ISLANDS: BIG, KAU, LAN, MAUI, MOLO, OAHU

Plate 50c

Short-finned Pilot Whale
Globicephala macrorhynchus
Nu'ao

ID: Mid-sized black whale with light patch on belly between flippers; bulbous, rounded head; flippers 15% to 20% of body length; largish broad dorsal fin placed slightly forward of mid-back; to 6 m (20 ft).

HABITAT: Off-shore and coastal waters, although not usually very close to shore; often seen in large groups, 30 to 100+ whales.

ISLANDS: BIG, KAU, LAN, MAUI, MOLO, OAHU

Plate 50d

Sperm Whale
Physeter macrocephalus
Palaoa

ID: Huge grayish, bluish gray, or brownish whale with blunt rectangular head; sometimes whitish around mouth and with lighter or whitish belly; small hump on back but no dorsal fin; wavy bumps along rear back; males to 18 m (60 ft), females to 12 m (40 ft).

HABITAT: Off-shore waters and open ocean; breaches (jumps from water) occasionally; a deep-sea whale.

ISLANDS: BIG

Note: Endangered, CITES Appendix I and USA ESA listed.

Plate 50e

Pygmy Sperm Whale
Kogia breviceps
Kohola

ID: Small whale, dark gray above, with light gray to whitish belly; light semi-circular mark between eye and flipper; blunt, squarish head; dorsal fin small and rearwards-curving, placed slightly behind mid-back; to 3.7 m (12 ft).

HABITAT: Off-shore waters.

ISLANDS: BIG, OAHU

b False Killer Whale
e Pygmy Sperm Whale
c Short-finned Pilot Whale
a Humpback Whale
d Sperm Whale

Plate 51a

Pacific Green Sea Turtle
Chelonia mydas agassizii
Honu

ID: A medium-sized sea turtle with blackish, gray, greenish, or brown heart-shaped back, often with yellowish spots or streaks; whitish, yellow, or orangish underneath; males' front legs each have one large, curved claw; males with longer, thicker tail than females; name refers to greenish body fat; to about 1.2 m (4 ft).

HABITAT: Ocean waters off main island coasts; feeds in shallow water; lays eggs on beaches in Northwestern Hawaiian Islands and other sites in the tropical and subtropical Pacific.

ISLANDS: MOLO, MID

ORIGIN: Hawaiian native

Note: Endangered over some of its range, CITES Appendix I and USA ESA listed.

Plate 51b

Pacific Hawksbill Sea Turtle
Eretmochelys imbricata bissa
Honu'ea

ID: A small to mid-sized sea turtle; shield-shaped back mainly dark greenish brown; yellow underneath; head scales brown or black; jaws yellowish with dark markings; chin and throat yellow; two claws on each front leg; narrow head and tapering hooked "beak" give the species its name; to 90 cm (35 in).

HABITAT: Feeds in clear, shallow ocean water near rocks and reefs, and also in shallow bays and lagoons; lays eggs on beaches.

ISLANDS: BIG, MAUI, MOLO

ORIGIN: Hawaiian native

Note: Endangered, CITES Appendix I and USA ESA listed.

Lengths given for fish are "standard lengths," the distance from the front of the mouth to the point where the tail appears to join the body; that is, tails are not included in the measurement.

Marine life text by Richard Francis

Plate 51c

Spotted Eagle Ray
Aetobatis narinari

This fairly common ray has white spots on a dark background. Quite active, it often leaps above the water surface. Often occurs in pairs (to 2.3 m, 7.5 ft)

Plate 51d

Manta Ray
Manta birostris

This large ray is often accompanied by remoras. Two flaplike projections on either side of the mouth aid in gathering plankton. The dorsal surface is dark and the belly white, usually with a dark blotch or two. The slow graceful "wingbeats" of this spectacular animal are mesmerizing.
(to 7 m, 23 ft)

a Pacific Green Sea Turtle

b Pacific Hawksbill Sea Turtle

d Manta Ray

c Spotted Eagle Ray

Plate 52a

Gray Reef Shark
Carcharhinus amblyrhynchos

This impressive shark can be identified by the black-edged caudal fin. Prefers clear water around coral reefs. Should be treated with respect. (to 1.8 m, 6 ft)

Plate 52b

Galápagos Shark
Carcharhinus galapagensis

This large shark is found around oceanic islands. There are no conspicuous markings, but look for a low ridge along the back between the two dorsal fins. This shark preys on sea lions in the Galápagos and should be treated with great respect. (3.5 m, 11.5 ft)

Plate 52c

Blacktip Reef Shark
Carcharhinus melanopterus

This common reef shark is often found in very shallow water. Prominent black markings on the tips of all fins, especially the first dorsal. (to 1.8 m, 6 ft)

Plate 52d

Whitetip Reef Shark
Triaenodon obesus

By day this shark can often be found resting under reef ledges or in caves. It is most active at night. Look for white tips on dorsal and caudal fins. (to 1.7 m, 5.75 ft)

Plate 52e

Scalloped Hammerhead Shark
Sphyrna lewini

Hammerheads are unmistakable. This is the only species that occurs in the Hawaiian islands. Most often seen off reef walls, sometimes in fairly large numbers. (to 4 m, 13 ft)

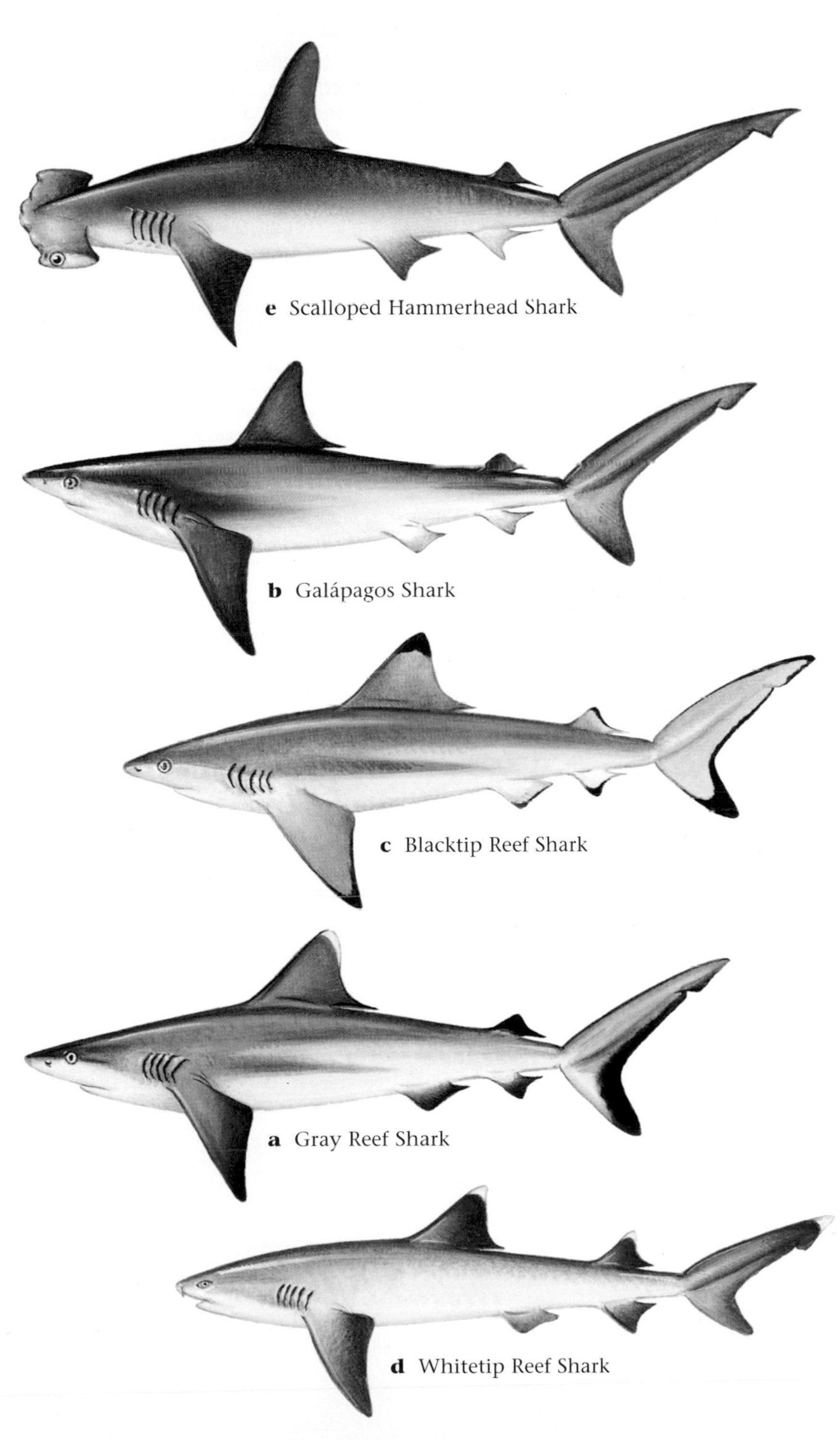

e Scalloped Hammerhead Shark

b Galápagos Shark

c Blacktip Reef Shark

a Gray Reef Shark

d Whitetip Reef Shark

Plate 53a

Snowflake Moray
Echidna nebulosa

This smallish moray is white, with small black spots and black bars extending from dorsal and ventral regions. You are more likely to find this species swimming in the open during the day than other morays. (to 71 cm, 28 in)

Plate 53b

Yellowmargin Moray
Gymnothorax flavimarginatus

Brown with yellow mottling. Look for a dark spot on gill opening. If it is out swimming, look for yellow lining on posterior fins. This is one of the most common morays in Hawaii and is easily tamed. (to 1.2 m, to 4 ft)

Plate 53c

Whitemouth Moray
Gymnothorax meleagris

This is the moray you are most likely to see in Hawaii. Numerous white spots on a dark background and striking white lining inside of mouth. (to 1 m, 40 in)

Plate 53d

Undulated Moray
Gymnothorax undulatus

Fairly common but not as easy to spot as the previous two morays. The background coloration is yellowish, but with dense irregular dark blotches of varying size. The head region has fewer and smaller blotches and hence is more yellow-green than the rest of the body. (to 1.1 m, 3.5 ft)

Plate 53e

Hawaiian Garden Eel
Gorgasia hawaiiensis

Look for aggregations of these eels in sandy areas between reef patches. Only a small portion of their grayish bodies extends above the surface and they disappear into the sand unless approached carefully. These plankton feeders prefer areas with some currents. (to 60 cm, 2 ft)

Plate 53f

Smallmouth Bonefish
Albula glossodonta

Prefers sandflats and lagoons. The body is silver with black highlights. Its upper jaw extends well beyond its lower jaw, like a freshwater sucker. (to 71 cm, 28 in)

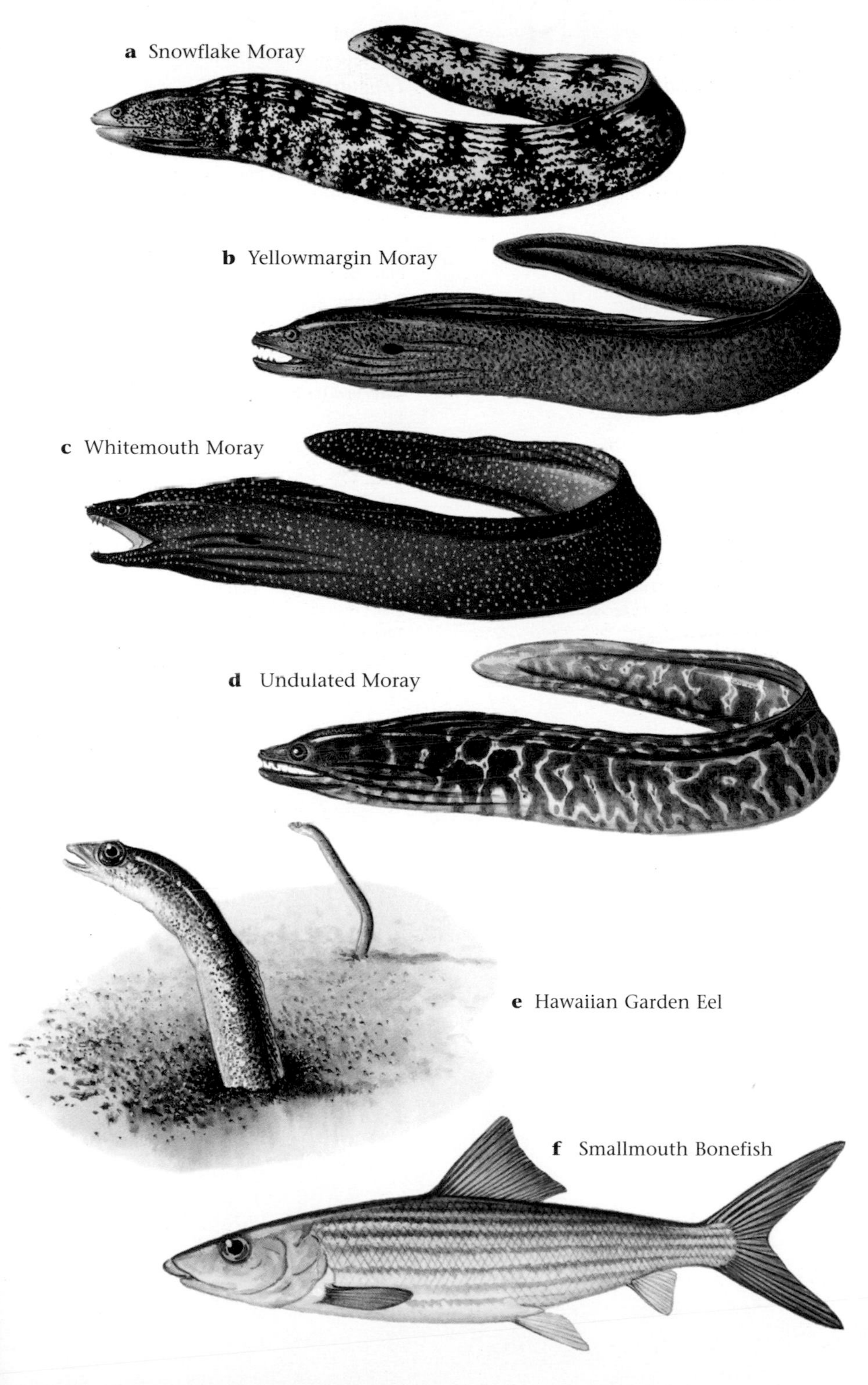
a Snowflake Moray
b Yellowmargin Moray
c Whitemouth Moray
d Undulated Moray
e Hawaiian Garden Eel
f Smallmouth Bonefish

Plate 54a

Milkfish
Chanos chanos

This silvery minnow-like fish was an important foodfish for the indigenous Hawaiians, who practiced aquaculture with this species in artificial ponds. (to 1.5 m, 5 ft)

Plate 54b

Reef Lizardfish
Synodus variegatus

This is one of the most common of the lizardfishes found in Hawaii and the most often seen. Variably colored with blotchy patterns which help it blend with coral rubble. (to 25 cm, 10 in)

Plate 54c

Hawaiian Freckled Frogfish
Antennarius drombus

Frogfishes are some of the most bizarre fishes to be found on reefs. All are well camouflaged and largely motionless. Easiest to spot when they move their eyes. This species is endemic to Hawaii. (to 11 cm, 4.5 in)

Plate 54d

Houndfish
Tylosurus crocodilus

You will find this elongate species hovering just below the surface. Countershaded: dark above, silvery below. The jaws extend forward to form a needlenose. This is the largest of the beaked fishes in Hawaii. (to 1.3 m, 53 in)

Plate 54e

Polynesian Halfbeak
Hemiramphus depauperatus

Another surface-hugging beaked fish. Distinguished by the red color at the tip of the lower jaw. (to 38 cm, 15 in)

Plate 54f

Spotfin Squirrelfish
Neoniphon sammara

Squirrelfishes are nocturnal. Usually some shade of red with large prominent eyes, and sharp spine in front of gill cover. During the day they can be found resting in reef crevices and caves. This species is more silvery than other squirrelfishes; characteristic dark spot on front of dorsal fin. (to 30 cm, 1 ft)

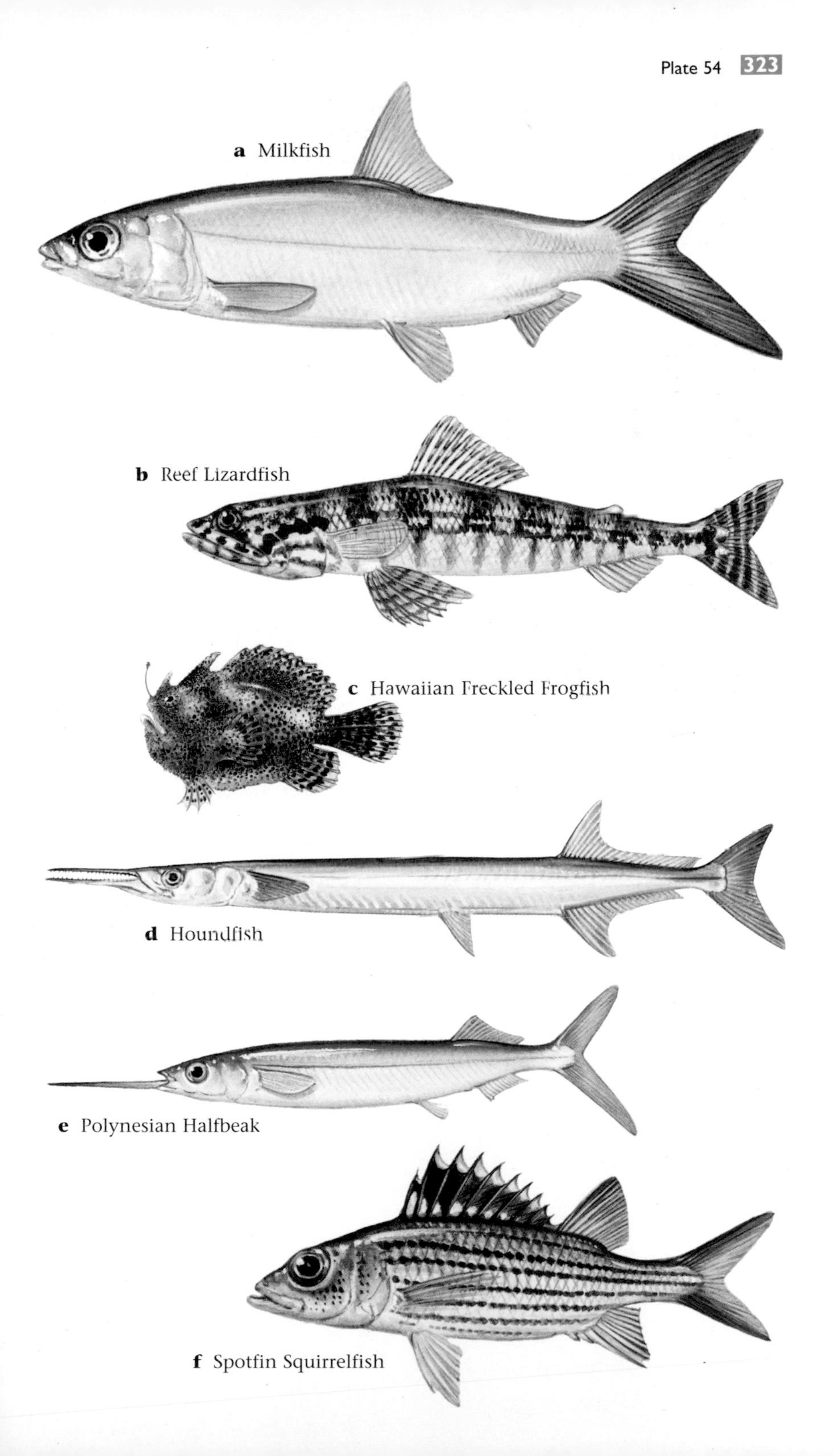
a Milkfish
b Reef Lizardfish
c Hawaiian Freckled Frogfish
d Houndfish
e Polynesian Halfbeak
f Spotfin Squirrelfish

Plate 55a

Peppered Squirrelfish
Sargocentron punctatissimum

Silvery background with red horizontal stripes. This species occurs in shallower water than most squirrelfishes; it can sometimes be found in tidepools. (to 13 cm, 5 in)

Plate 55b

Hawaiian Squirrelfish
Sargocentron xantherythrum

This endemic species is the most common squirrelfish in Hawaii. Red body with white horizontal stripes. (to 16 cm, 6.5 in)

Plate 55c

Trumpetfish
Aulostomus chinensis

This species will not be mistaken for any other. It occurs in two colors: yellow (the most common) and brown. Usually observed hanging head down in the water column. (to 76 cm, 30 in)

Plate 55d

Cornetfish
Fistularia commersonii

Similar in shape to the trumpetfish but much more slender. Most active at night, it can be seen resting on the bottom by day. (to 1.5 m, 5 ft)

Plate 55e

Redstripe Pipefish
Dunkerocampus baldwini

Pipefishes are closely related to seahorses (Family Syngnathidae) and, like seahorses, the males carry the developing young in a special brood pouch. All pipefishes are extremely slender and elongate. This species, which has a prominent longitudinal red stripe, is endemic to Hawaii. Look for it in crevices. (to 14 cm, 5.5 in)

Plate 55f

Hawaiian Lionfish
Dendrochirus barberi

Lionfishes and scorpionfishes belong to the same family as California rockfishes (family Scorpaenidae). All have venomous spines projecting from their fins. Lionfishes are well camouflaged and spend most of the time on the substrate waiting to ambush unsuspecting prey. This species is a Hawaiian endemic. It is distinguished by its red eyes. (to 91 cm, 3 ft)

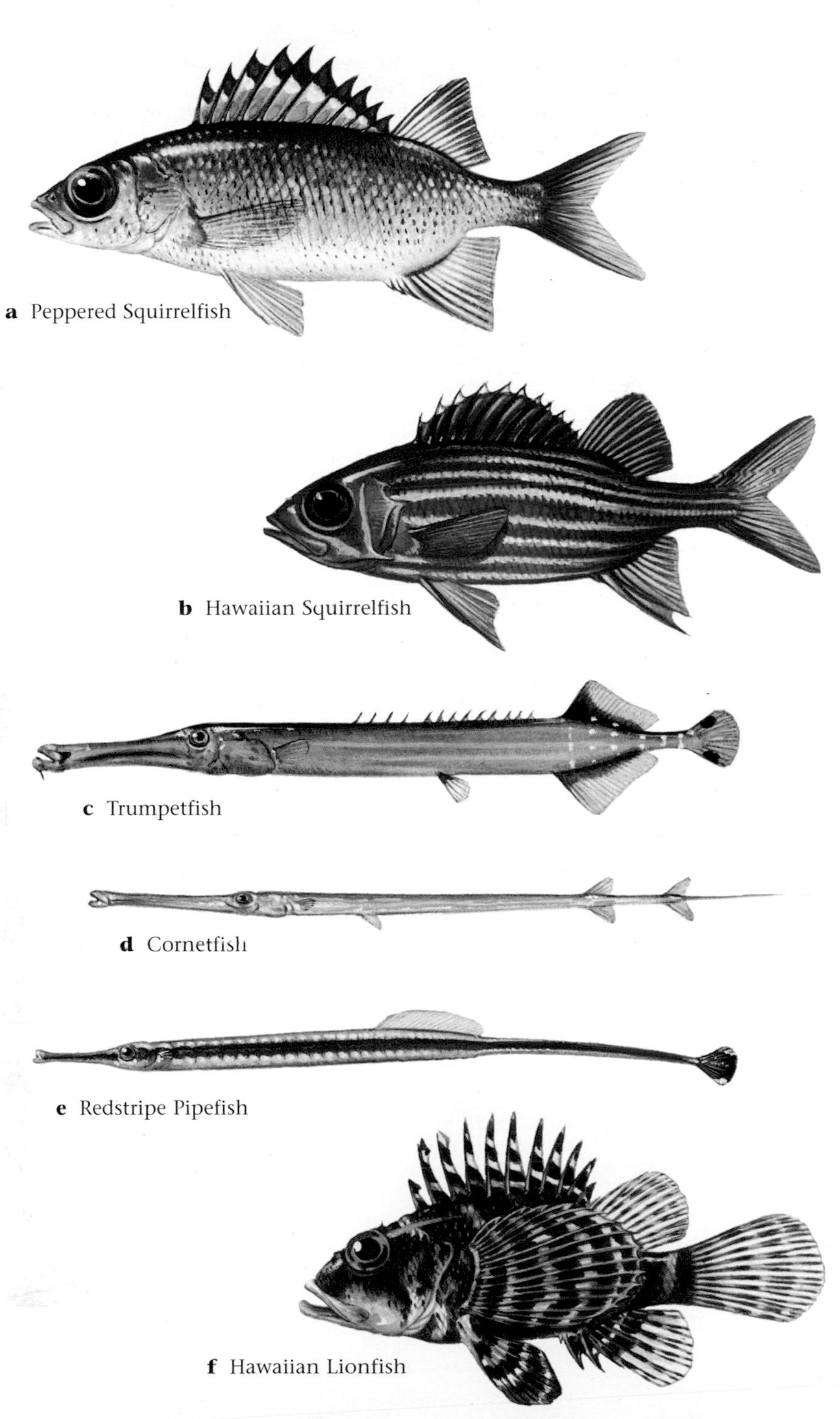

a Peppered Squirrelfish

b Hawaiian Squirrelfish

c Trumpetfish

d Cornetfish

e Redstripe Pipefish

f Hawaiian Lionfish

Plate 56a

Hawaiian Turkeyfish
Pterois sphex

This striking beauty has much longer spines than the lionfish. Brown and red bars on light background. During the day it tends to hover beneath ledges, often upside down. It is quite venomous, so don't stick your hand into reef recesses without looking first. Another Hawaiian endemic. (to 20 cm, 8 in)

Plate 56b

Titan Scorpionfish
Scorpaenopsis cacopsis

The spines on scorpionfishes are less prominent than those on lionfishes and turkeyfishes. This species is a mottled orange-red, with fleshy projections from body and head which give it a plant-like countenance. Endemic to Hawaii, this species is valued as a food fish. Unfortunately, it has become uncommon owing to spearfishing. Look for it in caves. (to 51 cm, 20 in)

Plate 56c

Hawaiian Bigeye
Priacanthus meeki

All members of this family (Priacanthidae) are laterally compressed with very large eyes. This species, a Hawaiian endemic, is generally some shade of red, but it can change its hue. (to 33 cm, 13 in)

Plate 56d

Arc-eye Hawkfish
Paracirrhites arcatus

Hawkfishes (Family Cirrhitidae) are smallish ambush predators and generally hug the substrate. Some are protogynous sex changers. This is one of the most common hawkfish species in Hawaii. It can be distinguished by the neon line arcing from below to above its eye. It occurs in two color phases, one with red background color and prominent white stripe, and the other a dark greenish-brown. (to 14 cm, 5.5 in)

Plate 56e

Hawaiian Morwong
Cheilodactylus vittatus

This easily identified fish is laterally compressed and deep-bodied in the front but tapering toward the tail. Striking black stripes on a white background. (to 41 cm, 16 in)

Plate 56f

Iridescent Cardinalfish
Apogon kallopterus

Cardinalfishes (Family Apogonidae) are nocturnal and often some shade of red. This small species, another Hawaiian endemic, is more likely to be found in shallow water than other members of the family. Its color is red but translucent, such that you can see its internal organs. (to 15 cm, 6 in)

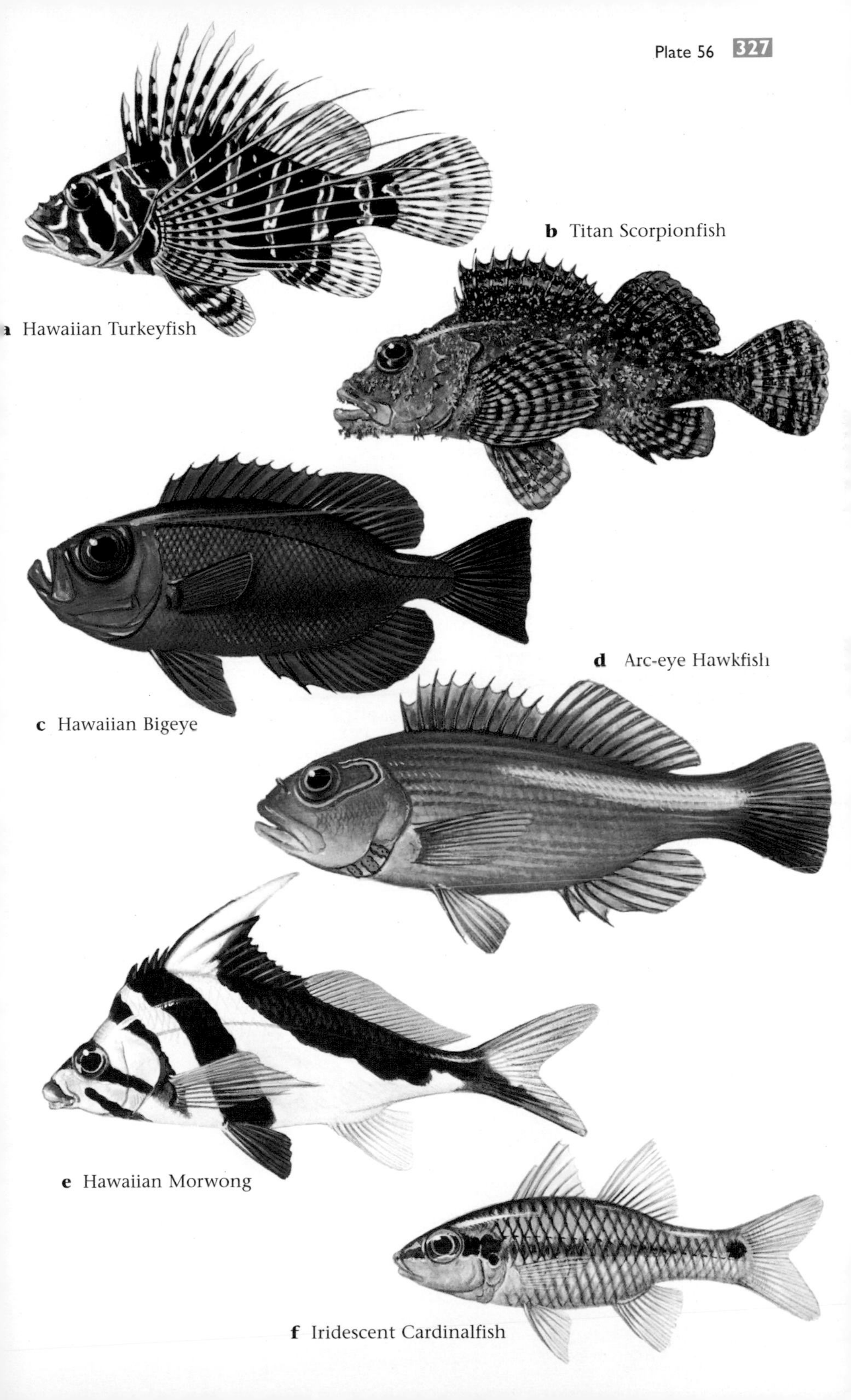

a Hawaiian Turkeyfish

b Titan Scorpionfish

c Hawaiian Bigeye

d Arc-eye Hawkfish

e Hawaiian Morwong

f Iridescent Cardinalfish

Plate 57a

Flagtail Tilefish
Malacanthus brevirostris

This species has a sloping head. The greenish gray body becomes lighter toward the bottom. It is quite elongate and slightly tapers toward the tail. Look for its characteristic inverted volcano-shaped burrows in sandy areas between reef patches. (to 30 cm, 12 in)

Plate 57b

Sharksucker
Echeneis naucrates

Like all remoras (Family Echeneidae) this species is usually found in the company of large sharks and mantas to which it attaches by means of a sucking disk on the top of its head. Elongate body with prominent longitudinal black stripe along midbody. (to 90 cm, 35 in)

Plate 57c

Island Jack
Carangoides orthogrammus

Jacks and other members of this family (Carangidae) are important food fishes. They prefer open water but make forays onto the reefs. They also come to cleaning stations (p. 194). All jacks are extremely strong swimmers. This species feeds more on benthic (bottom-dwelling) animals than most jacks, primarily on sandy substrates. The silvery blue body is typical of jacks, but look for ovoid yellow spots on sides that are characteristic of this species. (to 68 cm, 27 in)

Plate 57d

Giant Trevally
Caranx ignobilis

This large jack can weigh up to 66 kg (145 lb). Usually found in small schools off reef walls. Silvery with very small black spots that can be seen only at close range. (to 1.7 m, 5.5 ft)

Plate 57e

Bigeye Trevally
Caranx sexfasciatus

Named for its relatively large eye, this species is more fusiform (torpedo-shaped) than the giant trevally. The bigeye is nocturnal and tends to form large schools by day. (to 91 cm, 3 ft)

Plate 57f

Greater Amberjack
Seriola dumerili

This jack is distinguished from the rest by a dark band extending from the mouth, through the eye to the nape. (to 1.8 m, 6 ft)

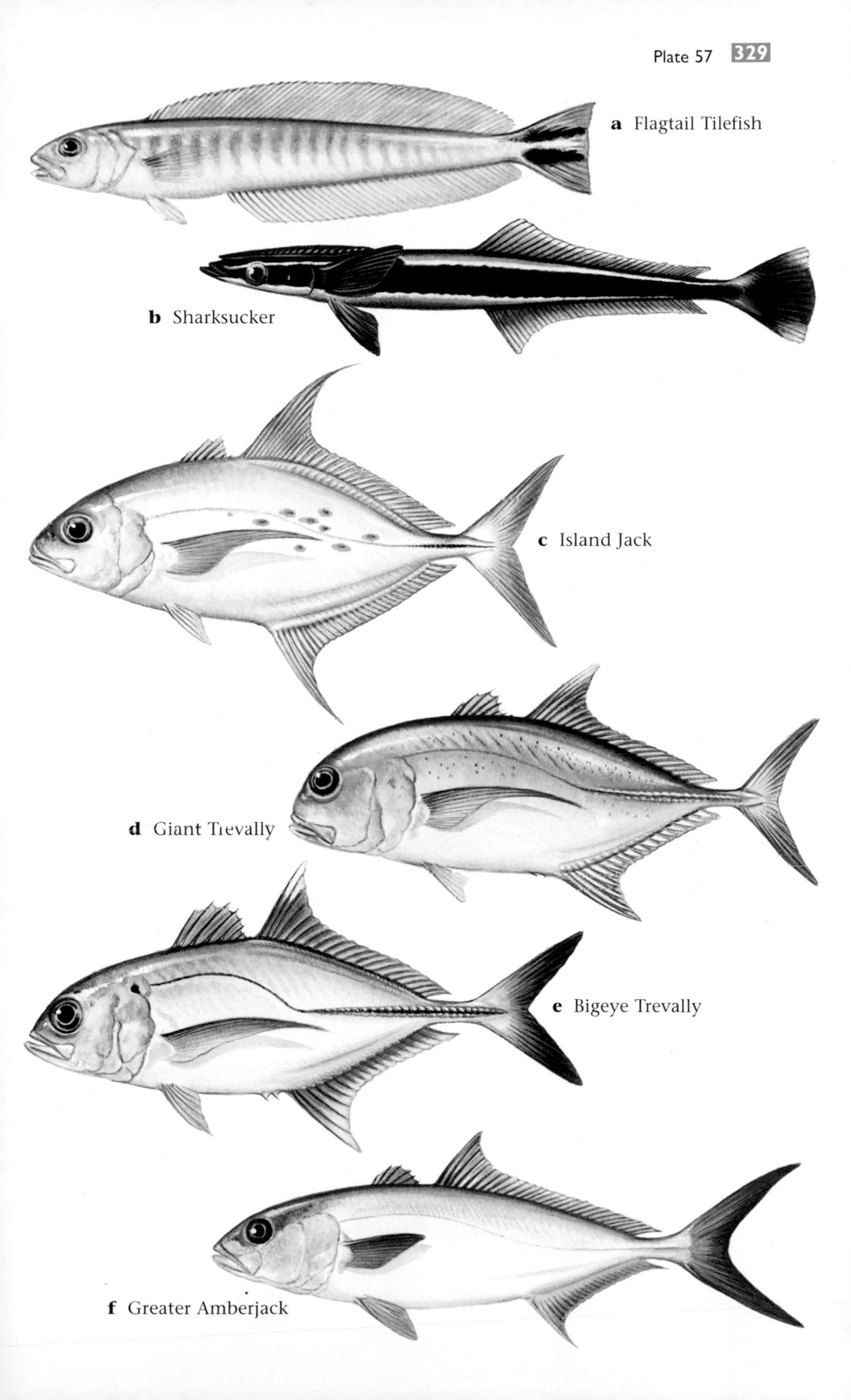
a Flagtail Tilefish
b Sharksucker
c Island Jack
d Giant Trevally
e Bigeye Trevally
f Greater Amberjack

Plate 58a

Yellowfin Goatfish
Mulloidichthys vanicolensis

Goatfishes (Family Mullidae) have catfish-like barbels extending from their chins. These are chemosensory organs that allow them to taste their surroundings in their quest for benthic prey in the sand or mud. When the prey (worms, brittle stars or mollusks) is located they root it out with their nose. This species feeds mostly at night but it can be readily seen by day in fairly stationary groups. The yellow midline stripe and yellow tailfin are characteristic. (to 38 cm, 15 in)

Plate 58b

Blue Goatfish
Parupeneus cyclostomus

This large goatfish differs from the rest in feeding primarily on other fishes. It has a long snout; wavy blue lines extend from its eyes. The body is blue to brown with a yellow saddle at the base of the tail. (to 51 cm, 20 in)

Plate 58c

Manybar Goatfish
Parupeneus multifasciatus

This is one of the most common goatfishes in Hawaii. It is variably colored but generally with whitish background overlayed by dark bars, especially toward the tail. (to 28 cm, 11 in)

Plate 58d

Gray Chub
Kyphosus bigibbus

Chubs are deep-bodied and almost perfectly symmetrical. Also called *rudderfishes*, they prefer exposed areas with hard substrates. This species feeds in groups above the reef. They can be seen overwhelming the territorial damselfishes and surgeonfishes by their size and numbers. This species is typical of its genus and fairly common. (to 60 cm, 2 ft)

Plate 58e

Threadfin Butterflyfish
Chaetodon auriga

Butteflyfishes are among the most spectacular reef inhabitants and readily spied by even the most casual fishwatcher. Their oval, laterally compressed bodies are colorfully marked with shades of yellow, black, white and blue, and they often possess a black eye-bar. Many species feed on live corals which they probe with their pincer-like snouts. Those that feed near the bottom tend to be territorial and the larger species occur in pairs that are mated for life. This species is white in front and yellow toward the tail. Two series of dark stripes extend at right angles from each other and diagonally relative to the body axis. A dark vertical stripe extends through the eye; and they have a characteristic black spot toward the rear of the dorsal fin, from which extends a filament-like projection. (to 20 cm, 8 in)

Plate 58f

Saddleback Butterflyfish
Chaetodon ephippium

One of the more common butterflyfishes in Hawaii, this beauty has a large black area extending from the base of the dorsal fin. It too has a filament extending from the dorsal fin. (to 23 cm, 9 in)

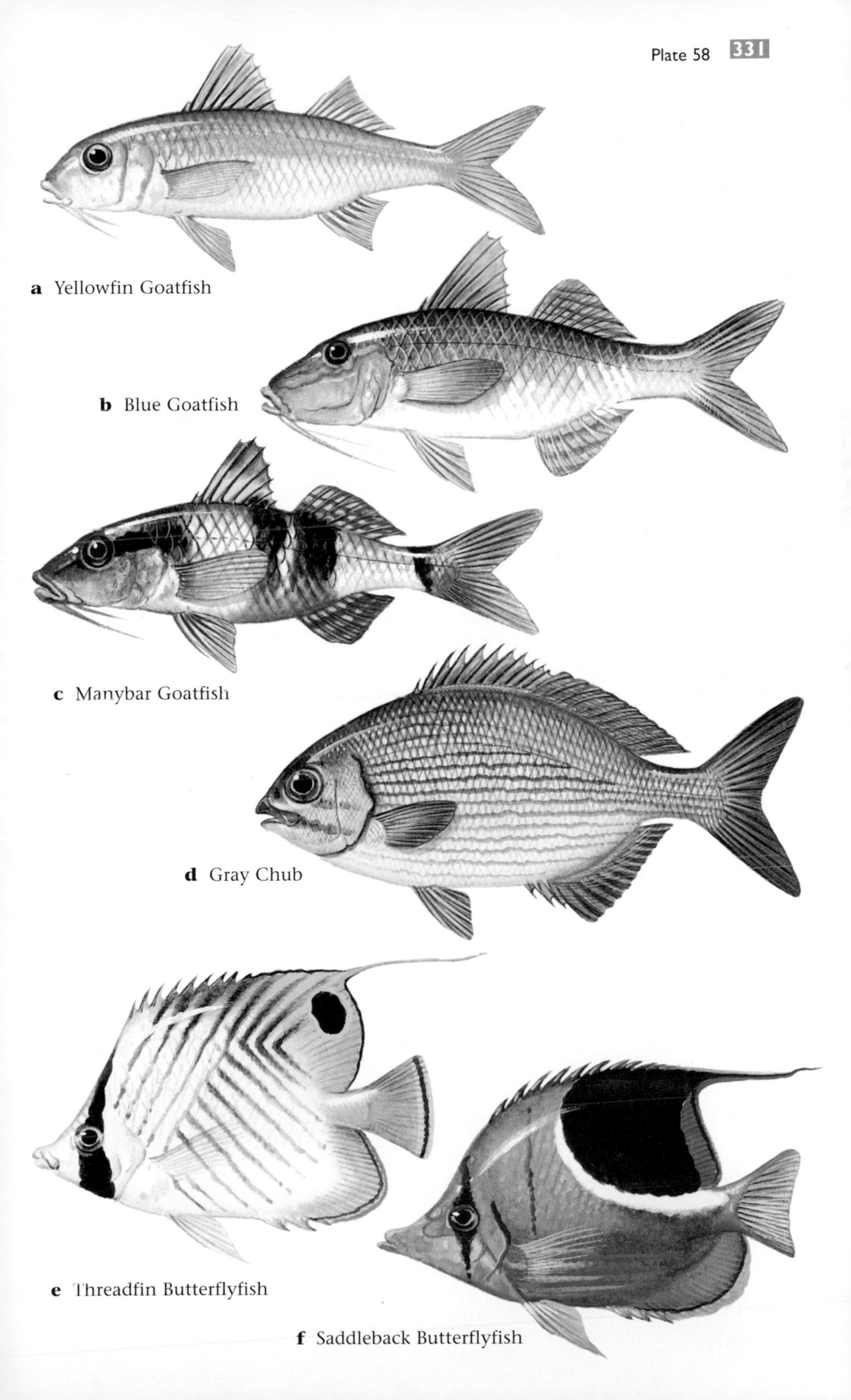

a Yellowfin Goatfish

b Blue Goatfish

c Manybar Goatfish

d Gray Chub

e Threadfin Butterflyfish

f Saddleback Butterflyfish

Plate 59a

Bluestripe Butterflyfish
Chaetodon fremblii

Yellow body with horizontal blue stripes. Dark spot on nape and a larger one near the base of the tail. This species is a Hawaiian endemic and is common in shallow water. (15 cm, 6 in)

Plate 59b

Lined Butterflyfish
Chaetodon lineolatus

Among the largest and most beautiful species of butterflyfishes. The body is white with narrow vertical black lines and a prominent black eyestripe. The dorsal, anal and caudal fins are yellow. The long snout is used to extract live coral. (to 30 cm, 12 in)

Plate 59c

Raccoon Butterflyfish
Chaetodon lunula

Named for the mask-like black band that covers its eyes, behind which is a white band. The body is yellow near the ventral area and dusky brown toward the nape. (to 20 cm, 8 in)

Plate 59d

Oval Butterflyfish
Chaetodon lunulatus (trifasciatus)

Perhaps the most beautiful fish in Hawaii, this butterfly is almost perfectly oval. It is not common. Almost always occurs in pairs. Prefers calm water such as lagoons. (to 14 cm, 5.5 in)

Plate 59e

Milletseed Butterflyfish
Chaetodon miliaris

Also called the Lemon Butterflyfish, this species is a Hawaiian endemic and also the most common butterflyfish in the area. Look for characteristic vertical rows of dark spots. This species feeds on zooplankton; it sometimes cleans larger fishes and sea turtles. (to 16 cm, 6.5 in)

Plate 59f

Ornate Butterflyfish
Chaetodon ornatissimus

There is nothing subtle about this large butterflyfish. Five black bars extend from tip of mouth to nape and one runs right through the eye. The bands are separated by yellow stripes. Orange stripes radiate from behind the gill. Another coral feeder that almost always is found in pairs. (to 20 cm, 8 in)

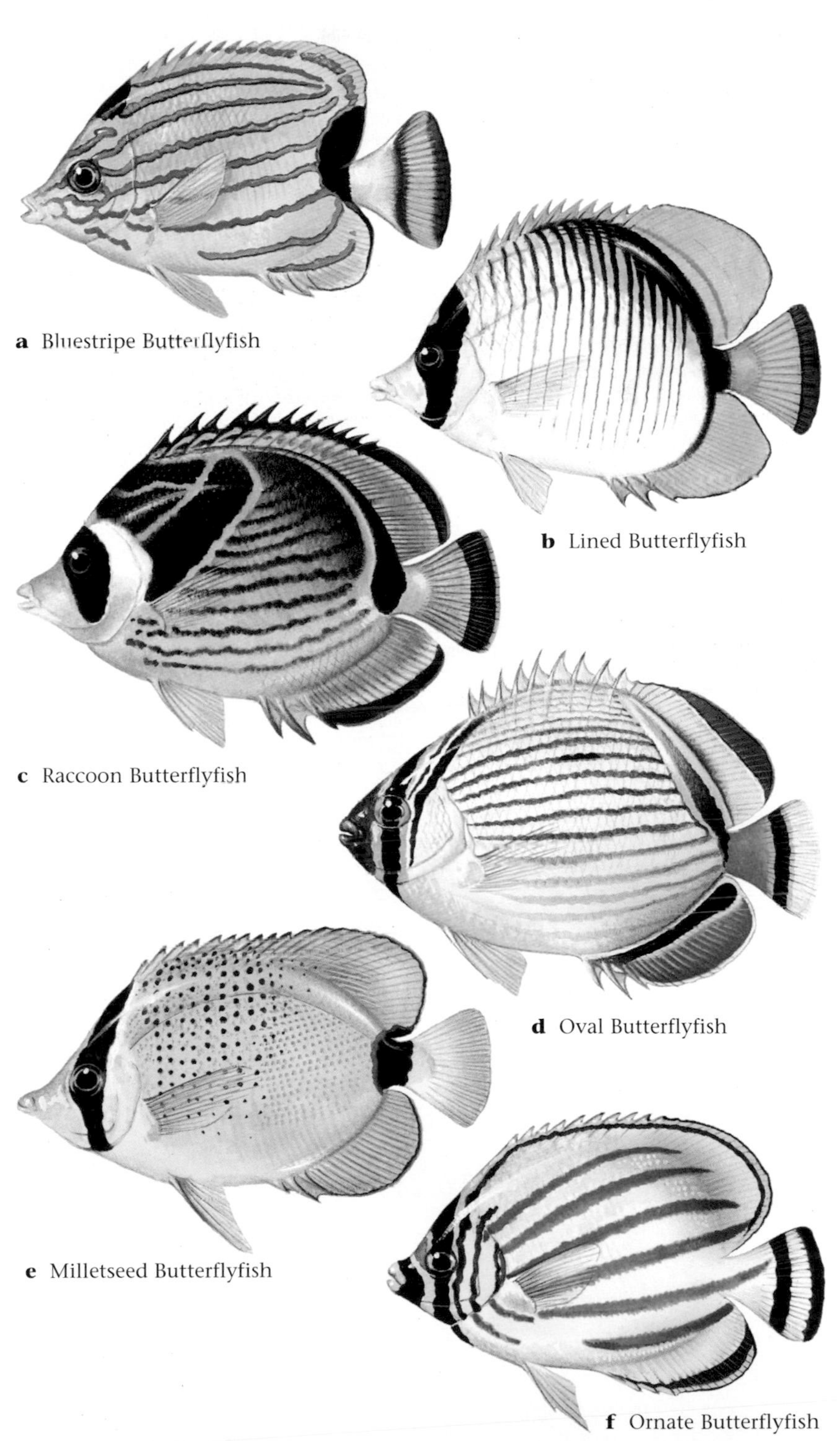

a Bluestripe Butterflyfish

b Lined Butterflyfish

c Raccoon Butterflyfish

d Oval Butterflyfish

e Milletseed Butterflyfish

f Ornate Butterflyfish

Plate 60a

Fourspot Butterflyfish
Chaetodon quadrimaculatus
Named for its four large white spots – two on each side – on dusky background. A prominent white stripe runs vertically just behind the eye. This species prefers the outer reef areas. Another coral feeder that occurs in pairs. (to 15 cm, 6 in)

Plate 60b

Reticulated Butterflyfish
Chaetodon reticulatus
You are most likely to find this uncommon species off the Big Island of Hawaii. Named for the finely cross-hatched black and yellow region on its sides. It occurs in pairs and feeds on coral. (to 18 cm, 7 in)

Plate 60c

Chevron Butterflyfish
Chaetodon trifascialis
This large species is uncommon in Hawaii. Named for its chevron-shaped black markings on a white background. Not at all shy. (to 18 cm, 7 in)

Plate 60d

Teardrop Butterflyfish
Chaetodon unimaculatus
The large teardrop shaped black spot is unique to this large butterflyfish. This coral-eater will also take worms and crustaceans. Almost always paired. (to 20 cm, 8 in)

Plate 60e

Longnose Butterflyfish
Forcipiger longirostris
Named for the long snout with which it probes for worms, fish eggs, and small crustaceans. The similar Forcepsfish (*Chaetodon flavissimus*) lacks the triangular black area on head and has a somewhat shorter snout. The longnose has a dark (chocolate-brown) color phase. (to 21 cm, 8.5 in)

Plate 60f

Pennantfish
Heniochus diphreutes
This species occurs in schools off the bottom. It is white with black bands. The dorsal fin extends as a long white filament. (to 20 cm, 8 in)

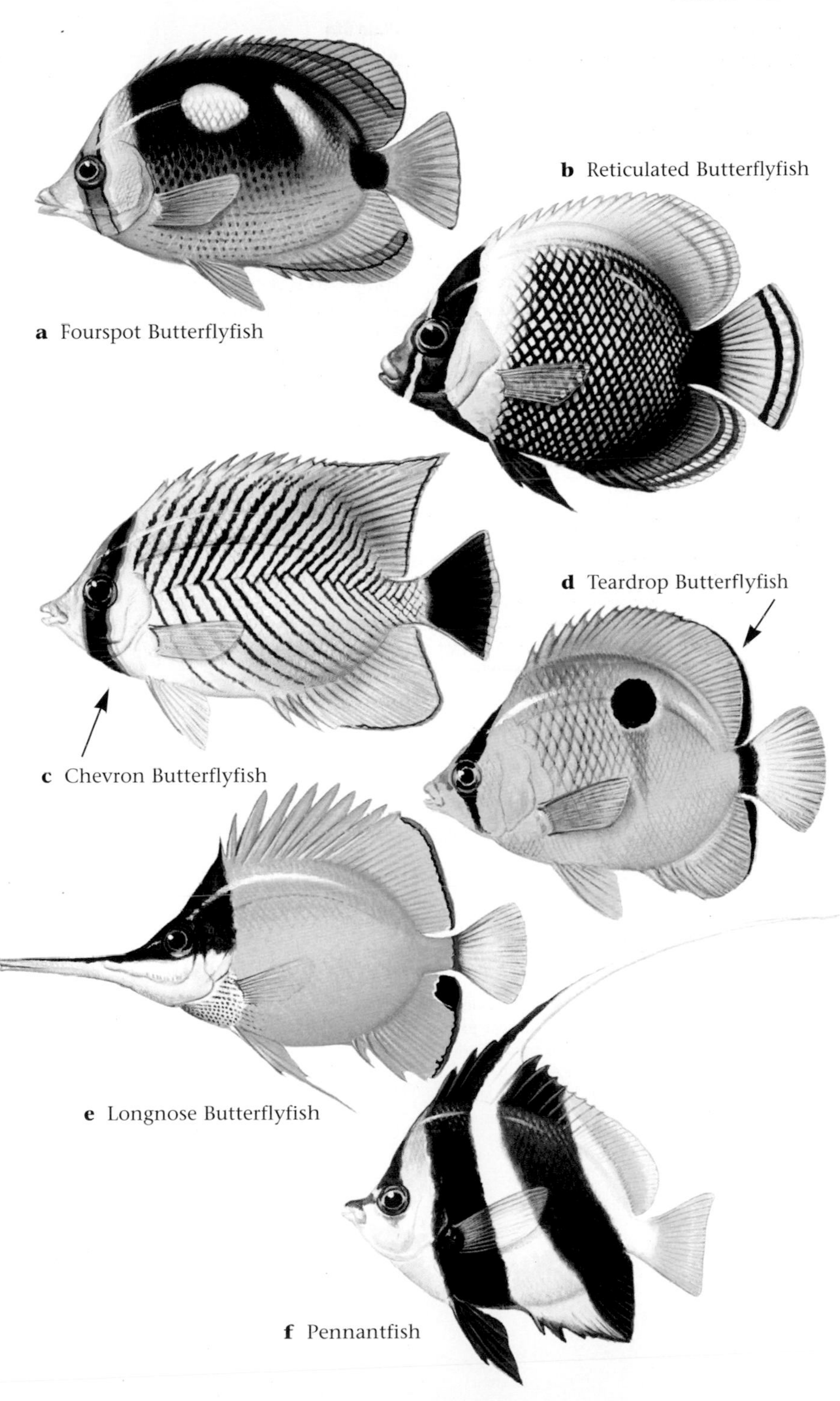
a Fourspot Butterflyfish
b Reticulated Butterflyfish
c Chevron Butterflyfish
d Teardrop Butterflyfish
e Longnose Butterflyfish
f Pennantfish

Plate 61a

Flame Angelfish
Centropyge loricula

Angelfishes (Family Pomacanthidae) are closely related to butterflyfishes and exhibit similar body shapes and mouths. Unlike butterflyfishes though, all angelfishes have a cheek spine in front of the gill cover. Angelfishes also undergo dramatic color changes as they develop; the adults and juveniles of a species do not look at all alike. Some and perhaps all angelfishes are protogynous sex changers (p. 192). Whereas butterflyfishes are well-represented in Hawaii, there are few angelfishes and none of the larger species found elsewhere in the Indo-Pacific. The Flame Angel is the most spectacular member of this family in Hawaii. Though gaudy, it is fairly secretive and not that easy to find. It is well worth the effort, however. (to 10 cm, 4 in)

Plate 61b

Potter's Angelfish
Centropyge potteri

This beauty is the most common angelfish in Hawaii and the most likely to be encountered while snorkeling. The irregular alternating orange and blue lines are characteristic. This is yet another species found only in the Hawaiian islands. (to 13 cm, 5 in)

Plate 61c

Hawaiian Sergeant
Abudefduf abdominalis

Damselfishes are among the most noticeable reef inhabitants. Members of this family also reward the observant diver with their interesting behavior. Most are highly territorial and quite pugnacious during the breeding season, often attacking fishes of much larger size. Like the angelfishes, damselfishes generally undergo dramatic color changes as they transition from the juvenile to the adult stage. Often the juveniles are more colorful than the adults. This species is extremely abundant and active. Five black bars on a gray-green background distinguish this Hawaiian endemic. (to 24 cm, 9.5 in)

Plate 61d

Agile Chromis
Chromis agilis

This rather drab damsel has a characteristic black spot at the base of the pectoral fins. It actively flits about the reef in pursuit of zooplankton. (to 11 cm, 4.5 in)

Plate 61e

Hawaiian Dascyllus
Dascyllus albisella

Members of this genus are protogynous sex changers (p. 192). This deep-bodied damsel is colored charcoal to black. Juveniles (shown) black with white spot on side and light spot on forehead. Endemic to Hawaii. (to 13 cm, 5 in)

Plate 61f

Chocolate-dip Chromis
Chromis hanui

The whimsical common name of this species says it all. The body is chocolate-brown and the tail (presumably by which it was held in the dipping process) is white. Endemic to Hawaii. (to 9 cm, 3.5 in)

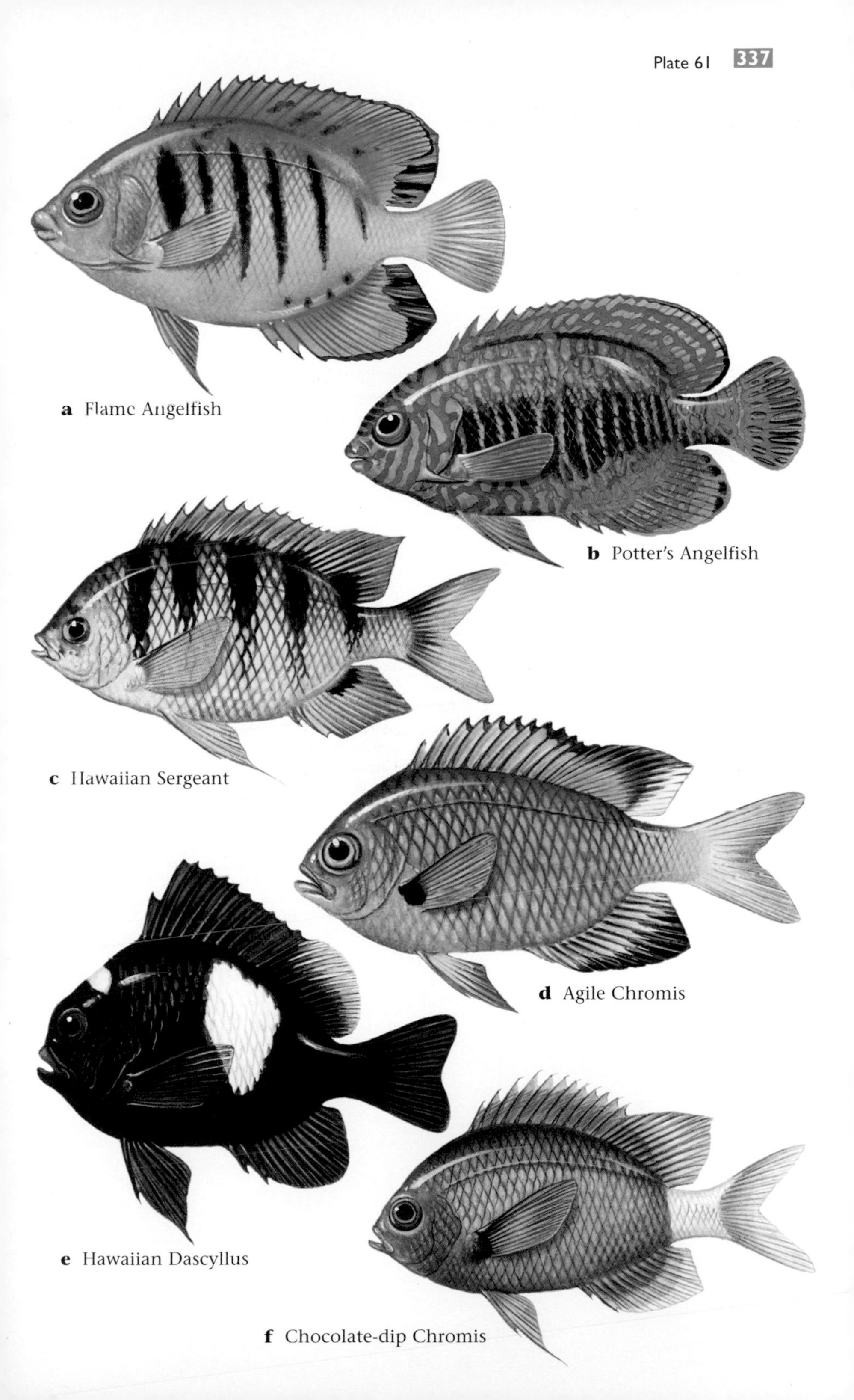

a Flame Angelfish

b Potter's Angelfish

c Hawaiian Sergeant

d Agile Chromis

e Hawaiian Dascyllus

f Chocolate-dip Chromis

Plate 62a

Blue-eyed Damselfish
Plectroglyphidodon johnstonianus

Look for blue eyes and blue markings at the tip of the dorsal fin. This damsel feeds partly on live coral. (to 11 cm, 4.5 in)

Plate 62b

Pacific Gregory
Stegastes fasciolatus

The body of this species is a drab gray, but the yellow iris stands out. Juveniles have a prominent yellow area at the base of the caudal fin, as well as an electric blue margin on the dorsal fin. (to 16 cm, 6.5 in)

Plate 62c

Pearl Wrasse
Anampses cuvier

The wrasses (Family Labridae) comprise an extremely large family and an important component of the reef fauna. Most, if not all, are protogynous sex changers (p. 192) and many undergo pronounced color changes during development. Wrasses have a characteristic jerky swimming motion, and primarily use their pectoral fins to propel themselves. Their bodies are generally dorsal ventrally compressed and often cigar shaped. The Pearl Wrasse is a Hawaiian endemic and prefers rocky (or lava) substrates. In contrast to most wrasses, the female is more colorful than the male. The males are typically olive-green, females dark with many white spots. (to 36 cm, 14 in)

Plate 62d

Hawaiian Hogfish
Bodianus bilunulatus

The Hawaiian population is considered a distinct subspecies of this broadly distributed Indo-Pacific fish. Males are a blotchy red-brown, females have a white body with horizontal brownish orange lines and a distinct black spot below the posterior portion of the dorsal fin. (to 51 cm, 20 in)

Plate 62e

Elegant Coris
Coris venusta

This common Hawaiian endemic can be readily found on shallow reefs. Males are green with red and blue bars in the head region; in females the red lines extend farther back. (to 19 cm, 7.5 in)

Plate 62f

Yellowtail Coris
Coris gaimard

This is one of the wrasses that undergoes a dramatic color change in the transition from the juvenile stage to adulthood. Juveniles (shown) are red with several white blotches outlined in black. Adult males and females are similar. Both have green bodies with small bright blue spots that become more numerous toward the tail, pinkish-orange heads with green lines radiating from the eyes, and yellow tails. (to 38 cm, 15 in)

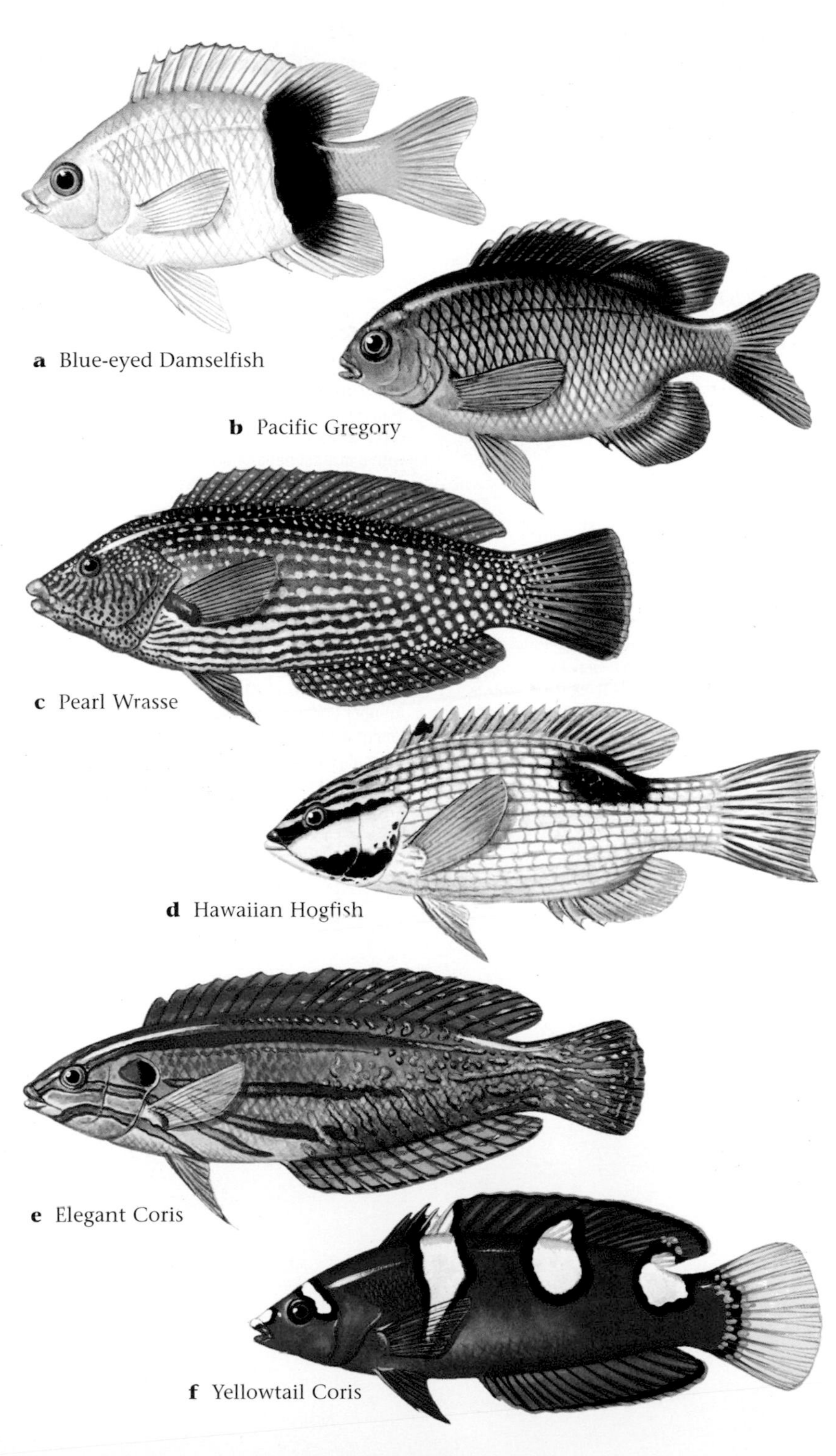

a Blue-eyed Damselfish

b Pacific Gregory

c Pearl Wrasse

d Hawaiian Hogfish

e Elegant Coris

f Yellowtail Coris

Plate 63a

Bird Wrasse
Gomphosus varius

This wrasse looks like something out of Dr. Seuss. The males are dark blue green; females exhibit much less color. (to 32 cm, 12.5 in)

Plate 63b

Ornate Wrasse
Halichoeres ornatissimus

Light red with green stripes on head and green spots on scales. This common wrasse roots around the bottom for crustaceans and mollusks. (to 15 cm, 6 in)

Plate 63c

Hawaiian Cleaner Wrasse
Labroides phthirophagus

This Hawaiian endemic is closely related to the Bluestreak Cleaner Wrasse (*Labroides dimidiatus*) found throughout the rest of the Pacific. The common names for both derive from their habit of cleaning larger fish species at traditional sites known as *cleaning stations*. They live in one-male groups. If the male dies, the most dominant female changes sex. Males are spectacularly colored with streaks of yellow, black, blue, and purple. (to 10 cm, 4 in)

Plate 63d

Rockmover Wrasse
Novaculichthys taeniourus

The juveniles of this species look like pieces of algae. Adults are dark with white spots on each scale and dark irregular lines radiating from the eyes. The common name reflects their penchant for moving pebbles and other objects with their mouths during their search for benthic prey. (to 30 cm, 12 in)

Plate 63e

Belted Wrasse
Stethojulis balteata

This Hawaiian endemic is quite common on inshore reefs. Terminal phase males are green with a distinct band of orange extending at midline from pectoral fin to near the base of the tail. Females and initial phase males are green with a yellow spot above pectoral fins. (to 15 cm, 6 in)

Plate 63f

Saddle Wrasse
Thalassoma duperry

Another endemic species, the Saddle Wrasse is one of the most common reef inhabitants. The head of terminal phase males is dark purple, the rest of the body green with a broad orange band around the pectoral fins. (to 25 cm, 10 in)

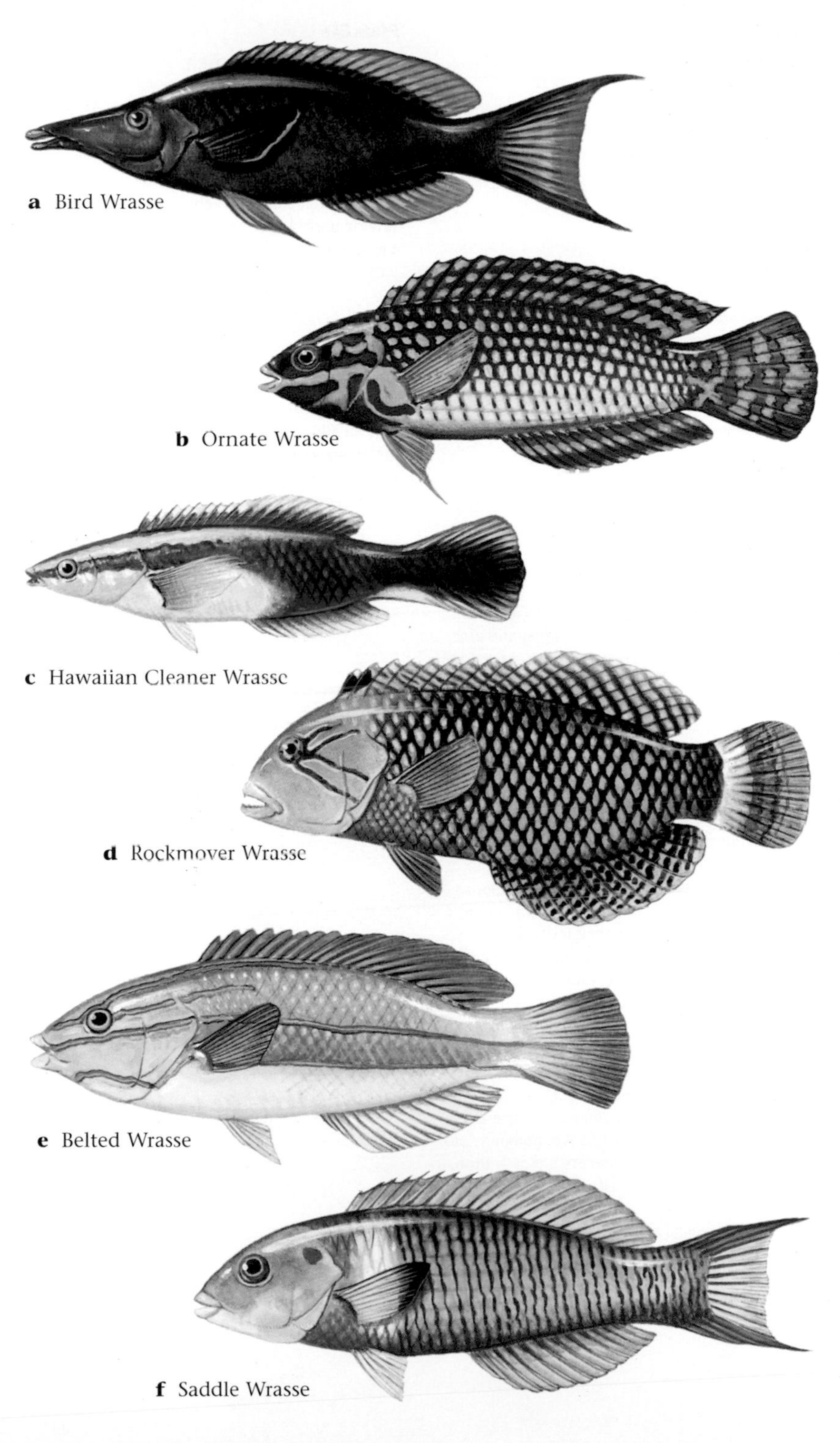

a Bird Wrasse

b Ornate Wrasse

c Hawaiian Cleaner Wrasse

d Rockmover Wrasse

e Belted Wrasse

f Saddle Wrasse

Plate 64a

Christmas Wrasse
Thalassoma trilobatum

In the terminal phase of this wrasse, the body is rose-colored with two rows of rectangular green. Initial phase fish are green with black mottling. (to 30 cm, 12 in)

Plate 64b

Psychedelic Wrasse
Anampses chrysocephalus

This Hawaiian endemic can be found on the deeper reefs. Males have bright orange heads with blue markings. Females are brown with a white spot on each scale. Both males and females have a white bar at the base of the tail. (to 18 cm, 7 in)

Plate 64c

Blackside Razorfish
Xyrichtys umbrilatus

Whitish gray with a large black blotch in the middle. In the juveniles of this endemic the first two dorsal spines project forward well beyond the head. (to 23 cm, 9 in)

Plate 64d

Stareye Parrotfish
Calotomus carolinus

Parrotfishes are so called because of their beak-like dentition formed by the fusion of several teeth. They use this structure to bite off chunks of coral from which they extract the living matter in their unique digestive system (they lack stomachs). The coral matrix is ground into a fine powder which they excrete. Parrotfish excrement is a major constituent of all those beautiful tropical beaches. While biting and crunching the coral, parrotfishes make sounds that are audible underwater. Probably all parrotfishes are protogynous sex changers (p. 192), and, like the wrasses, many undergo dramatic changes in coloration which may or may not accompany sex change. The initial phase fish are typically drab, while the terminal phase fish – usually males – can be quite strikingly colored. The Stareye Parrotfish is typical in this respect: initial phase fish are mottled gray, white and pink, but the terminal phase is blue-green with pinkish lines radiating from the eyes. (to 51 cm, 20 in)

Plate 64e

Spectacled Parrotfish
Chlorurus perspicillatus

This large parrotfish is endemic to Hawaii, where it is fairly common. The initial phase fish, which sometimes form large groups, are dusky brown with red fins and a white bar at the base of the tail. Terminal phase fish are blue-green, tending toward lavender in the head region. A distinct darker bar runs across the forehead. (to 61 cm, 2 ft)

Plate 64f

Palenose Parrotfish
Scarus psittacus

This smallish parrotfish is common in Hawaii. The initial phase is dusky brown; the terminal phase is green with pink-edged scales and pink cheeks and chin. (to 30 cm, 12 in)

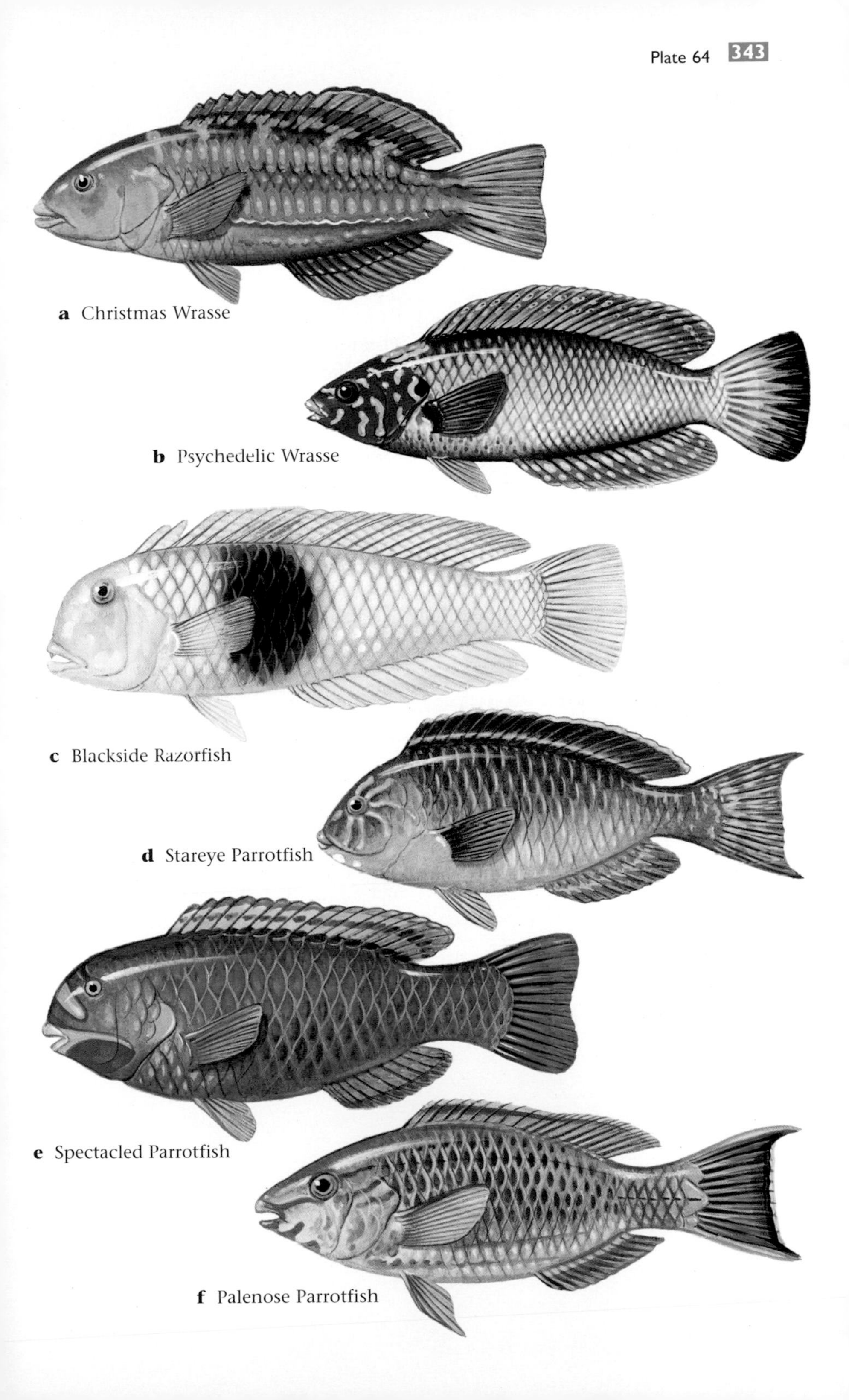

a Christmas Wrasse

b Psychedelic Wrasse

c Blackside Razorfish

d Stareye Parrotfish

e Spectacled Parrotfish

f Palenose Parrotfish

Plate 65a

Redlip Parrotfish
Scarus rubroviolaceus

Initial phase fish have reddish brown head and dusky gray body. Terminal phase males have the distinctive pink line above the upper lip; their bodies are green with purplish fins. (to 30 cm, 28 in)

Plate 65b

Scarface Blenny
Cirripectes vanderbilti

Blennies (Family Blenniidae) are small bottom-dwelling fish in which the pelvic fins are in front of the pectoral fins. This blenny is endemic to Hawaii, and it is one of the species you are most likely to encounter. Dark brown with a blunt reddish head on which can usually be found bright red small lines and spots. (to 11 cm, 4.5 in)

Plate 65c

Shortbodied Blenny
Exallias brevis

Males and females of this blenny look quite different. Males have many closely packed brown-red spots forming reddish blotches toward the tail. Females lack the red coloration. Deeper bodied than most blenny species. (to 15, cm, 6 in)

Plate 65d

Zebra Rockskipper
Istiblennius zebra

This Hawaiian endemic can be found in very shallow water, including tide pools. The color is uniform but quite variable. Look for the fleshy projections on the head. This blenny is named for its ability to leap from one tide pool to another. (to 18 cm, 7 in)

Plate 65e

Hawaiian Sand Goby
Coryphopterus sp.

Gobies comprise the largest fish family but there are relatively few species in Hawaii. Moreover, gobies tend to be small and inconspicuous, so they are overlooked by all but the most observant divers. This endemic species is a case in point. It is tiny and almost translucent, so it is hard to spot against its sandy backdrop. (to 6 cm, 2.5 in)

Plate 65f

Hawaiian Shrimp Goby
Psilogobius mainlandi

Another endemic, this species is named for its habit of cohabitating with shrimp. Prefers protected areas where silt has built up. Shown guardinng a shrimp burrow. (to 6 cm, 2.3 in)

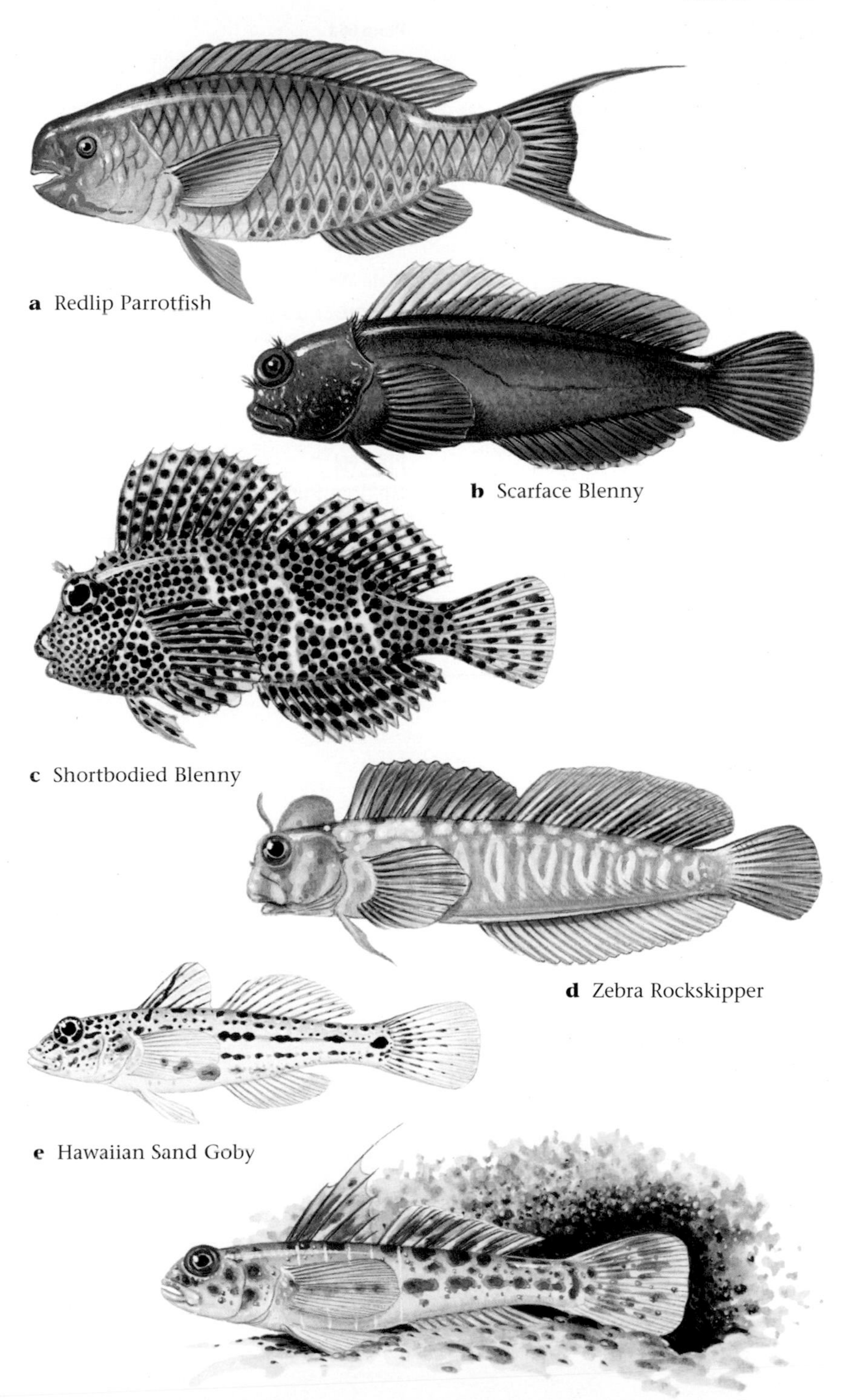

a Redlip Parrotfish

b Scarface Blenny

c Shortbodied Blenny

d Zebra Rockskipper

e Hawaiian Sand Goby

f Hawaiian Shrimp Goby

Plate 66a

Moorish Idol
Zanclus cornutus

The sole member of its family, the Moorish Idol requires no description. Perhaps the single most spectacular reef inhabitant. This species will usually be found in pairs, moving in a deliberate and stately manner about the reef as they probe the coral with their narrow mouths. (to 20 cm, 8 in)

Plate 66b

Achilles Tang
Acanthurus achilles

Hawaii is well endowed with surgeonfishes (Family Acanthuridae). This is one of the more common and conspicuous members of the clan. Strikingly colored, it has a black body with a bright orange area near its tail. The fins are rimmed with white. (to 25 cm, 10 in)

Plate 66c

Ringtail Surgeonfish
Acanthurus blochii

You are more likely to find this large surgeonfish over sandy areas than on the reef itself. It has a dark greenish body with bluish purple median fins, and a white swatch at the base of the tail. (to 43 cm, 17 in)

Plate 66d

Whitespotted Surgeonfish
Acanthurus guttatus

Look for this fish in the surge zone. Often forms small schools. The posterior half of the body has white spots, while two vertical white stripes adorn the head region. This species is also more deep-bodied than most surgeonfishes. (to 28 cm, 11 in)

Plate 66e

Whitebar Surgeonfish
Acanthurus leucopareius

Look for the white stripe behind the eye. The body is colored a light greenish gray. (to 25 cm, 10 in)

Plate 66f

Goldrim Surgeonfish
Acanthurus nigricans

Your best chance of seeing this uncommon species is off the Big Island. It is black with bright yellow streaks separating the body from the dorsal and anal fins. The tail is white with a yellow stripe. (to 22 cm, 8.5 in)

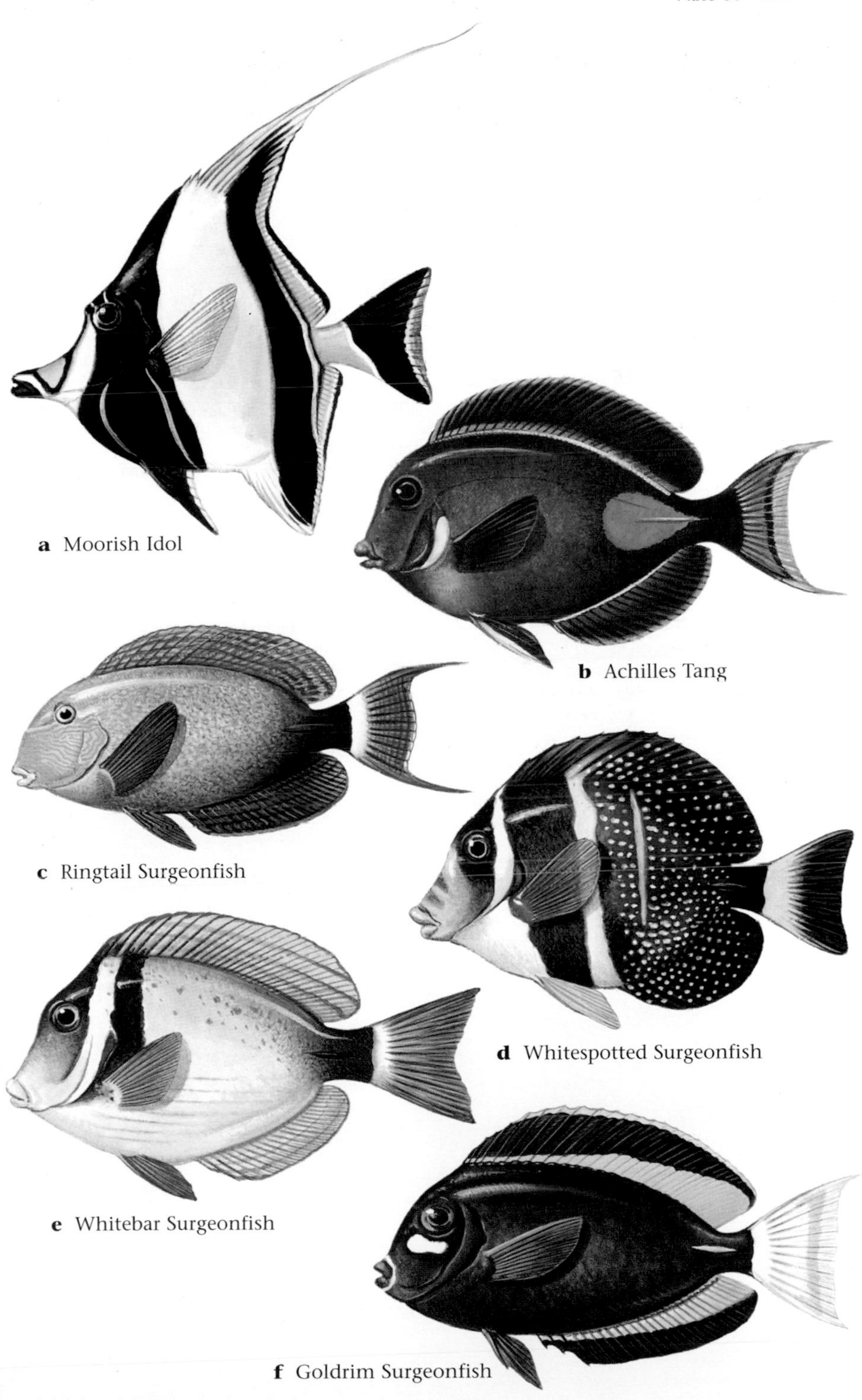

a Moorish Idol

b Achilles Tang

c Ringtail Surgeonfish

d Whitespotted Surgeonfish

e Whitebar Surgeonfish

f Goldrim Surgeonfish

Plate 67a

Brown Surgeonfish
Acanthurus nigrofuscus

This rather drab surgeonfish is quite common. The body is tan with lavender tones; yellowish spots on head; and black spots at the base of the dorsal and anal fins near the tail. (to 20 cm, 8 in)

Plate 67b

Orangeband Surgeonfish
Acanthurus olivaceus

This is one of the more distinctively colored reef fish. Its body is tan in the head half and dark brown in the tail half; a striking horizontal orange streak extends from behind the eye toward the tail. Its mouth is situated more toward the ventral than in most surgeonfish. Juveniles are bright yellow and can be confused with adult yellow tangs. (to 30 cm, 12 in)

Plate 67c

Convict Surgeonfish
Acanthurus triostegus

This extremely common fish is named for its vertical black stripes on a grayish background. It often occurs in large aggregations. (to 25 cm, 10 in)

Plate 67d

Goldring Surgeonfish
Ctenochaetus strigosus

This one is named for the yellow area around the eye. Numerous thin light blue lines run horizontally across a dark brown background. (to 18 cm, 7 in)

Plate 67e

Bluespine Unicornfish
Naso unicornis

The unicornfish are a subgroup of surgeonfish with forward-directed projections on their head. This common species has fairly modest "horns" by unicornfish standards, but they are quite noticeable none the less. Like other members of the family, it eats primarily algae, and it will come into quite shallow water to get it. (to 69 cm, 27 in)

Plate 67f

Yellow Tang
Zebrasoma flavescens

This is one of the first fish you will notice when you enter the water. It is more common in Hawaii than elsewhere in the tropical Pacific. (to 20 cm, 8 in)

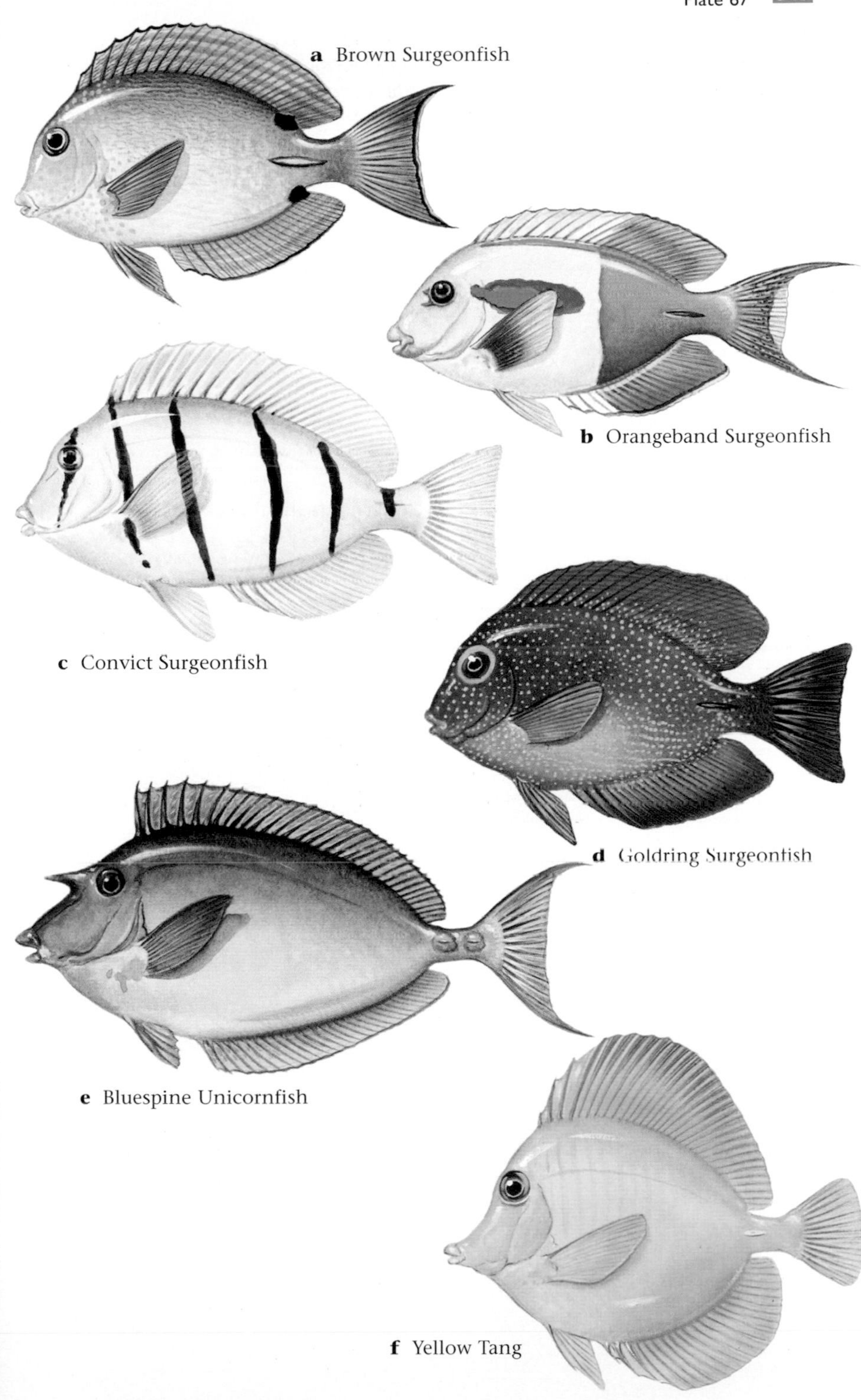
a Brown Surgeonfish
b Orangeband Surgeonfish
c Convict Surgeonfish
d Goldring Surgeonfish
e Bluespine Unicornfish
f Yellow Tang

Plate 68a

Sailfin Tang
Zebrasoma veliferum

This surgeonfish likes to be near where the waves are breaking on the reef. When alarmed it raises its large dorsal and anal fins, seeming to double its body size. Dark brown bars alternate with light gray. (to 39 cm, 15.5 in)

Plate 68b

Great Barracuda
Sphyraena barracuda

It looks much more menacing than it is. This silvery predator usually hangs in the water unless disturbed, its dramatic underbite revealing an impressive array of needle-like teeth. The smaller, younger fish prefer to hang around mangroves. (to 1.7 m, 5.5 ft)

Plate 68c

Black Durgeon
Melichthys niger

Triggerfish (Family Balistidae) are named for their ability to lock their first dorsal spine in the erect position with the second dorsal spine. Inside their relatively small mouths are some impressive chisel-like teeth, which they use to munch sea urchins and other invertebrates with hard exteriors. They have a distinctive way of swimming by means of undulations of the dorsal and anal fins. The females generally guard the eggs, and when nesting they can be quite aggressive; hence they should be treated with respect. This triggerfish feeds mainly on zooplankton in the water column and tends to form aggregations, unlike most other members of the family. From a distance it appears all black, with white lines at the base of the dorsal and anal fins. Up close you can discern densely packed black lines on a dark blue-green background. (to 32 cm, 12.5 in)

Plate 68d

Lagoon Triggerfish
Rhinecanthus aculeatus

As its common name implies, this species prefers calm water. The yellow line extending from the mouth is distinctive in this fish, which resembles a piece of abstract art. (to 30 cm, 12 in)

Plate 68e

Reef Triggerfish
Rhinecanthus rectangulus

This beauty is also often referred to as the Picasso Fish or Picasso Trigger, because of its resemblance to the paintings of that artist inspired by African masks. (to 25 cm, 10 in)

Plate 68f

Lei Triggerfish
Sufflamen bursa

This species is common in Hawaii. Look for the thin white line extending from the mouth to the anal fin, and the thicker yellow lines, one through the eye and the other behind. (to 22 cm, 8.5 in)

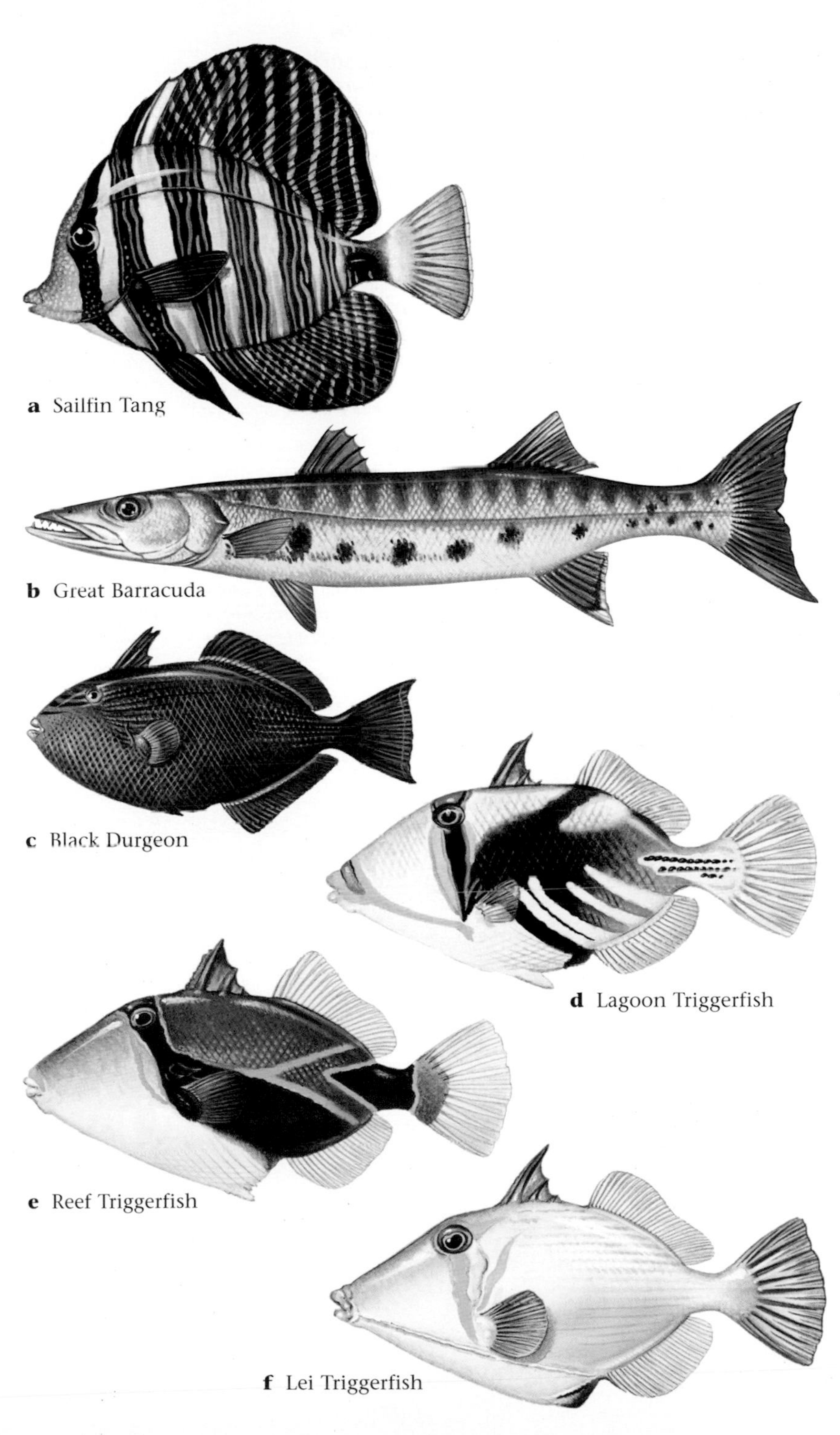

a Sailfin Tang

b Great Barracuda

c Black Durgeon

d Lagoon Triggerfish

e Reef Triggerfish

f Lei Triggerfish

Plate 69a

Scrawled Filefish
Aluterus scriptus

This fish comports itself as if it were ill, its tail hanging down, and its dorsal and anal fins held close to the body. The body color is grayish blue and it is covered with dark blue spots in an irregular array. (to 76 cm, 30 in)

Plate 69b

Fantail Filefish
Pervagor spilosoma

The first dorsal spine is especially long on this Hawaiian endemic. Look for the orange tail. This species is noteworthy because of its boom and bust cycles. In some years they are about the most common reef animal and in others they are hard to find. (to 18 cm, 7 in)

Plate 69c

Spotted Boxfish
Ostracion meleagris

Hawaii has its own unique subspecies. Like the puffers and porcupinefishes, trunkfishes primarily use their pectoral fins to propel themselves. The sexes have dramatically different coloration in this species. Males are black on top with white spots, and blue on the side with black spots. Females are black with white spots. (to 15 cm, 6 in)

Plate 69d

Spotted Puffer
Arothron meleagris

Also referred to as the Guineafowl Puffer, especially in the Eastern Pacific. There are two distinct color types: one dark with white spots and the other yellow all over. The yellow form is rare in Hawaii. (to 34 cm, 13.5 in)

Plate 69e

Crown Toby
Canthigaster coronata

This small puffer is off-white with brown bands on top, and yellow or blue spots scattered all over. It prefers areas of rubble. (to 13 cm, 5 in)

Plate 69f

Spiny Balloonfish
Diodon holocanthus

This nocturnal fish is the one that inflates to form a spiny ball. It has a whitish belly and various shades of brown elsewhere. Note the large eyes. (to 38 cm, 15 in)

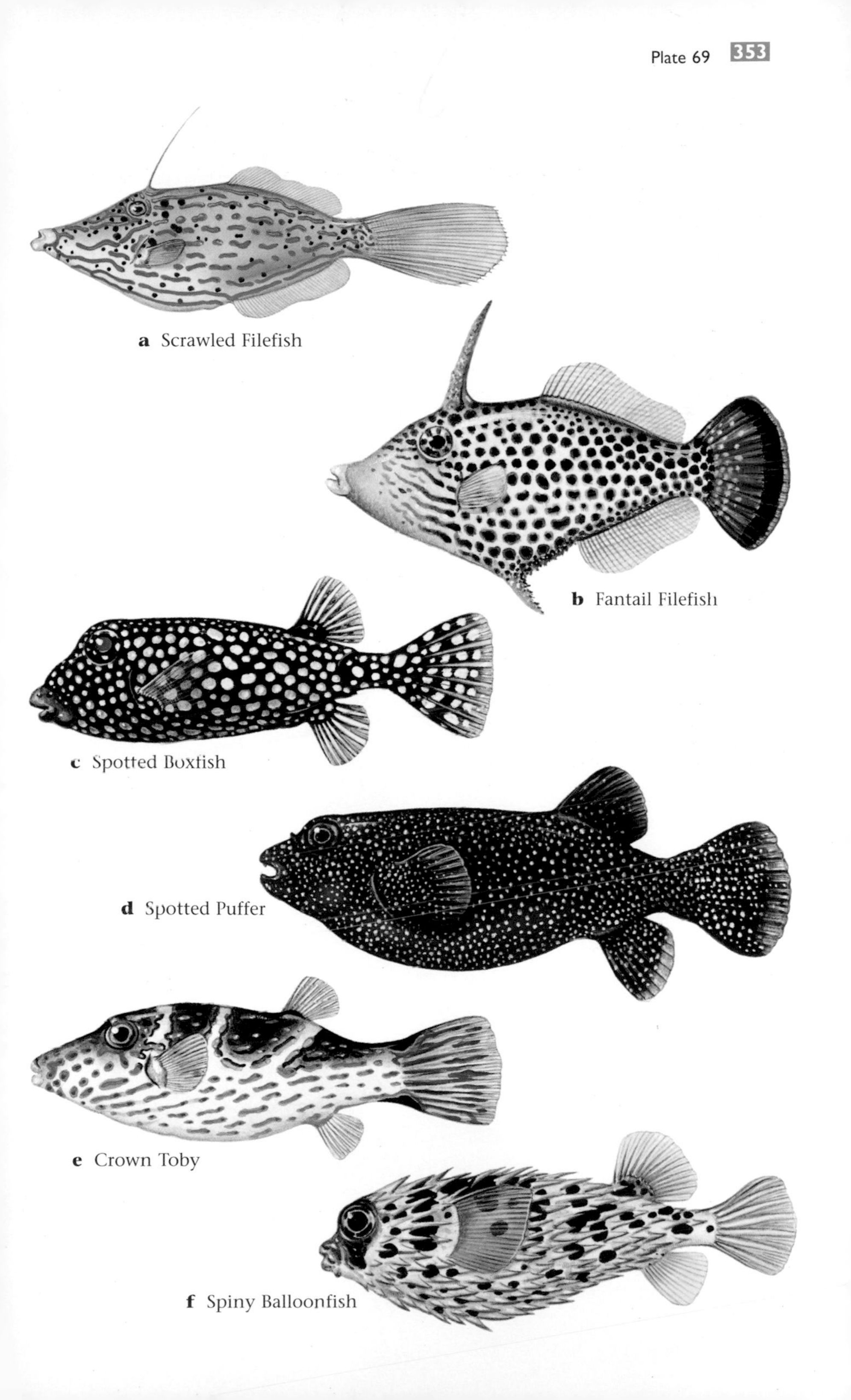

a Scrawled Filefish

b Fantail Filefish

c Spotted Boxfish

d Spotted Puffer

e Crown Toby

f Spiny Balloonfish

Plate 70a
Clathrina sp.
A shiny sponge with long white tubes. Usually found in sheltered areas such as crevices.

Plate 70b
Ralpharia sp.
This species prefers very shallow water. It has two sets of tentacles: an outer spreading group and a shorter group held close together.

Plate 70c
Fire Coral
Millepora
An encrusting, sometimes branching, group of species that inflict very nasty burning stings when touched.

Plate 70d
Carijoa sp.
Members of this genus of octocoral are white but are often associated with sponges that alter their coloration. It forms mats of long cylindrical branches that taper toward the top.

Plate 70e
Dendronephythya (Spongodes)
Members of this coral genus form attractive, prickly, rounded colonies.

Plate 70f
Sea Whip
Ctenocella sp.
Colonies consist of long whip-like extensions. Prefers deeper parts of reef, especially slopes.

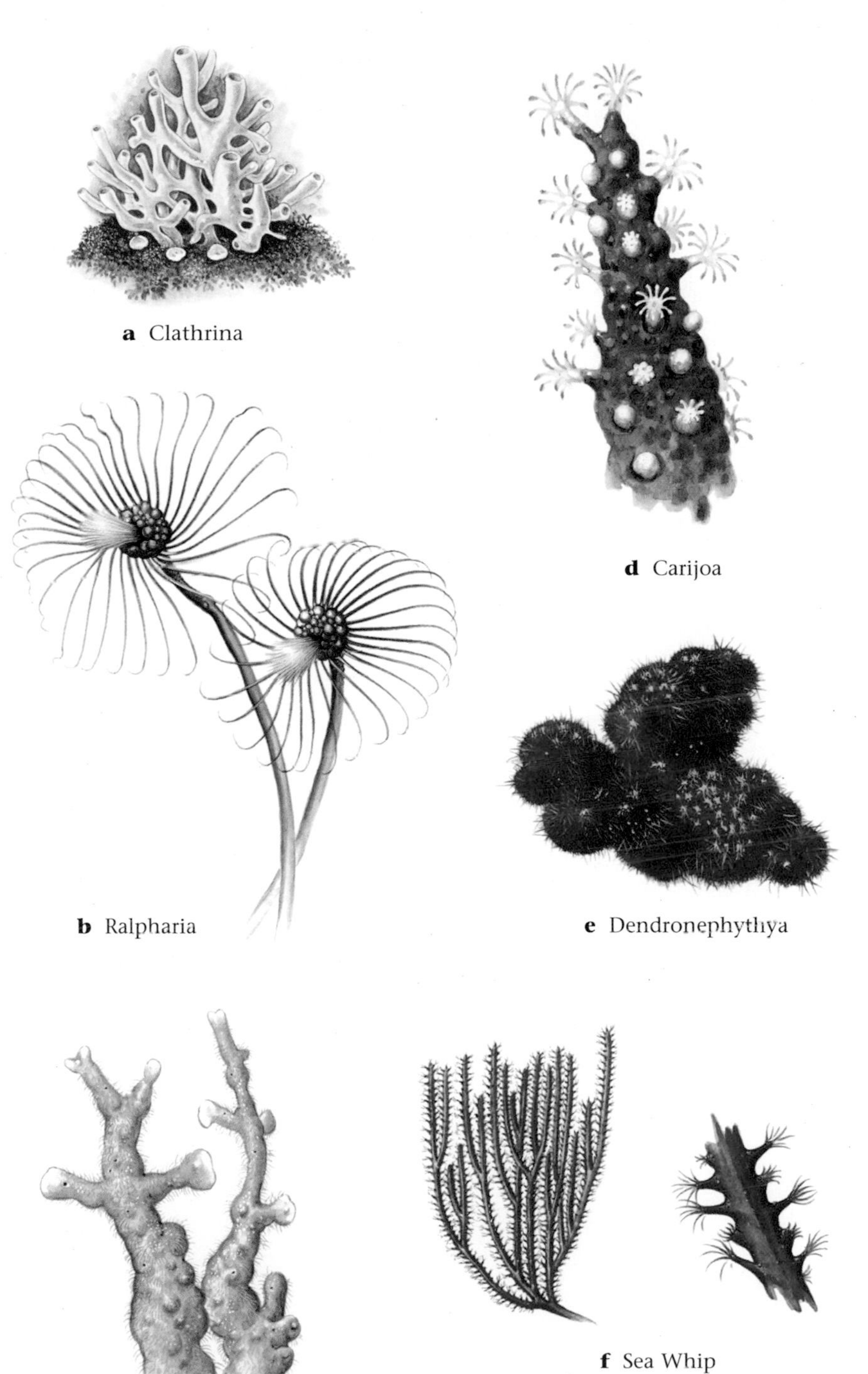
a Clathrina
b Ralpharia
c Fire Coral
d Carijoa
e Dendronephythya
f Sea Whip

Plate 71a

Crab Claw (or Pom Pom) Anemone
Triactis sp.

This very small cream-colored anemone is often held in each claw of the crab *Lybia tesselata* (as shown in illustration). It is considered a mutualistic relationship.

Plate 71b

Pocillopora eydouxi

This coral species is fairly common in Hawaii but it prefers areas with lots of water movement. Forms round heads with compact branches of uniform size.

Plate 71c

Porites sp.

This mounding coral forms lumpy shallow domes. It prefers calm waters and can be particularly abundant in lagoons.

Plate 71d

Fungia sp.

Mushroom shaped coral with ridges radiating from the center.

Plate 71e

Bushy Black Coral
Antipathes

It's only black when it's dead. Living coral is brownish orange. Very plant-like in appearance: bushy and densely branched.

Plate 71f

Upside-down Jellyfish
Cassiopea andromeda

Look for this jellyfish in calm water over sand. It hangs upside down in order to absorb the most sunlight for its symbiotic algae.

a Crab Claw Anemone

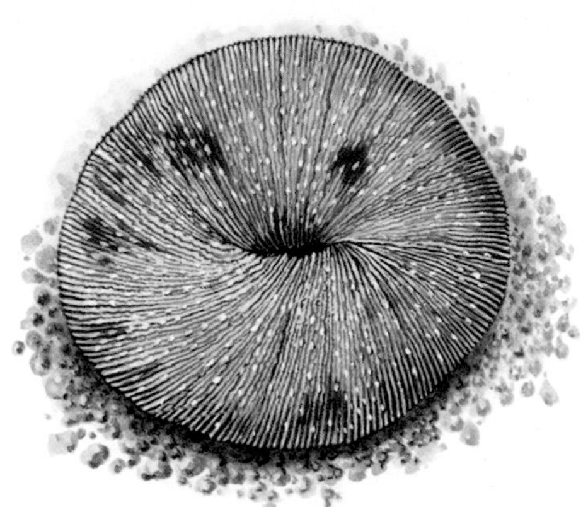

d Fungia

f Upside-down Jellyfish

b *Pocillopora eydouxi*

e Bushy Black Coral

c Porites

Plate 72a
Phrikoceros sp.
This flatworm is endemic to Hawaii. The body is brown with white spots of variable size.

Plate 72b
Pseudobiceros gratus
This flatworm has an undulating margin. The body is a creamy color over which run three longitudinal brown stripes. This species is common on the top of reefs.

Plate 72c
Pseudoceros dimidiatus
The striking color of this species suggests that, like nudibranchs, it is toxic. The most common form has wide yellow stripes alternating with wide black stripes. Also look for the orange margin.

Plate 72d
Pseudoceros ferrugineus
This beauty has a yellow margin; the body has shades of lavender and burgundy. The bright coloration indicates that it is toxic.

Plate 72e

Spaghetti Worm
Loimia medusa
Most of the body of this worm is always hidden, but the distinctive thin white tentacles are easy to spot.

Plate 72f
Pherecardia striata
This is a type of fireworm, known for their painful stings. It is covered with white hairy structures called *setae*, and if you get close enough, you can see a series of thin pink and white stripes running longitudinally along the back.

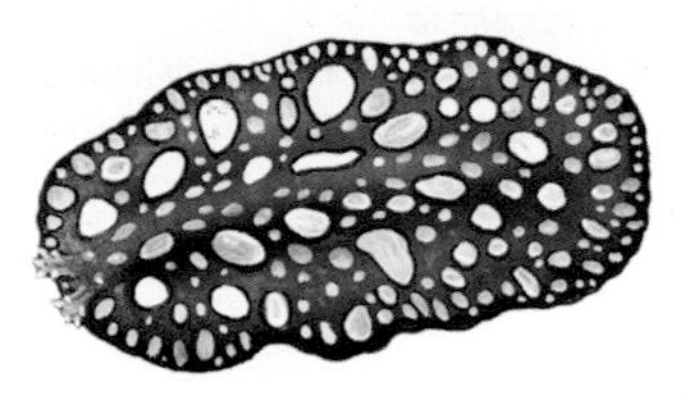

a Phrikoceros

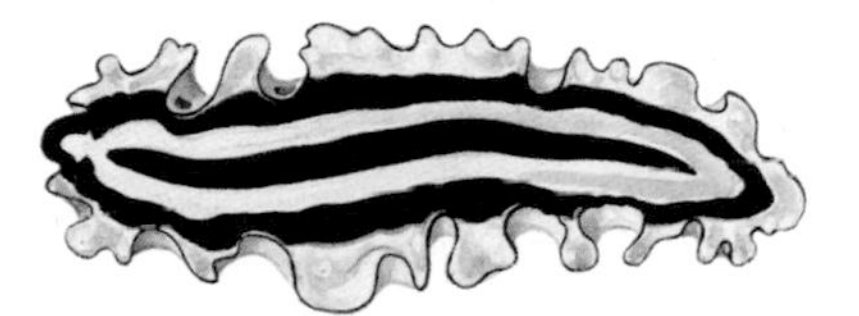

b Pseudobiceros gratus

c Pseudoceros dimidiatus

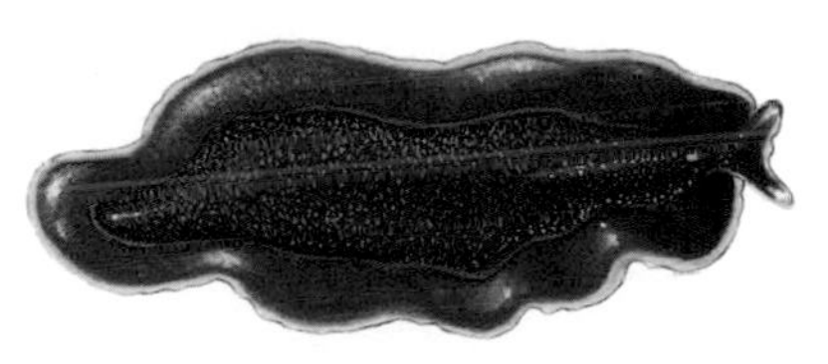

d Pseudoceros ferrugineus

e Spaghetti Worm

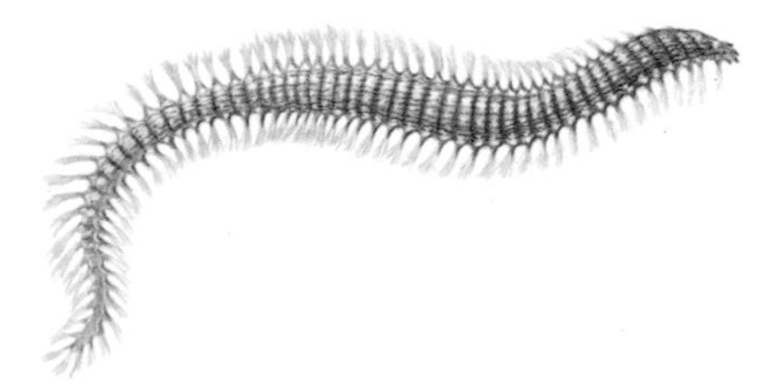

f Pherecardia striata

Plate 73a

Filograna implexa

This colonial worm is found in many habitats, including wharf pilings. The relatively small tentacles are typically red near the base and white at the tips.

Plate 73b

Christmas Tree Worm

Spirobranchus giganteus

Variable in color, but always with two spirals of tentacles. It lives in tubes on living coral.

Plate 73c

Opihi Limpet

Cellana sandwicensis

This Hawaiian endemic is always dark colored with ribs that are quite pronounced. Opihi was an important source of food for native Hawaiians.

Plate 73d

Titiscania limacina

The adults of this species lack a shell but their white defensive glands are visible.

Plate 73e

Horned Helmet

Cassis cornuta

This common helmet is also the largest. It feeds on echinoderms, including the notorious Crown-of-thorns (Plate 82). Look for it on sand inside the outer barrier reefs.

Plate 73f

Moon snail

Polinices mammilla

The glossy shell is white and so is the animal inside. Moon snails feed on other mollusks. They can be found on sand or on mudflats in shallow water.

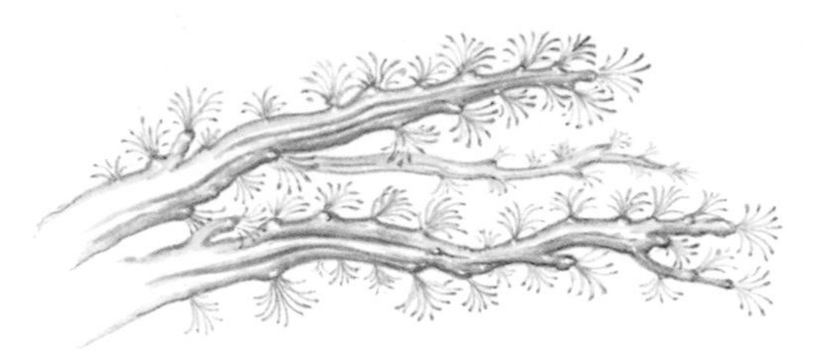

a Filograna implexa

e Horned Helmet

d Titiscania limacina

f Moon Snail

b Christmas Tree Worm

c Opihi Limpet

Plate 74a

Gold Ringed Cowrie
Cypraea annulus

This cowrie has a greenish color with a bright yellow ring. It is very common in shallow water.

Plate 74b

Honey Cowrie
Cypraea helvola

A very common cowrie, this species is reddish brown with many white spots. Prefers shallow water.

Plate 74c

Mole Cowrie
Cypraea talpa

The shell is brown and gold banded; the mantle is black with many tiny white spots. Look for this cowrie underneath coral heads.

Plate 74d

Tiger Cowrie
Cypraea tigris

This species frequents a number of habitats but it is probably most common in coral rubble. It has a white shell with black spots. It is one of the larger cowries.

Plate 74e

Pacific Deer Cowrie
Cypraea vitellus

The shell of this cowrie is brown with white spots. The papillae on the mantle are particularly large.

Plate 74f

Triton's Trumpet
Charonia tritonis

This is one of the largest sea snails and an important predator on the Crown-of-thorns (Plate 82). Look for it on sandy areas between reef patches.

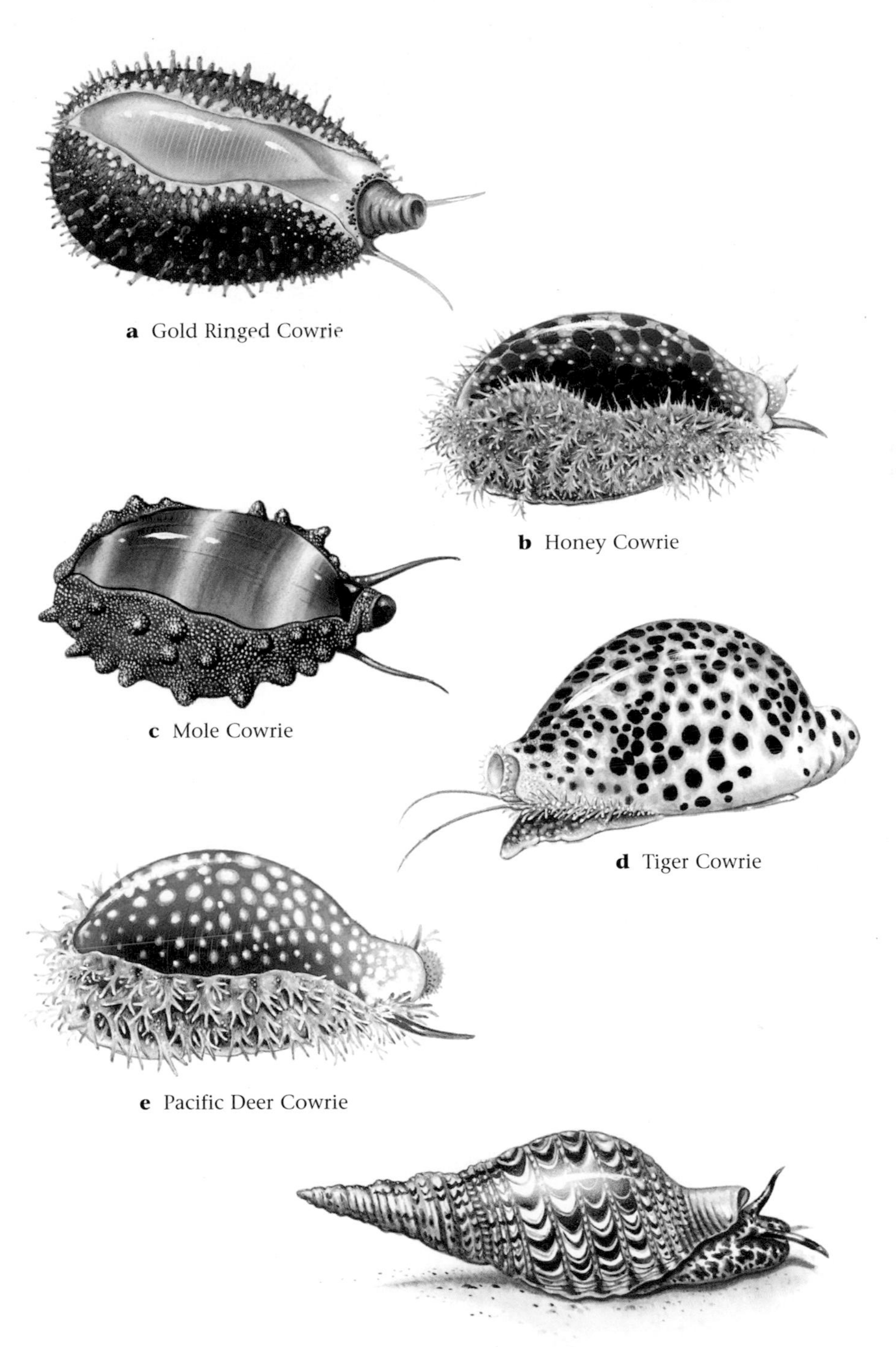

a Gold Ringed Cowrie

b Honey Cowrie

c Mole Cowrie

d Tiger Cowrie

e Pacific Deer Cowrie

f Triton's Trumpet

Plate 75a
Nassarius papillosus
Another very large gastropod, it is the color of honey. The beaded shell is also characteristic.

Plate 75b
Apple Snail
Malea pomum
This nocturnal snail prefers shallow silty water. Its color is light brown with whitish spots.

Plate 75c
Pope's Miter
Mitra papalis
This large miter has orange spots on a cream colored shell. Unlike most miters, this species prefers sandy habitats.

Plate 75d
Marble Cone
Conus marmoreus
Like many cones, this species has many triangular markings, but it has a darker shell than most other members of the genus. Cones are predators and this one prefers to eat other cone snails.

Plate 75e
Textile Cone
Conus textile
A favorite among shell collectors, this species is extremely venomous. The venom, which it uses to kill small fish, can also kill ecotravellers. It is most common in coral rubble in shallow water.

Plate 75f
Sundial
Architectonica perspectiva
This beautiful gastropod has a characteristic coiled form. It likes sandy areas in shallow water.

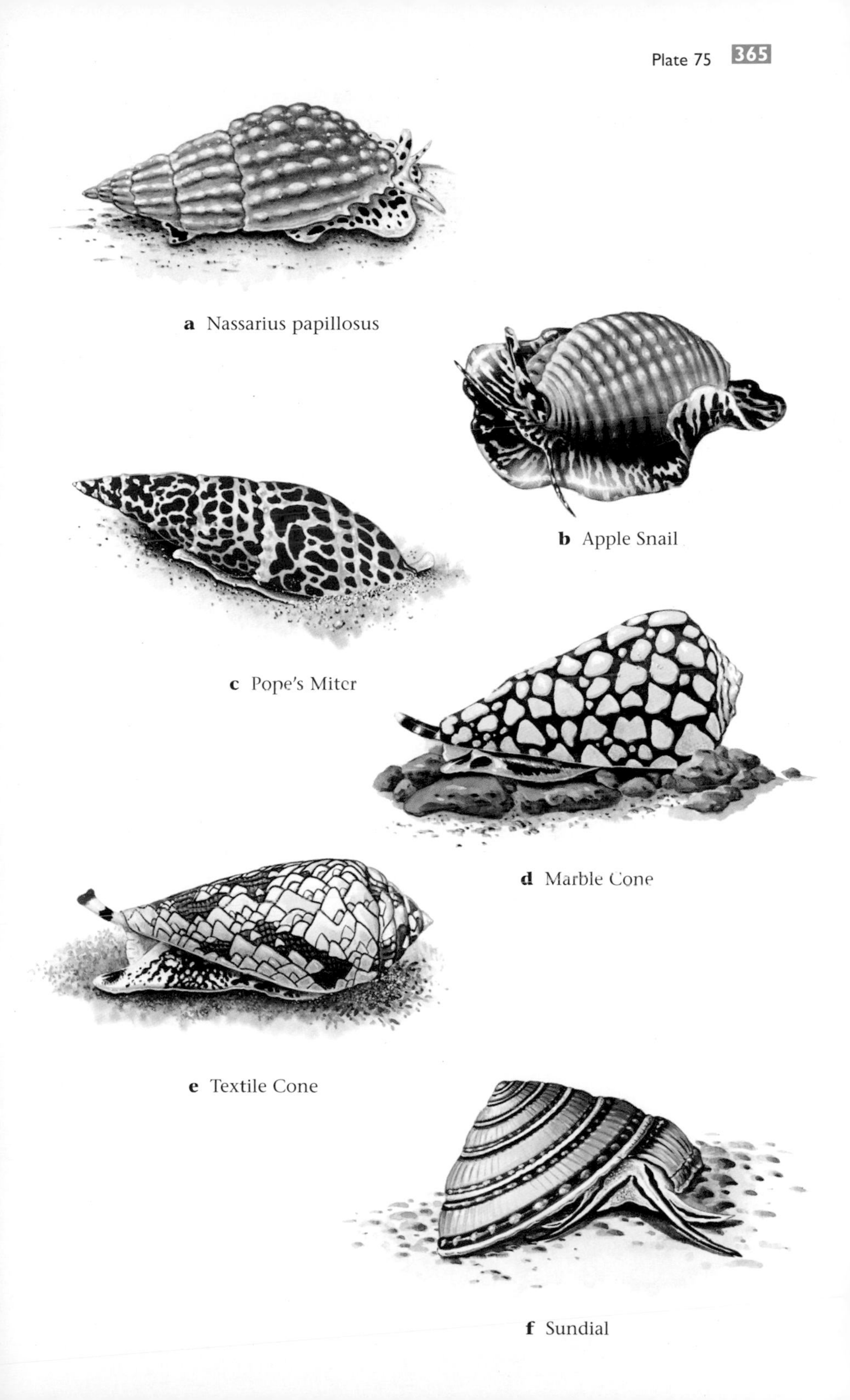

a Nassarius papillosus

b Apple Snail

c Pope's Miter

d Marble Cone

e Textile Cone

f Sundial

Plate 76a
Hydatina physis
A smallish white shell on a large reddish pink body. It feeds on marine worms.

Plate 76b
Bubble Snail
Bulla vernicosa
These small brown snails are algae grazers. They burrow in muddy silt.

Plate 76c
Chelidonura hirundinina
This delicate little nudibranch rewards close inspection. It is often blue-green (but color varies) with black lines forming a T on the head. It prefers shallow water over either sand or rocks.

Plate 76d
Aplysia dactylomela
This common species of sea hare lives in shallow water where algae is abundant. This species has characteristic large dark rings on its mantle.

Plate 76e
Elysia ornata
This sea slug has a green body with orange and black margins.

Plate 76f
Umbraculum umbraculum
Look for this sea slug in tide pools or shallow reefs. It is bright orange with round white papillae projecting from its back.

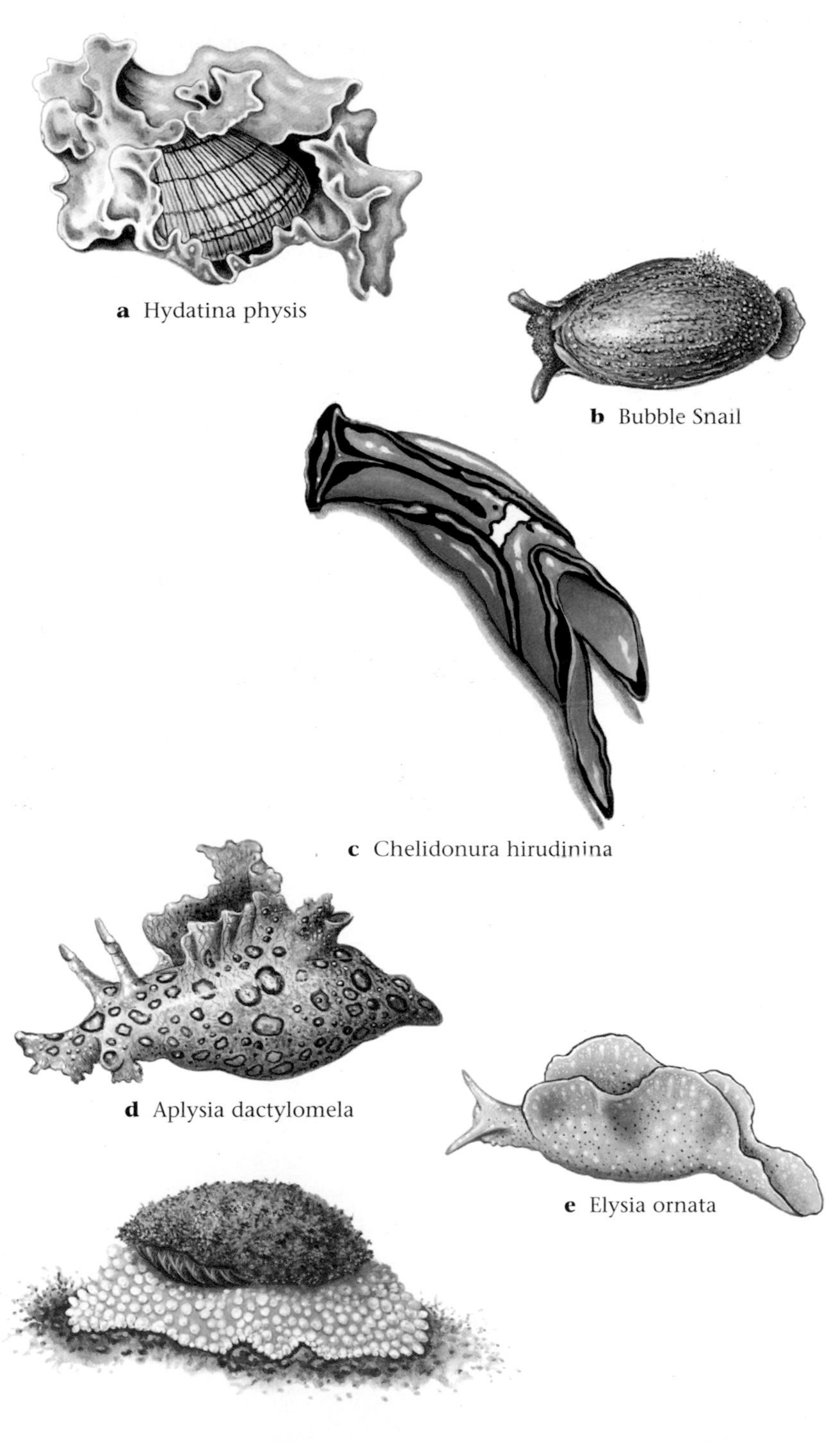

a Hydatina physis

b Bubble Snail

c Chelidonura hirudinina

d Aplysia dactylomela

e Elysia ornata

f Umbraculum umbraculum

Plate 77a

Berthella martensi

This sea slug can literally come apart before your eyes. When disturbed, it sheds all or part of its mantle, which consists of three parts. The color is quite variable.

Plate 77b

Spanish Dancer

Hexabranchus sanguineus

This species is named for the undulating movements it makes while swimming through the water. Always some shade of red, it is the largest nudibranch in the world.

Plate 77c

Phyllidiella pustulosa

This nudibranch has a black body with pink tubercles. One of the most common species of nudibranch throughout the Indo-Pacific.

Plate 77d

Phyllidiopsis sphingis

A cream-colored body with blue areas at the margins, as well as numerous black lines of various orientations. It prefers reef slopes where it can be found out in the open.

Plate 77e

Pteraeolidia ianthina

Members of this group of nudibranchs lack gills. They respire by means of numerous projections known as *cerata*, which also contain branches of the digestive system. Nematocysts from cnidarian prey are transported to these structures, which discourages would-be predators. This elongate species is covered with cerata. It comes in various shades of blue and green.

Plate 77f

Common Pearl Oyster

Pinctada margaritifera

This bivalve attaches to coral rubble and rocks. It can be recognized by the jagged teeth around the aperture.

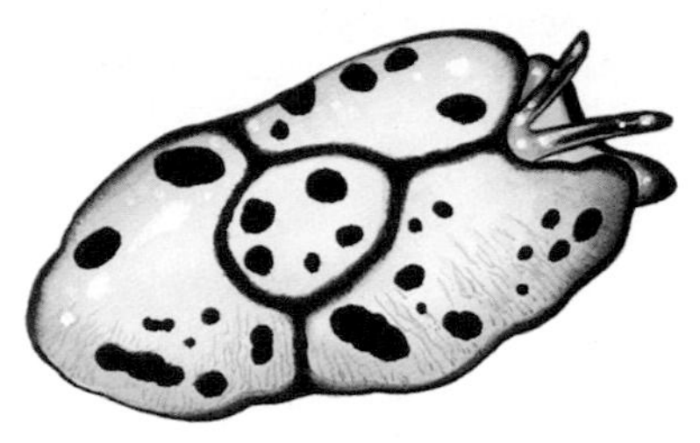

a Berthella martensi

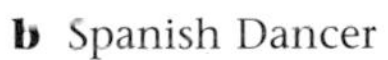

b Spanish Dancer

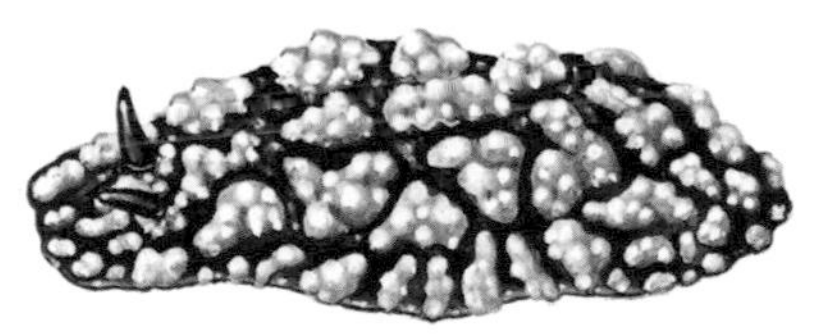

c Phyllidiella pustulosa

d Phyllidiopsis sphingis

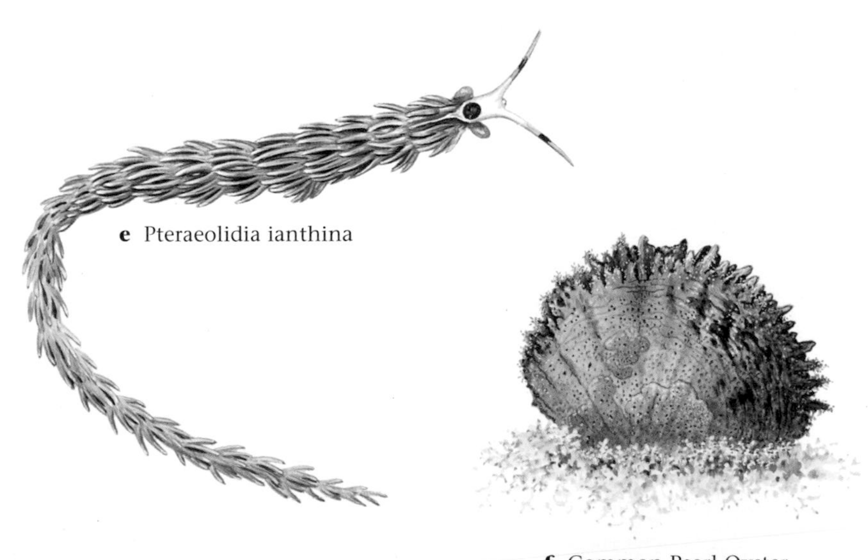

e Pteraeolidia ianthina

f Common Pearl Oyster

Plate 78a

Cat's Tongue Thorny Oyster
Spondylus linguaefelis
This deeper-water oyster has a thick covering of long spines.

Plate 78b

Bigfin Reef Squid
Sepioteuthis lessoniana
This is the squid you are most likely to encounter on the reef, especially at its edges. The fins of this species extend much farther down the mantle than in most squids. The color, as in most squids, is variable and changeable.

Plate 78c

Octopus
Octopus cyanea
This is a relatively large octopus which is out and about during the day. It also prefers relatively shallow water favored by snorkelers and scuba divers. Look for a black spot surrounded by a thin black ring at the base of the arm web.

Plate 78d

Odontodactylus scyllarus
This mantis shrimp is a real beauty. It has a bright green body, a blue head and eyes, and bright red appendages. It prefers sand and rubble and feeds on a variety of other crustaceans, mollusks and worms at night.

Plate 78e

Stegopontonia commensalis
This shrimp lives among sea urchin spines (as shown), for which its long flattened body is well suited. Dark purple to black with three thin white stripes running the length of the body.

Plate 78f

Alpheus paracrinitus
This shrimp can be readily identified by its characteristic orange and white banding.

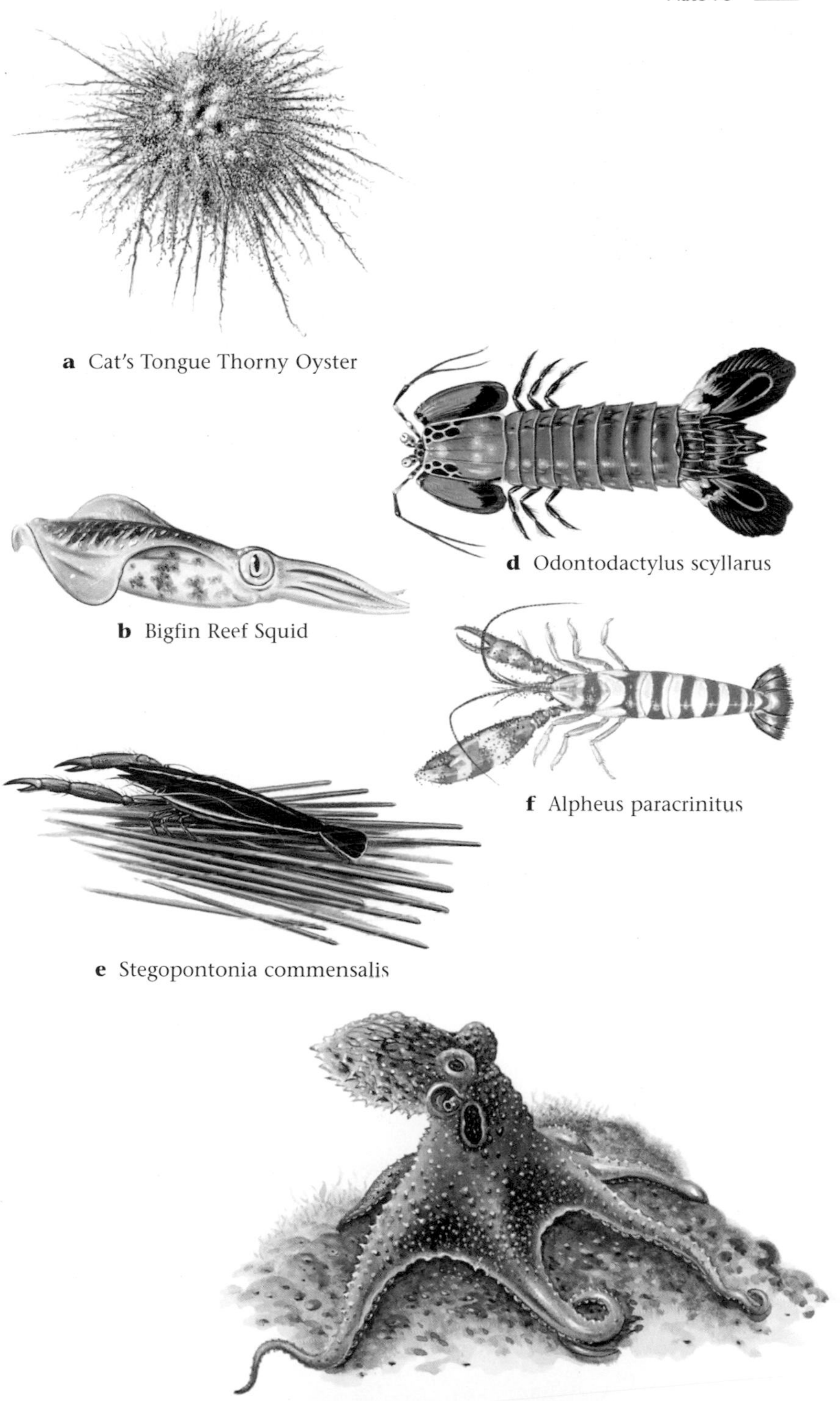

a Cat's Tongue Thorny Oyster

d Odontodactylus scyllarus

b Bigfin Reef Squid

f Alpheus paracrinitus

e Stegopontonia commensalis

c Octopus

Plate 79a

Banded Coral Shrimp
Stenopus hispidus

Cleaner shrimp are often strikingly colored, ostensibly to attract clients to their cleaning stations. This species has a thin body with red and white bands, extending to the claws. When a fish arrives at the station, the shrimp palpates it with its antennae; this causes the fish to relax so that the shrimp can crawl all over its body looking for ectoparasites.

Plate 79b

Fountain Shrimp
Stenopus pyrosonotus

This species, though less colorful than the preceding, is also thought to be a cleaner. A red stripe runs down the top of the otherwise transparent abdomen.

Plate 79c

Humpback Cleaner Shrimp
Lysmata amboinensis

This species belongs to a group known as humpback shrimp and it is also a cleaner. Its body is more robust than the previous two. The orange body is adorned with a broad red stripe that runs down the back, inside of which is a thin white stripe.

Plate 79d

Bumblebee Shrimp
Gnathophyllum americanum

This blunt-headed shrimp has one large claw that is almost as long as its body, and a shorter one on the other side, like a fiddler crab. Its body is white with black and brown bands.

Plate 79e

Harlequin Shrimp
Hymenocera picta

This shrimp can be easily identified by its bold markings consisting of red, burgundy or purple spots on a white body. It preys on starfish, which it consumes from the tip of the arm toward the central mouth. In this way it keeps the starfish alive for an extended period so that the shrimp can dine at its leisure.

Plate 79f

Hairy Lobster
Enoplometopus occidentalis

This is one of the so-called soft lobsters, though it does have two spiny ridges on its carapace. It has a white body covered with red spots. The legs and antennae are light orange.

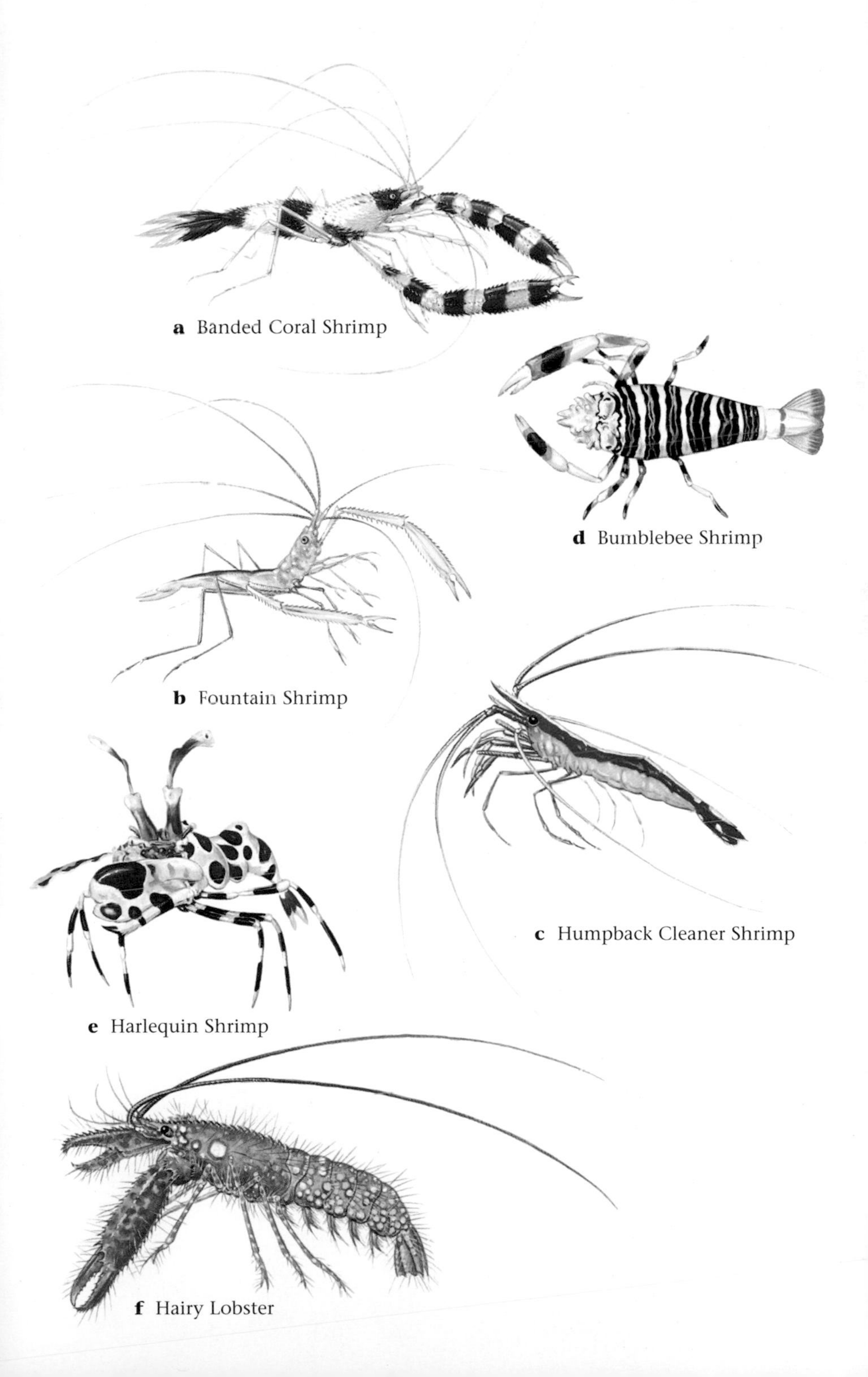

a Banded Coral Shrimp

d Bumblebee Shrimp

b Fountain Shrimp

c Humpback Cleaner Shrimp

e Harlequin Shrimp

f Hairy Lobster

Plate 80a

Long-handed Lobster

Justitia longimanus

This spiny lobster is quite colorful. It has distinctive red markings over a yellow background color. It can be found in caves on reef edges.

Plate 80b

Mole Lobster

Palinurellus wieneckii

This small spiny lobster lacks the long spines characteristic of the group. Instead it is covered with numerous tiny spines. It is entirely orange in color.

Plate 80c

Panulirus marginatus

This spiny lobster is endemic to Hawaii. Each abdominal segment has a white line across it; its carapace has white spots, and its legs and antennae are black.

Plate 80d

Regal Slipper Lobster

Arctides regalis

Slipper lobsters have much flatter bodies than spiny lobsters. This species, a Hawaiian endemic, is one of the most colorful species of this group. The body is brick red with an assortment of black markings. Look for it in caves, which is where small groups tend to congregate.

Plate 80e

Trizopagurus strigatus

This hermit crab is always found in cone shells (as shown). Its bright yellow and red stripes make it easy to identify.

Plate 80f

Mole Crab

Emerita pacifica

Mole crabs are burrowers, as the common name implies. They are usually buried in the sand so they are not commonly seen. They are always found in sandy areas in which they can disappear with stunning speed, tail-first.

a Long-handed Lobster

d Regal Slipper Lobster

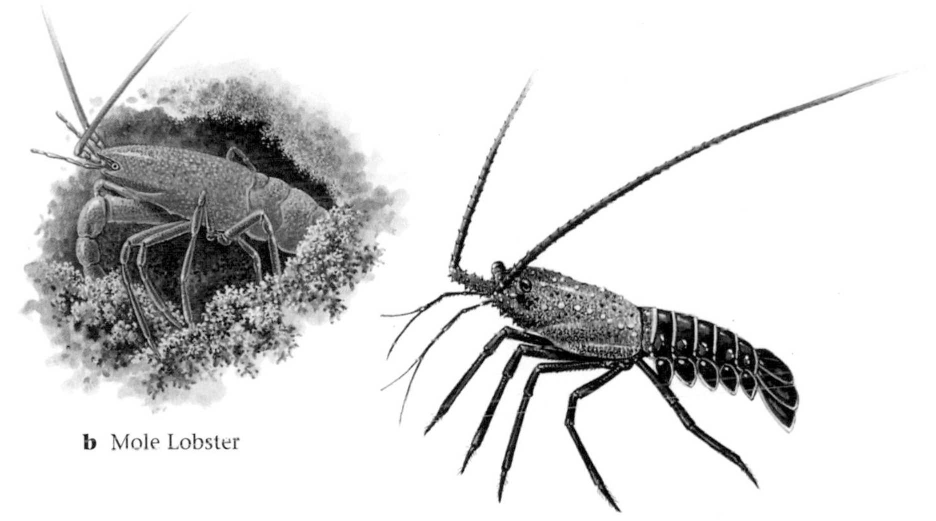

b Mole Lobster

c Panulirus marginatus

e Trizopagurus strigatus

f Mole Crab

Plate 81a

Arrowhead Crab
Huenia heraldica

This is one of the decorator crabs, so-called because they adorn themselves with anemones, algae, tunicates and anything else they can apply to themselves. This habit renders them quite cryptic. This species, though, is less given to decoration. It has a long triangular green carapace.

Plate 81b

Red-legged Swimming Crab
Charybdis erythrodactyla

Some crabs do swim, and this species is one of them. It has a reddish body with blue blotches. Also look for the five large lateral spines.

Plate 81c

Charybdis hawaiiensis

This swimming crab is tan to red in color. Outside of Hawaii it is found only in the Society Islands and Tuamotu.

Plate 81d

Carpilius maculatus

This handsome crab has a gray-tan body with several large brown spots. It hangs out in rocky areas.

Plate 81e

Pompom Crab
Lybia edmondsoni

This fascinating crab is endemic to Hawaii. Its name derives from the small anemones with which it adorns its two claws. The anemones seem to help ward off predators, because the crab waves its pompoms when threatened. Its body is salmon-colored, with fine white lines running in various directions. Also look for maroon bands on its legs.

Plate 81f

Astropecten polycanthus

This sea star is brown with white spines running along the edge of each arm. It is found on sandy bottoms, where it preys on bivalves.

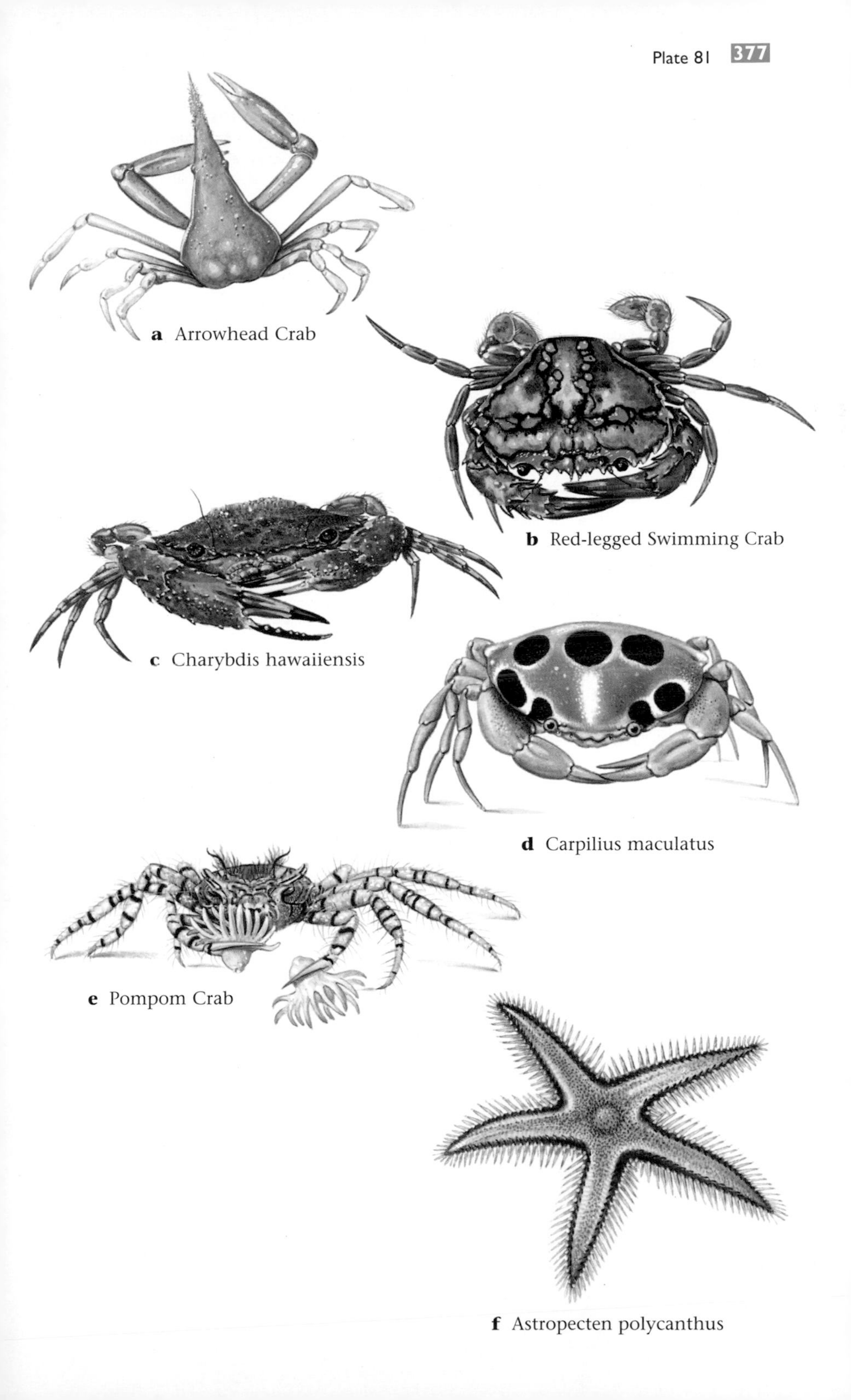

a Arrowhead Crab

b Red-legged Swimming Crab

c Charybdis hawaiiensis

d Carpilius maculatus

e Pompom Crab

f Astropecten polycanthus

Plate 82a
Pentaceraster cumingi
This stout sea star prefers rocky areas or reefs. It has short, thick arms. The body, which is reddish orange, is adorned with thick spines that are lighter in color.

Plate 82b
Linckia multiflora
This variably colored sea star is common on shallow reefs. It is usually mottled with red, blue and yellow hues.

Plate 82c
Crown-of-thorns
Acanthaster planci
This notorious sea star is no longer considered quite the demon it was once thought to be. But local population explosions of this species have resulted in the decimation of coral reefs. Its formidable spines make it easy to identify.

Plate 82d
Prionocidaris hawaiiensis
This sea urchin is endemic to Hawaii, where it resides in the reef's interstices. Its spines are often covered with algae, perhaps for camouflage.

Plate 82e
Echinothrix calamaris
This urchin has fairly well spaced tubular spines. Look for it in shallow reefs, but be careful because its spines can inflict painful wounds.

Plate 82f
Echinothrix diadema
The spines of this urchin are closed at the tip and pointed. They are also densely packed. This urchin is more likely to be found out in the open than most others.

Plate 82g
Tripneustes gratilla
This is one of the more common urchins. It prefers calm shallow water such as lagoons. It frequently covers itself with debris (as shown), presumably for camouflage. The color is quite variable; the spines are arranged in distinct bands.

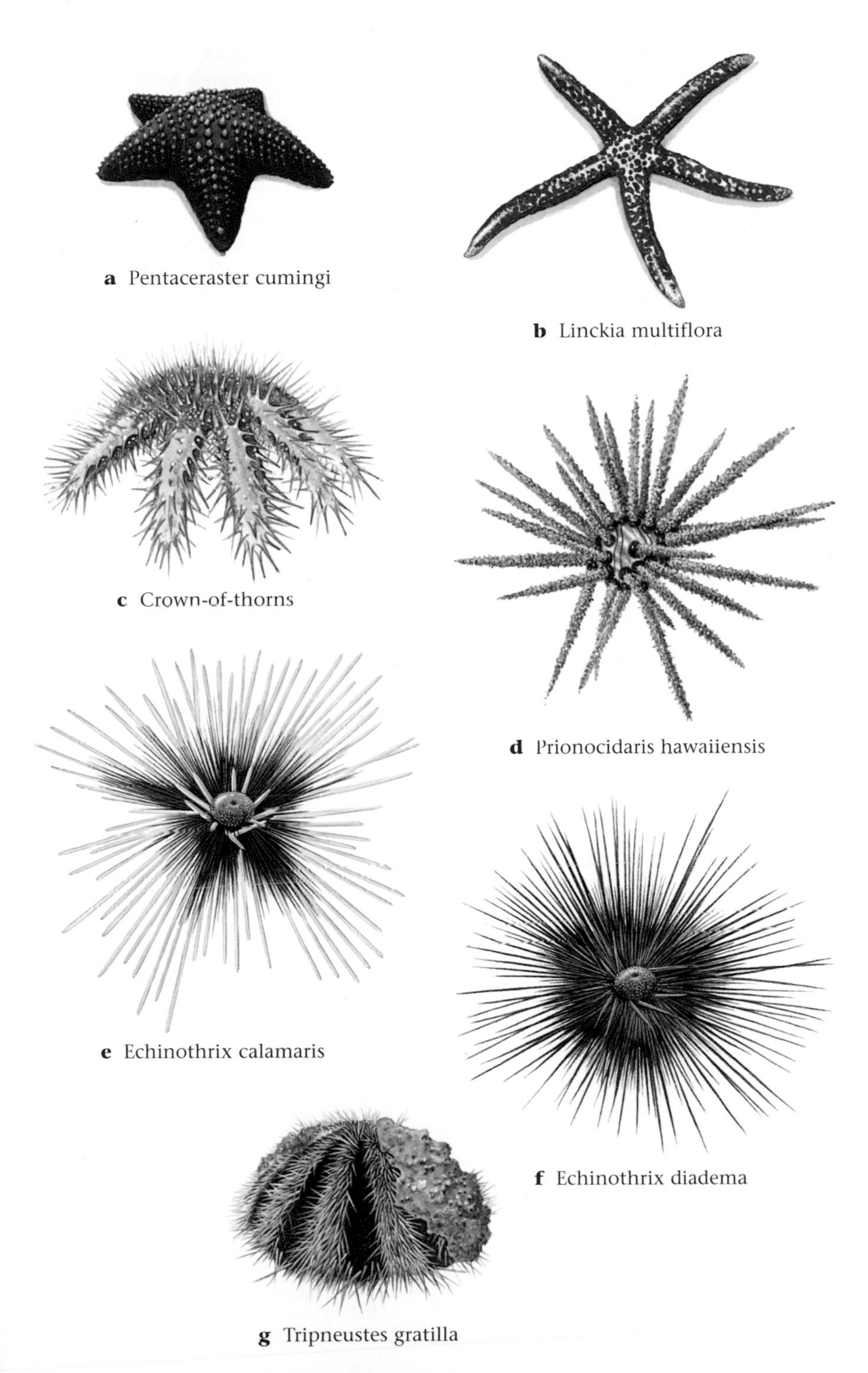

a Pentaceraster cumingi

b Linckia multiflora

c Crown-of-thorns

d Prionocidaris hawaiiensis

e Echinothrix calamaris

f Echinothrix diadema

g Tripneustes gratilla

Plate 83a
Echinometra mathaei
This is a stout-spined urchin. Spine color varies, but they are often reddish with a white ring around the base of each. It is found on very shallow reefs.

Plate 83b
Slate Pencil Urchin
Heterocentrotus mammillatus
The extremely thick, blunt spines are what distinguishes this sea urchin. The color is extremely variable. It prefers shallow reefs; look for it in the crevices.

Plate 83c
Echinoneus cyclostomus
This is one of the heart urchins, which, along with the sand dollars, lack radial symmetry (they are not perfectly round). This small species is oval-shaped and covered with short spines. It buries itself in sand in shallow water.

Plate 83d
Bohadschia paradoxa
This sea cucumber likes sandy habitats in shallow water. One of the fatter species, it is brown with fairly short papillae.

Plate 83e
Holothuria atra
This sausage-shaped sea cucumber is black, but it often covers itself with sand. One of the most common species, it sometimes forms aggregations in shallow sandy areas.

Plate 83f
Stichopus chloronotus
This sea cucumber prefers coral rubble and rocky areas. Its dark green body is adorned with long black papillae with orange tips.

Plate 83g
Botryllus sp.
These colonial tunicates form lumpy amorphous masses. All members of the genus are variably colored. The siphons are sometimes shaped like volcanoes.

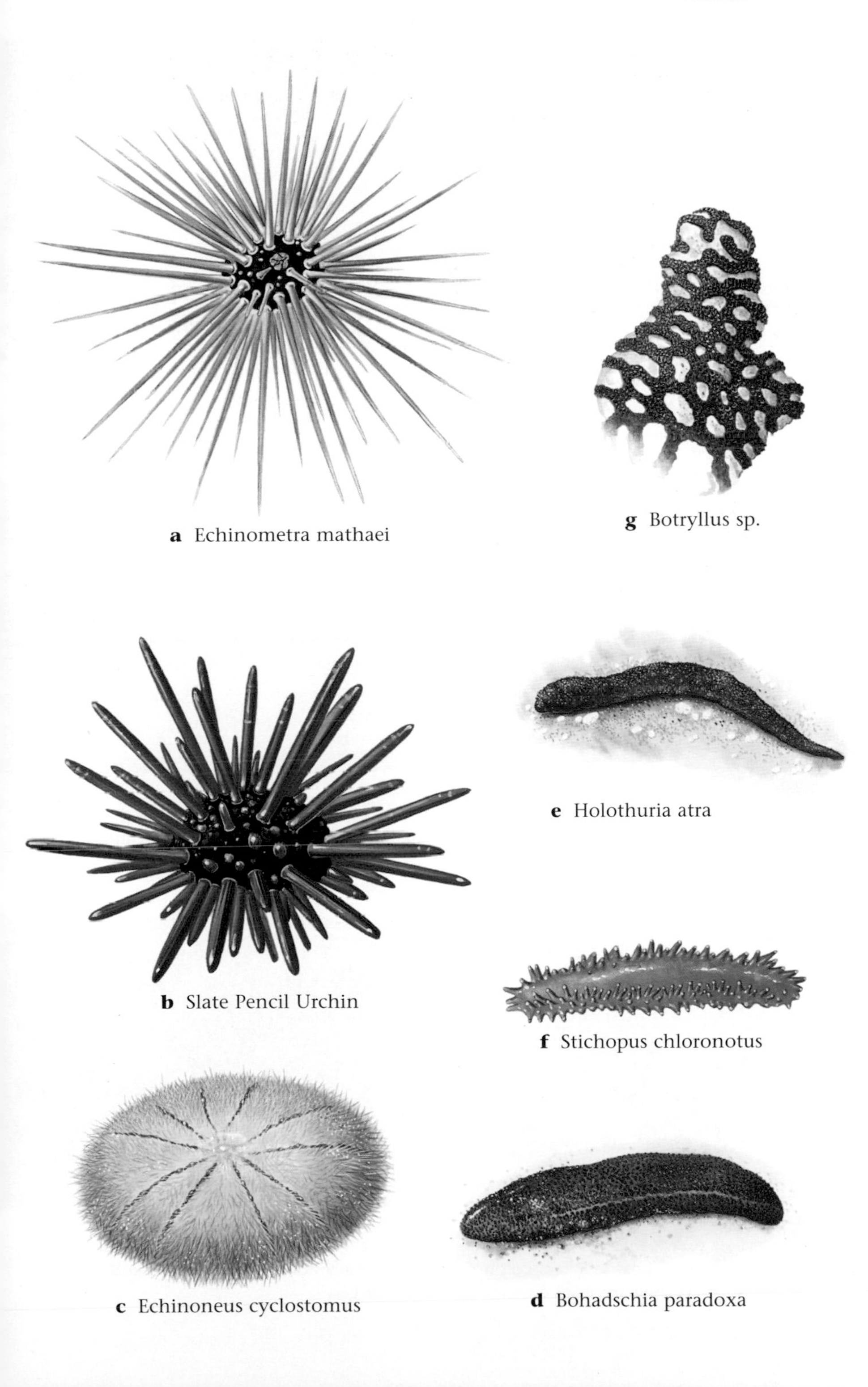

a Echinometra mathaei

g Botryllus sp.

b Slate Pencil Urchin

e Holothuria atra

f Stichopus chloronotus

c Echinoneus cyclostomus

d Bohadschia paradoxa

Plate 84a
Hawaiian picture wing fruit fly (*Drosophila grimshawi*) on rotting tree bark. Each *Drosophila* species specializes on different parts of different rotting plants.

Plate 84b
Casebearing moth (*Hyposmocoma* sp.) caterpillar crawling on bark. When disturbed, the caterpillar retreats into its homemade shell.

Plate 84c
Lava flow cricket (*Caconemobius fori*) searching for windblown debris. These crickets colonize lava flows as soon as they are cool enough to walk on.

Plate 84d
Lava flow wolf spider (*Lycosa* sp.) carrying empty egg sac (below) and hundreds of babies (on her back).

Plate 84e
Wekiu bug (*Nysius wekiuicola*) feeding on a dead blowfly by sticking its beak into the fly's leg. Found only in the summit region of the Big Island's Mauna Kea.

Plate 84f
Aa bug (*Nysius aa*) perched on rough aa lava. Found only above treeline on the Big Island's Mauna Loa.

Plate 85a
Big Island cave cixiid planthopper (*Oliarus polyphemus*) on lava tube wall. Adaptations to the cave environment include reduced eyes and pale coloration.

Plate 85d
Predacious caterpillar (*Eupithecia staurophragma*) attacking a cricket. The attack is initiated by the prey touching the caterpillar's rear end.

Plate 85b
Maui lava tube cricket (*Caconemobius* sp.) foraging on lava rocks. Adaptations to the cave environment include reduced eyes and pale coloration.

Plate 85e
Predacious caterpillar (*Eupithecia orichloris*) eating a freshly caught fly. The meal usually starts with the head.

Plate 85c
Cave water-treader (*Cavaticovelia aaa*) walking on lava tube wall. Aaa is the Hawaiian word for lava tube.

Plate 85f
Happyface spider (*Theridion grallator*) displaying its clown-faced abdomen. Patterns range from plain yellow to full black circles.

Plate 86a
Koa bug (*Coleotichus blackburniae*) nymph on koa plant. The red and black warning coloration indicates that this insect tastes bad.

Plate 86b
Koa bug (*Coleotichus blackburniae*) adult on koa leaf. Young adults are mostly red and become more green as they get older.

Plate 86c
Hawaiian damselfly (*Megalagrion calliphyas*) perched on a stick. This species breeds in pools of water.

Plate 86d
Kamehameha butterfly (*Vanessa tameamea*) alighted on a leaf. One of two endemic butterflies in Hawaii.

Plate 86e
Blackburn's little blue butterfly (*Udara blackburni*) alighted on aalii plant. This endemic species has iridescent green scales on the underside of its wings.

Plate 86f
Monarch (*Danaus plexippus*) caterpillar on milkweed. Albino adults are not uncommon in some areas of Hawaii.

WCS Conservation Work in North America

A Century of Leadership

Since its creation as the New York Zoological Society in 1895, the Wildlife Conservation Society (WCS) has played a central role in the North American conservation movement. In 1897, WCS carried out the first survey of Alaskan wildlife, which led to the passage of legal limits on hunting. In 1905, WCS General Director William Hornaday initiated the campaign to save the bison from extinction through captive breeding and successful reintroduction in reserves across the American West. In 1912, Hornaday drafted the first legislation protecting migratory birds and their habitats. Throughout this past century, WCS has supported pioneering field studies of key species such as bighorn sheep, black-footed ferrets, grizzly bears, mountain lions and bald eagles, and helped to create more than 30 U.S. parks and reserves, including the Arctic National Wildlife Refuge, and the Olympic and Wind Cave National Parks. As it enters its second century, WCS continues to build on this North American tradition by informing and inspiring people to care about native wildlife and ecosystems.

North American Field Projects

Projects in the Western United States

One of the main goals of the WCS North America program is re-establishing extirpated species and their roles in the wider landscape. In many parts of the country, particularly the lower 48 states, large predators have been eliminated from many ecosystems, causing other animal populations to increase unchecked. One consequence of this scenario is that the flora of a wilderness area can become devastated through over-grazing, depriving smaller animals of needed habitats.

To better understand the dynamics of a complete ecosystem, WCS conducts studies in Alaska and the northern Rockies to learn more about predator–prey relationships for ecosystem management. Research has found that predator-free environments create behavioral changes in prey species such as the moose. Comparisons between ecosystems where predators are present and absent can provide insights about wildlife management, particularly with respect to carnivore recovery programs. WCS's century-long affiliation with the American bison continues in the southern Greater Yellowstone ecosystem. Researchers currently

Bighorn sheep rams, Glacier National Park. © S. Morse

Moose near Grand Teton Mountains. © J. Berger

are monitoring the health of bison and elk populations to determine the effects of the bacterial disease brucellosis on reproduction and ways to limit its potential transmission to cattle on nearby ranches.

Whereas large predators like wolves and bears have disappeared from most of the lower 48 states, some smaller carnivores have managed to hang on – primarily by staying out of sight. Recent WCS research revealed that the elusive lynx is present in Montana, Idaho, Washington and Oregon, sometimes in places where the medium-sized cat was long thought to have disappeared. These lynx were detected by means of non-intrusive scent traps, which collect hairs for DNA analysis. Accurate information on lynx distribution is important; the animal is slated for classification as an endangered species, a listing which would affect how large areas of forest are managed. Another finding of the study is that lynx favor

Caged Bison about to be released during early 1900s species re-introduction program. © WCS

Researcher observing Bison behavior. © J. Berger

a mosaic of mature and partially open wooded areas, where its preferred prey, the snowshoe hare, is found. In other words, logging interests could be balanced with wildlife conservation in lynx habitat.

Maintaining the integrity of biological communities is a major issue in other parts of North America as well. With its variety of animal life and breathtaking scenery, Alaska has become one of the most popular destinations for eco-tourists. But human activity, however well-intentioned, can become disruptive and even harmful to wildlife if not properly managed. The Chilkat River valley, located in southeast Alaska, is well-known for its vast population of bald eagles. The eagles

Lynx. © J. Weaver

Grizzly Bear. © J. Berger

and other animals of the area – bears, wolves, coyotes, wolverines, otters and numerous bird species – attract visitors in ever-growing numbers. WCS researchers report that a local commercial raft company in the region brought 12,000 visitors through the eagle's main feeding area in 1997, as opposed to a few hundred per year in the early 1980s. A rise in commercial and residential development on private land near the valley threatens to fragment the environment. Hoping to find a balance between eco-tourism and ecological health, WCS has initiated a wildlife monitoring program to gather information on the impact of human activity on the bald eagle and to limit disturbances in areas critical to the bird's survival.

Other WCS projects have a broader focus; knowing how an ecosystem functions as a whole often will determine if a particular wildlife management strategy succeeds or fails. The Ponderosa pine forests of the western U.S. are a good example of how management can sometimes go astray. WCS researchers have determined that the suppression of forest fires which would otherwise naturally occur, and the long-term logging of large trees, results in an understory forest of small, densely packed trees. This change in the structure of these forests affects many wildlife species such as spotted owls and their ground-dwelling prey, that historically depend on more open habitat. Properly controlled, periodic fires can open and invigorate forest environments, thus helping the animals which depend on them. WCS also studies the role woodpeckers play in engineering decaying trees into cavity-bearing snags. Many songbirds and small mammals depend on tree cavities, and so depend on the activities of woodpeckers in coniferous forests.

Ecological restoration efforts also have been applied to riparian habitats on western range lands. When left unattended, livestock animals such as cattle can destroy stream function and vegetation. This leads to a decline in both water quality and quantity, and in neotropical migratory birds such as the yellow warbler and the willow flycatcher. By carefully managing livestock grazing, landowners can improve water conditions and restore biodiversity on their lands.

Projects in the Northeastern United States

WCS projects in the northeastern U.S. also focus on understanding the role and relationships of predators to their environments. Once indigenous to the eastern U.S., the gray wolf was forced from the area by human pressures such as hunting. If re-established in the northeastern United States, wolves could help regulate the numbers of white-tailed deer and beaver, both of which can cause damage to the environment in the absence of predators.

Over the past few years, WCS researchers have explored the possibility of bringing the wolf back to the northeast. One study investigated the feasibility of a biological corridor, through which wolves in Canadian regions such as Quebec move down into both Maine and possibly New York state's Adirondack Park. However, given the gauntlet of manmade obstacles such as highways and other developments which the animals must run in order to repopulate former territory, a release program may stand a better chance of establishing a stable population. And because interactions between humans and wolves would almost certainly occur, education of the public would be key in the success of any such project. By providing this kind of information to decision makers, WCS plays a leading role in ensuring that science – not emotion – steers this important issue.

Another WCS effort concentrates on the challenges that the Adirondack Park itself presents to conservation efforts. As the flagship of WCS's northeast forest program, the Adirondack Park contains a rich mixture of forests, lakes and wetlands – homes for numerous species of animals and plants. Nearly as numerous are the competing interests for the park's resources. The 10,000-square-

Gray Flycatcher. © S. Zack

Timber Wolf. © S. Morse

mile park borders many municipalities and is subjected to conservation mandates from many layers of local and state agencies. Conflicts between public interests and private landowners are a constant concern to park managers. Since 57% of the park is privately owned, activities such as fence construction and logging could seriously fragment the environment and adversely affect wildlife. And ironically, park visitors wishing to experience and protect nature are increasing in such numbers that the public lands are being degraded.

In an attempt to balance the interests of disparate stakeholders and groups, WCS researchers are working to build an effective conflict-resolution model for community-based conservation. The program regards human populations as a key component of the natural system, and therefore just as important as that system's flora and fauna. Targeted research on socio-economic and political factors can be used to complement ecological knowledge, paving the way for a more integrated method of wildlife management. Issues such as a potential exodus of logging companies from the region to more profitable areas could result in economic dislocations for local communities, followed by the sale and subdivision of forest lands. And activities such as hunting, fishing and trapping could become detrimental to the Adirondack ecosystem if not properly managed.

Another key part of the management plan is active cooperation among the park's stakeholders. Every year, an informal body called the Oswegatchie Roundtable brings together a diverse collection of stakeholders – including private landowners, logging companies, local conservation groups, elected officials and other residents – to discuss important park issues. Participants can share objective research provided by WCS and other organizations and subsequently make more informed decisions in balancing the park's ecology with economics. Information about issues such as the impact of logging on biodiversity and the widespread

Sugar Maple sapling. © J. Jenkins

failure of hardwood forest to regenerate on both public and private grounds can be examined by all affected groups before decisions are reached.

Roundtable meetings also have helped identify key areas for study and management, prompting such WCS projects as the first comprehensive study of the beaver in the Adirondack region. WCS activities, such as applied research on forest ecosystems, key species and commercial activities, will continue to form the basis of debates on those issues, helping to inform and improve the stewardship of the park's resources in the process.

Tri-state Area

While many WCS projects deal with animals and environments of distant locations, the ecosystems which surround New York City are considered equally worthy subjects of study and conservation. Of particular interest to WCS researchers is the frontier between suburban communities and areas which are still undeveloped. Surprising as it may seem, in an area of nearly 20 million people, there still exists a diverse assemblage of beautiful rivers, forests and valleys, home to turtles, butterflies, eagles, bobcats – even the occasional black bear.

Unfortunately, the area's wildlife is threatened by a variety of human pressures such as roads, parking lots, drainage ditches and subdivisions, all of which can render the environment unsuitable for many species. To preserve wildlife in the New York metro area, WCS has initiated the Metropolitan Conservation Alliance (MCA). Since its inception in 1997, the MCA has launched several programs to monitor and protect areas such as New York state's Great Swamp, which lies approximately 75 miles north of Times Square, and the Wallkill Valley, which bisects the New York/New Jersey border.

The MCA works to educate, inform and galvanize public support for the continued protection of wildlife areas in the region. Townships surrounding the Great Swamp are playing an increasing role in protecting the area from urban sprawl and development. In the Wallkill Valley, nearly 40 communities are creating a plan that will focus development away from delicate ecosystems, while promoting recreational uses of their natural resources such as hiking and canoeing. Of particular importance to maintaining biodiversity is protecting the area's wetlands. Species such as the bog turtle and the mole salamander, the focus of two different WCS projects, are extremely sensitive to over-development and must be vigorously protected.

Since many of the New York metro area residents are as yet unaware of the area's natural wealth, education is critical not only for local councilmen, legislators and elected officials, but also for the leaders of tomorrow. By crafting effective classroom curricula to promote awareness of local ecosystems, students of the area's school systems will grow up mindful of the natural richness of the tri-state region and hopefully endeavor to restore and protect it.

Beaver pond. © L. Graham

Bog Turtle

You Can Make A Difference — Your Membership Helps Save Animals And Their Habitats

Join today! Your membership contribution supports our conservation projects around the world and here at home. For over 100 years, we've been dedicated to saving endangered species — including tigers, snow leopards, African and Asian elephants, gorillas and thousands of other animals.

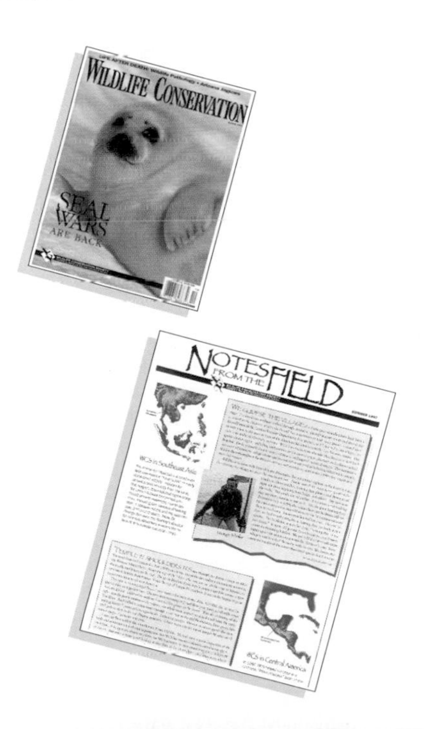

FOR A COMPLETE LIST OF BENEFITS PLEASE TURN THE PAGE . . .

YES! I want to help save rare and endangered animals. Here is my membership contribution of:

- ☐ \$25 ☐ \$35 ☐ \$75 ☐ \$150
- ☐ I cannot join at this time but enclosed is my contribution of \$ ________.

Full name (Adult #1) ____________________

Address ____________________

City ________ State ________ Zip ________ Daytime Phone (____) ________

For Family Plus and Conservation members only:

Full name Adult #2 ____________________

Full name Adult #3 ____________________ #Children ages 2–18 ____

Your employer may have a matching gift program that can double, even triple, your contribution. Be sure to check with your Personnel or Community Relations Office.

A35Ø99

Species Index

Note: Plate references give plate numbers and are in **Bold** type

General Index